Y0-BJP-121

Annual Reports in
COMPUTATIONAL CHEMISTRY

VOLUME **3**

Annual Reports in
COMPUTATIONAL CHEMISTRY

VOLUME 3

Edited by

DAVID C. SPELLMEYER
Nodality, Inc.
7000 Shoreline Court, Suite 250
South San Francisco, CA 94080
USA

RALPH A. WHEELER
Department of Chemistry & Biochemistry
University of Oklahoma
620 Parrington Oval, Room 208
Norman, OK 73019
USA

ELSEVIER

Amsterdam • Boston • Heidelberg • London • New York • Oxford
Paris • San Diego • San Francisco • Singapore • Sydney • Tokyo

Elsevier
Radarweg 29, PO Box 211, 1000 AE Amsterdam, The Netherlands
Linacre House, Jordan Hill, Oxford OX2 8DP, UK

First edition 2007

Copyright © 2007 Elsevier B.V. All rights reserved

No part of this publication may be reproduced, stored in a retrieval system or transmitted in any form or by any means electronic, mechanical, photocopying, recording or otherwise without the prior written permission of the publisher

Permissions may be sought directly from Elsevier's Science & Technology Rights Department in Oxford, UK: phone (+44) (0) 1865 843830; fax (+44) (0) 1865 853333; email: permissions@elsevier.com. Alternatively you can submit your request online by visiting the Elsevier web site at http://elsevier.com/locate/permissions, and selecting *Obtaining permission to use Elsevier material*

Notice
No responsibility is assumed by the publisher for any injury and/or damage to persons or property as a matter of products liability, negligence or otherwise, or from any use or operation of any methods, products, instructions or ideas contained in the material herein. Because of rapid advances in the medical sciences, in particular, independent verification of diagnoses and drug dosages should be made

Library of Congress Cataloging-in-Publication Data
A catalog record for this book is available from the Library of Congress

British Library Cataloguing in Publication Data
A catalogue record for this book is available from the British Library

ISBN: 978-0-444-53088-2
ISSN: 1574-1400

For information on all North-Holland publications
visit our website at books.elsevier.com

Printed and bound in the USA
07 08 09 10 11 10 9 8 7 6 5 4 3 2 1

Working together to grow libraries in developing countries

www.elsevier.com | www.bookaid.org | www.sabre.org

ELSEVIER   BOOK AID International   Sabre Foundation

CONTENTS

Contributors ix
Preface xiii

Section 1: Simulation Methodologies (Section Editor: Carlos Simmerling)

1. **Molecular Simulations of pH-Mediated Biological Processes** 3
   Jana Khandogin and Charles L. Brooks

   1. Introduction 3
   2. Static Structure Based pK_a Prediction Methods 4
   3. Molecular Dynamics Coupled with Acid–Base Titration 5
   4. Applications of CPHMD 7
   5. Summary and Outlook 10
   . Acknowledgement 11
   . References 11

2. **Extending Atomistic Time Scale Simulations by Optimization of the Action** 15
   A.S. Clarke, S.M. Hamm and A.E. Cárdenas

   1. Introduction 15
   2. Applications 22
   3. Conclusions 25
   . Acknowledgements 26
   . References 27

3. **Fishing for Functional Motions with Elastic Network Models** 31
   A.J. Rader

   1. Introduction 31
   2. Background 32
   3. Identification of Functional Motions 33
   4. Detailed EN Models 34
   5. Merging EN Models and MD Simulations 36
   6. Summary & Future Prospects 36
   . References 37

4. **Alchemical Free Energy Calculations: Ready for Prime Time?** 41
   Michael R. Shirts, David L. Mobley and John D. Chodera

   1. Introduction 41
   2. Background 42
   3. Equilibrium Methods 43

4.	Nonequilibrium Methods	45
5.	Intermediate States	46
6.	Sampling	48
7.	Applications	50
8.	Conclusion	51
.	Acknowledgements	53
.	References	53

*Section 2: Biological and Biophysical Applications
(Section Editor: Heather Carlson)*

5. Linear Quantitative Structure–Activity Relationships for the Interaction of Small Molecules with Human Cytochrome P450 Isoenzymes   **63**
Thomas Fox and Jan M. Kriegl

1.	Introduction	63
2.	The Cytochrome P450 Superfamily	64
3.	Methodological Overview	66
4.	Applications	68
5.	Discussion and Outlook	73
.	References	75

Section 3: Chemical Education (Section Editor: Theresa Zielinski)

6. Observations on Crystallographic Education   **85**
Phillip E. Fanwick

1.	Introduction	85
2.	Objectives for Teaching Crystallography	86
3.	Bragg's Law	89
4.	Relating Crystallography to Chemistry	90
5.	Creativity	96
6.	Conclusions	98
.	References	98

7. Achieving a Holistic Web in the Chemistry Curriculum   **99**
Henry S. Rzepa

.	Introduction: The Impact of the Web on the Chemistry Curriculum	99
.	Background: The Trend Towards an Accumulation of Acrobat	101
1.	The Properties of a PDF Collection	103
2.	Formal Metadata Based Approaches	106
3.	The Concept of Document Re-Use	111
4.	Data as the Intel Inside	113
5.	Towards the Holistic Approach: The Podcast	118
6.	The Wiki	121
.	Conclusion	131
.	References	132

Section 4: Materials and Polymers (Section Editor: Jeffry Madura)

8. The Role of Long-Time Correlation in Dissipative Adsorbate Dynamics on Metal Surfaces   137
Jeremy M. Moix and Rigoberto Hernandez

1.	Introduction	137
2.	Classical Surface Diffusion	140
3.	Molecular Dynamics Simulations and Projective Models	143
4.	Summary and Conclusions	146
.	Acknowledgements	147
.	References	147

Section 5: Quantum Chemistry (Section Editor: T. Daniel Crawford)

9. An Active Database Approach to Complete Rotational–Vibrational Spectra of Small Molecules   155
Attila G. Császár, Gábor Czakó, Tibor Furtenbacher and Edit Mátyus

1.	Introduction	155
2.	Nonadiabatic Computations—Where Theory Delivers	158
3.	MARVEL—An Active Database Approach	158
4.	Electronic structure computations	160
5.	Variational Nuclear Motion Computations	165
6.	Outlook	169
.	Acknowledgement	169
.	References	169

10. The Effective Fragment Potential: A General Method for Predicting Intermolecular Interactions   177
Mark S. Gordon, Lyudmilla Slipchenko, Hui Li and Jan H. Jensen

1.	Introduction	177
2.	EFP2 Theory	178
3.	Example Applications	183
4.	Summary and Future Developments	190
.	Acknowledgements	191
.	References	191

11. Gaussian Basis Sets Exhibiting Systematic Convergence to the Complete Basis Set Limit   195
Kirk A. Peterson

1.	Introduction	195
2.	Correlation Consistent Basis Sets: A Review	196
3.	Recent Advances in Correlation Consistent Basis Sets	200
4.	Conclusions	203
.	Acknowledgements	203
.	References	203

Section 6: Emerging Technologies (Section Editor: Wendy Cornell)

12. Principles of G-Protein Coupled Receptor Modeling for Drug Discovery   209
Irache Visiers

1.	Introduction	209
2.	Homology Models of Rhodopsin-Like GPCRS	211
3.	Ab Initio Methods	219
4.	Modeling the Activated State	220
5.	Conclusions	222
.	Acknowledgement	222
.	References	222

Index   229
Cumulative Index Vols 1–2   233

CONTRIBUTORS

Charles L. Brooks, III
Department of Molecular Biology, TPC6, The Scripps Research Institute, 10550 North Torrey Pines Road, La Jolla, California 92037, USA

A.E. Cárdenas
Department of Chemistry, University of South Florida, Tampa, Fl 33620, USA

John D. Chodera
Department of Chemistry, Stanford University, Stanford, CA 94305, USA

A.S. Clarke
Department of Chemistry, University of South Florida, Tampa, Fl 33620, USA

Attila G. Császár
Laboratory of Molecular Spectroscopy, Institute of Chemistry, Eötvös University, P.O. Box 32, H-1518 Budapest 112, Hungary

Gábor Czakó
Laboratory of Molecular Spectroscopy, Institute of Chemistry, Eötvös University, P.O. Box 32, H-1518 Budapest 112, Hungary

Phillip E. Fanwick
Department of Chemistry, Purdue University, West Lafayette, IN 47907, USA

Thomas Fox
Computational Chemistry, Department of Lead Discovery, Boehringer Ingelheim Pharma GmbH & Co. KG, 88397 Biberach, Germany

Tibor Furtenbacher
Laboratory of Molecular Spectroscopy, Institute of Chemistry, Eötvös University, P.O. Box 32, H-1518 Budapest 112, Hungary

Mark S. Gordon
Department of Chemistry, Iowa State University and Ames Laboratory, Ames, IA 50011, USA

S.M. Hamm
Department of Chemistry, University of South Florida, Tampa, Fl 33620, USA

Rigoberto Hernandez
Center for Computational Molecular Science and Technology, School of Chemistry and Biochemistry, Georgia Institute of Technology, Atlanta, GA 30332-0400, USA

Jan H. Jensen
Department of Chemistry, Universitetsparken 5, University of Copenhagen, 2100 Copenhagen, Denmark

Jana Khandogin
Department of Molecular Biology, TPC6, The Scripps Research Institute, 10550 North Torrey Pines Road, La Jolla, California 92037, USA

Jan M. Kriegl
Computational Chemistry, Department of Lead Discovery, Boehringer Ingelheim Pharma GmbH & Co. KG, 88397 Biberach, Germany

Hui Li
Department of Chemistry, University of Nebraska, Lincoln, NE 68588, USA

Edit Mátyus
Laboratory of Molecular Spectroscopy, Institute of Chemistry, Eötvös University, P.O. Box 32, H-1518 Budapest 112, Hungary

David L. Mobley
Department of Pharmaceutical Chemistry, University of California, San Francisco, CA 94143, USA

Jeremy M. Moix
Center for Computational Molecular Science and Technology, School of Chemistry and Biochemistry, Georgia Institute of Technology, Atlanta, GA 30332-0400, USA

Kirk A. Peterson
Department of Chemistry, Washington State University, Pullman, WA 99164-4630, USA

A.J. Rader
Department of Physics, Indiana University-Purdue University Indianapolis, Indianapolis, IN 46202, USA

Henry S. Rzepa
Department of Chemistry, Imperial College London, South Kensington Campus, London, SW7 2AZ, UK

Michael R. Shirts
Department of Chemistry, Columbia University, New York, NY 10027, USA

Lyudmilla Slipchenko
Department of Chemistry, Iowa State University and Ames Laboratory, Ames, IA 50011, USA

Irache Visiers
Millennium Pharmaceuticals, Department of Computational Chemistry, 40 Landsdowne Street, Cambridge, MA 02139, USA

PREFACE

Annual Reports in Computational Chemistry (ARCC) focuses on providing timely reviews of topics important to researchers in the field of computational chemistry. The *ARCC* is published and distributed by Elsevier and is sponsored by the Division of Computers in Chemistry (COMP) of the American Chemical Society. Members in good standing of the COMP Division receive a copy of the *ARCC* as part of their membership benefits. We are pleased that both Volumes 1 and 2 have received a very positive response from our readers. The COMP Executive Committee expects to deliver future volumes of the *ARCC* that build on the solid contributors of our first three volumes. To ensure that you receive future installments of this series, please join the Division as described on the COMP website at http://membership.acs.org/c/Comp/index.html.

In Volume 3, our Section Editors have assembled 12 contributions in six sections. Topics covered include Simulation Methodologies (Carlos Simmerling), Biological and Biophysical Applications (Heather Carlson), Chemical Education (Theresa Zielinski), Materials and Polymers (Jeffry Madura), Quantum Chemistry (T. Daniel Crawford), and Emerging Technologies (Wendy Cornell). With Volume 3, we extend the practice of cumulative indexing of both the current and past editions in order to provide easy identification of past reports and topics.

As was the case with our previous volumes, the *Annual Reports in Computational Chemistry* has been assembled entirely by volunteers in order to produce a high-quality scientific publication at the lowest cost possible. The Editors extend our gratitude to the many people who have given their time to make this edition of the *Annual Reports in Computational Chemistry* possible. The authors of each of this year's contributions and the Section Editors have graciously dedicated significant amounts of their time to make this Volume successful. This year's edition could not have been assembled without the help of Deirdre Clark, of Elsevier. Thank you one and all for your hard work, your time, and your contributions.

We hope that you will find this edition to be interesting and valuable. We are actively planning the fourth volume and are soliciting input from our readers about future topics. Please contact either of us with your suggestions and/or to volunteer to be a contributor.

Sincerely,

David Spellmeyer and Ralph Wheeler, Editors

Section 1
Simulation Methodologies

Section Editor: Carlos Simmerling

Center for Structural Biology
Stony Brook University
Stony Brook, NY 11794
USA

CHAPTER 1

Molecular Simulations of pH-Mediated Biological Processes

Jana Khandogin[*] and Charles L. Brooks, III[*]

Contents		
	1. Introduction	3
	2. Static Structure Based pK_a Prediction Methods	4
	3. Molecular Dynamics Coupled with Acid–Base Titration	5
	3.1 Discrete protonation state methods	5
	3.2 Continuous protonation state methods	6
	4. Applications of CPHMD	7
	4.1 De novo prediction of pK_a values in proteins	7
	4.2 Exploring pH-coupled conformational processes	8
	5. Summary and Outlook	10
	Acknowledgement	11
	References	11

1. INTRODUCTION

Molecular simulations are beginning to capture physical realism at increasing resolution, with the exponential growth in computing power and rapid advances in theory, model and algorithm development. These simulation experiments allow us to explore many more questions regarding the molecular mechanisms of cellular machinery and offer valuable complements to wet lab observations. One of these areas is related to the role of solution acidity, or pH, in mediating biological processes occurring in various cellular environments, where pH ranges between 4.5 and 8. For example, electron transfer in cell respiration is coupled to proton translocation across membranes [1]. The resulting proton gradient is utilized in ATP synthesis [2]. The ion-exchange function of the integral membrane, Na^+/H^+ antiporter, relies on a conformational change elicited by pH [3]. Protein misfolding and amyloid fibril formation are often induced or facilitated by a change in

---

[*] Department of Molecular Biology, TPC6, The Scripps Research Institute, 10550 North Torrey Pines Road, La Jolla, California 92037, USA

the environmental pH [4]. Finally, pH plays a major role in the catalytic activity of many enzymes [5].

Solution pH modulates electrostatic interactions in biomolecules by shifting the equilibrium between protonated and deprotonated states of ionizable groups. The observable, pK_a or p$K_{1/2}$, describes the pH condition under which the protonated and deprotonated states of a titratable group exist in equal populations. Although the pK_a value of a titratable side chain in an isolated amino acid, often referred to as the standard or model compound pK_a, is well known, its value can shift substantially in a protein environment. Thus, the determination of pK_a shifts with respect to the standard values is a central problem in studying pH-mediated biological processes [6,7]. Theoretical methods for pK_a predictions have emerged over the past decade. The first class of methods relies on a high-resolution structure and deduces pK_a shifts from calculations of electrostatic energies using an implicit solvent model and two distinct dielectric constants, one for the protein and one for the solvent [8]. In the second class of methods, the free energy of deprotonation is evaluated during molecular dynamics evolution of a biomolecule in an explicit or implicit solvent representation [9]. The latter methods, referred to as pH-coupled molecular dynamics, directly account for the coupling between the ionization process and protein relaxation, thereby offering a more physical and hence, in principle, more accurate solution to the pK_a prediction problem. Most importantly, these methods can be applied to elucidate the aforementioned pH-mediated conformational processes, many of which can not be directly explored by conventional molecular simulations where the protonation states are pre-assigned and kept fixed at specified pH [10]. Examples include simulations under a pH condition where protonation states are unknown, fractionally occupied or perturbed by conformational transitions. Recently, a pH-coupled molecular simulation study demonstrated that the equilibrium properties of a pentapeptide can not be reproduced by linear combinations of properties obtained from simulations with fixed protonation states [11].

This report is mainly concerned with the recent development of a pH-coupled molecular dynamics technique, continuous constant pH molecular dynamics (CPHMD), [12,13] and its applications to *first principles* pK_a calculations [14] and pH-coupled biological phenomena [15]. To put this into context, we will also briefly survey other existing methods for pK_a predictions and pH-coupled conformational dynamics, but we refer interested readers to a recent preliminary review of this area for further details [9]. Finally, we will discuss opportunities for further development of the CPHMD method and our ongoing application studies.

2. STATIC STRUCTURE BASED pK_a PREDICTION METHODS

In the past decade, several groups have advanced the methods based on Poisson–Boltzmann (PB) electrostatics calculations to predict pK_a's for multiple ionizable groups in proteins [16–18]. One major problem in these methods arises from the lack of a rigorous treatment for protein conformational flexibility. Various strategies have been devised to offset this problem. An effective dielectric constant, typically between 4 and 20 [18], is assigned for the solute (protein), to implicitly account for the polarizability of the protein interior [19]. The effects due to local rearrangement of polar side chains are implicitly taken into account by increasing the dielectric constant for residues with high B factors or multiple rotamers [20] in an *ad hoc* fashion, by including multiple side chain conformations [21,22], or by using an average structure from a short-time molecular dynamics trajectory [23]. Another continuum electrostatics calculation based method utilizes a sigmoidally screened Coulomb potential, which incorporates a hydrophobicity scale to account for the dielectric microenvironments around titratable groups [24]. More recently, purely empirical methods have emerged that make use of statistically observed relationships between pK_a shifts and hydrogen bonding, charge–charge interactions, electrostatic potential, and desolvation effects [25,26]. Undoubtedly, the static structure based pK_a prediction methods, using continuum electrostatics or purely empirical calculations, have provided insights into the relationship between protein structure and protonation equilibria. However, the computed pK_a values are highly sensitive to the input structure [20]; and these methods can not be applied to obtain pK_a's where the ionization process is accompanied by a large conformational transition.

3. MOLECULAR DYNAMICS COUPLED WITH ACID–BASE TITRATION

3.1 Discrete protonation state methods

To avoid the conceptual problem associated with the use of a protein dielectric constant and to allow protein structural reorganization during the ionization process of side chains, a number of methods have been developed that explicitly take into account the dielectric property of the protein in a microscopic framework. Early work along these lines include a grand canonical ensemble based method for coupling dynamics and acid–base equilibrium [27], and a method based on a protein dipoles/Langevin dipoles model in all-atom molecular dynamics (MD) simulations [28]. More recently, discrete and continuous protonation state techniques have been developed. In the discrete approaches, also called stochastic titration methods [29], molecular dynamics is periodically interrupted by a Monte Carlo (MC) sampling of protonation states. These approaches mainly differ in the solvent model (explicit or implicit) used for propagation of dynamics and evaluation of protonation state energies [29–32]. Two potential drawbacks arise from the discontinuous energy and force which may lead to instability in the molecular dynamics trajectory; and the additional computational cost for energy evaluations at each MC trial. A recent implementation using a generalized Born (GB) implicit solvent model for both dynamics and protonation state samplings was tested through 1-ns simulations of hen egg lysozyme at various pH [31]. A major problem was the incorrect prediction of protonation states when the pH deviates from the expected pK_a value by more than 2 units [31].

3.2 Continuous protonation state methods

3.2.1 The acidostat method

In the continuous protonation state methods, a set of continuous titration coordinates, λ_i, bound between 0 and 1, that describe the progress toward the protonated or deprotonated state of titrating groups, are evolved simultaneously with the atomic degrees of freedom. The acidostat method [33] propagates λ_i towards the equilibrium values, which represent the extent of deprotonation for titrating sites, via a first-order coupling scheme in analogy to Berendsen's thermostat [34] in explicit solvent simulations. This method has been tested in the pK_a calculation of small amines [33] and the pH-dependent helix stability of decalysine [35]. However, due to severe convergence problems, applications to pK_a calculations and pH-coupled conformational dynamics in proteins have not been explored.

3.2.2 The CPHMD method

An alternative formalism to the acidostat method, continuous constant pH molecular dynamics (CPHMD), has recently been developed [12,13]. In this approach, the titration coordinates, defined as $\lambda_i = sin^2(\theta_i)$, are propagated simultaneously with the conformational degrees of freedom, in GB implicit solvent simulations. An extended Hamiltonian formalism in analogy to the λ dynamics technique that was developed for free energy calculations [36], is used to propagate the titration coordinates. The end-point states of λ ($\lambda \approx 1$ or 0) represent the deprotonated or protonated states, respectively. Thus, unlike the acidostat method, the extent of deprotonation is computed from the relative population of the state with $\lambda \approx 1$. In CPHMD, an extended Hamiltonian is defined as the sum of a hybrid Hamiltonian, kinetic energy of θ, which has fictitious mass similar to a heavy atom, and a biasing potential.

$$H^{extend}(\{r_i\},\{\theta_i\}) = H^{hybrid}(\{r_i\},\{\theta_i\}) + \sum_i \frac{m_i}{2}\dot{\theta}_i^2 + U^*(\{\theta_i\}).$$

In the hybrid Hamiltonian, the charge states and van der Waals parameters involving titrating side chains are linearly attenuated by the values of λ, thereby enabling a coupling between conformational and protonation dynamics. The biasing potential is defined as

$$U^*(\{\theta_i\}) = \sum_i (-U^{barrier}(\theta_i) - U^{model}(\theta_i) + U^{pH}(\theta_i))$$

where $U^{barrier}(\theta_i)$ is a harmonic barrier potential that equally lowers the energies of end-point states, thus minimizing the residence time of states with intermediate λ values (unphysical states). $U^{model}(\theta_i)$ is an analytic function that describes the potential of mean force (PMF) for deprotonation of a model compound along the titration coordinate. $U^{pH}(\theta_i)$ represents the change in the deprotonation free energy of the model compound due to a deviation of pH from the equilibrium value at which both protonation states are equally populated. Thus, dynamics of the titrating coordinates in a protein simulation is driven by the difference between simulation pH and standard pK_a's, and between the nonbonded environment in a protein and that in a model compound.

Simultaneous titration at two competing sites, such as the Nδ and Nϵ in histidine, is a scenario that was not dealt with in the original formulation of the CPHMD method [12], since titration coordinates are independent from each other, as in λ dynamics [36]. This deficiency resulted in qualitative disagreement with experiments in the calculation of pK_a's for histidines [12], the ionization state of which often plays a key role in protein stability [37] and enzyme catalysis [38]. To overcome this problem, a two-dimensional λ dynamics technique was developed, in which a second variable, x, representing a tautomeric interconversion coordinate, is introduced for each residue with two competing titrating sites [13]. This strategy can also be applied to the titration of carboxyl groups, since the exchange of the carboxylate oxygens via rotation around the terminal C−C bond is slow on a MD timescale. Thus, for the titrating histidine and carboxyl groups, charge states and van der Waals interactions are linearly attenuated by both λ and x, allowing the protonation process of competing titrating sites to be modeled on an equal footing. In principle, the two-dimensional λ dynamics technique can be extended to devise a multiple-site titration model to describe the titration of the amino group in lysine. However, the necessity for developing such a sophisticated model is unclear. The ionization equilibrium of lysine is typically not considered in studies under cellular (pH = 4.5–8) or physiological (pH $\approx$ 7.5) conditions because its known pK_a's in proteins are above 10 [39]. Further details of the formalism and methodologies of the CPHMD method are given in references [12,13].

4. APPLICATIONS OF CPHMD

4.1 De novo prediction of pK_a values in proteins

Convergence is a major bottleneck in bringing computed pK_a's into quantitative agreement with experiment using pH-coupled molecular dynamics, which directly links protonation with protein conformational fluctuations as noted in the previous review [9]. De novo prediction of protein pK_a's in explicit solvent is practically unattainable using either discrete [30] or continuous protonation state methods [35], given the current CPU power. However, implicit solvent representations have the potential to be applied in the first-principles folding of peptides or mini-proteins, an area that has begun to be explored by GB implicit solvent models [40–43]. Even so, random errors for the computed pK_a's using GB models have been observed to be as large as 0.5 units in 4-ns single-temperature CPHMD titration simulations of blocked histidine or aspartic acid. Since pK_a is computed from sampling protonation states that are dependent on local conformational variations, convergence in the calculation of pK_a's can be greatly improved by making use of advanced conformational sampling protocols such as replica-exchange (REX) [44]. Indeed, the magnitude of random errors was reduced to below 0.16 units in model compound titration simulations by combining CPHMD with the REX sampling protocol using the same total simulation time of 4 ns [14].

The accuracy of pK_a predictions using pH-coupled molecular dynamics is intimately linked to that of the underlying solvent model as well. Overstabilization of salt bridges in GB models [45] has led to the systematic underestimation of pK_a's for carboxyl residues that interact with positively charged groups [13]. Recently, significant improvement has been achieved in REX-CPHMD titration simulations of 10 benchmark proteins [14] by using an optimized set of GB input atomic radii, which were obtained through calculations of the interaction PMF between polar amino acid groups and attention to conformational equilibria of both helical and β-hairpin peptides [43]. Implicit solvent models employing a dielectric boundary model based on the van der Waals surface, which neglects solvent excluded volumes, underestimate desolvation energies of buried groups as well as buried charge–charge interactions. These effects manifest themselves as pronounced deviations from experiment in the pK_a calculation for buried residues [14].

Effects due to low salt concentrations in the GB model can be accounted for by scaling the solvent dielectric constant by a function, $e^{\kappa r}$, where κ is the inverse of the Debye–Hückel screening length, to approximate the ionic-strength dependence of electrostatic energies in the PB calculation [46]. Although the extent of salt dependence recovered by this crude treatment is still unclear; and may be somewhat larger than that from the finite-difference linearized PB calculations [46], REX-CPHMD titration simulations that incorporate the approximate screening function yield pK_a results in better agreement with experiments that were conducted in an ionic strength of 100–300 mM for 9 out of 10 proteins [14].

Using REX-CPHMD simulations, de novo prediction of protein pK_a's has reached a quantitative level. A recent benchmark study demonstrated that 1-ns REX-CPHMD simulations give results with an RMS error below 1 pK_a unit for a stringent test set of 10 proteins that exhibit anomalously large pK_a shifts in carboxyl and histidine residues [14]. The ionization equilibria of these residues are often involved in the catalytic activity of enzymes [47,48] and proton-driven biological processes [4,3,1]. Although the efficiency of REX-CPHMD simulations can be further enhanced by optimizing the temperature range and the total number of replica's in the REX protocol, and by coupling both temperature and pH in a two-dimensional version of the REX protocol, this [14] and other on-going studies clearly indicate that de novo pK_a calculations by REX-CPHMD will soon become a routine task.

4.2 Exploring pH-coupled conformational processes

Because of the quantitative accuracy in predicting protein ionization equilibria and the advantage of being only marginally slower than standard GB simulations, which are significantly faster compared to explicit solvent simulations, CPHMD has become a powerful tool for gaining atomistic insights into a host of pH-dependent conformational phenomena in biology.

Many globular proteins are maximally folded at a pH that minimizes the total net charge, or electrostatic repulsions among ionizable groups [49]. Although molecular mechanisms governing the pH-dependent folding behavior can be probed by spectroscopic measurements combined with amino acid mutations, much detailed information is inaccessible. For example, a bell-shaped pH profile of helix content for the C-peptide from ribonuclease A has been known for a long time [50]. However, despite intense experimental efforts [50–52], questions regarding the composition of the conformational equilibrium and the specific interactions responsible for the pH dependence remained elusive. A recent de novo folding simulation study of the C-peptide using REX-CPHMD reveals not only a pH-dependent total helix content in agreement with experiment, but also a pH-dependent conformational equilibrium consisting of unfolded and partially folded states of various helical lengths [15]. Most importantly, this study was able to provide a direct correlation between electrostatic interactions and stability of different helical conformations, thus offering insights into the mechanism of pH-modulated helix-coil transitions occurring in many proteins [15].

REX-CPHMD simulations have also been applied to explore the formation of folding intermediate states involving ionizable side chains. Recent solution NMR data has revealed a sparsely populated intermediate of the villin headpiece domain in which the N-terminal subdomain is largely random but the C-terminal subdomain remains in a native-like fold [37]. Interestingly, H41 in this intermediate state titrates at a pH value of 5.6, more than 1.5 units higher than that of the native state [37]. REX-CPHMD simulations [53] initiated from an X-ray crystal and a minimized average solution NMR structure gave rise to two conformational states that have distinct titration behavior for H41, similar to the measurements for the native and intermediate states. Together with the dynamic and structural properties, the simulation data suggest that the state derived from the solution NMR structure resembles a putative intermediate that has a largely unfolded N-terminal subdomain. Moreover, the formation of the putative intermediate was found to be the result of the loss of a hydrogen-bonded network centered at H41. The disruption of the same hydrogen bond network was found in the acid-induced unfolding of the N-terminal subdomain in a REX-CPHMD simulation initiated from the X-ray crystal structure. Thus, this work put forth a proposal that the hydrogen-bonded network and not the protonation state of H41 is a prerequisite for the folding in the villin headpiece domain.

Finally, we conclude the review of recent applications of CPHMD noting that REX-CPHMD simulations can be utilized to explore the pH-dependent molecular mechanism of protein misfolding and aggregation phenomena. Unlike the C-peptide and many globular proteins, the amyloid β peptides from Alzheimer's disease (Aβ) exhibit an inverse bell-shaped pH profile of helix content in TFE solution [54]. A recent work (JK and CLB, manuscript submitted) reveals that both Aβ(1–28), which contains all of the ionizable residues in the full-length amyloid β peptide, and Aβ(10–42) display low helical propensity in a pH-dependent manner similar to that observed in TFE, in sharp contrast to the prediction made by the empirical helix predictors such as AGADIR [55]. This study also confirmed the significant preference for β-turn formation in the region between D23 and N27, similar to the one proposed for Aβ(21–30) based on solution NMR data [56]. A structural loop or bend in a similar region was observed in the fibril model of the full-length Aβ based on solid-state NMR data as well [57]. Finally, solvent exposure of the central hydrophobic cluster (L17–A21) in both Aβ fragments reaches a maximum at pH 6 relative to other pH values, consistent with the pH range where the largest rate of Aβ aggregation was observed [59]. Taken together, the simulation data suggest a pH-dependent aggregation mechanism that rationalizes the pH-dependent fibril formation in Aβ(1–28) [58] as well as the pH-dependent kinetics of the initial aggregation step in Aβ(1–40) and Aβ(1–42) [59].

5. SUMMARY AND OUTLOOK

Conformational dynamics and acid–base equilibria are microscopically coupled. However, traditional molecular dynamics simulations are performed with fixed protonation states. The development of CPHMD has improved the physical realism in molecular simulations. It has enabled, for the first time, quantitative pK_a prediction for biological macromolecules on a first-principles level, thereby eliminating the need for *ad hoc* assignment of a protein dielectric constant and for a high-resolution structure as in the traditional Poisson–Boltzmann based approaches. Despite this success, the accuracy of the CPHMD method can be further improved through the continuing development of the underlying GB implicit solvent model and its parameterization. The underestimation of pK_a shifts for deeply buried residues may be reduced by using a molecular surface that accounts for the solvent excluded volume. A more elaborate scheme relative to the current approximate function [14] may be considered to better capture salt screening effects at the Debye–Hückel level. Another effect that may play an important role in further improving the accuracy for pK_a determinations is polarization. In this regard, the CPHMD technique can be combined with methods of treating polarization such as the fluctuating charge model [60], for which an all-atom force field for protein simulations has been recently developed [61,62]. Since the solvation energy is coupled with titration coordinates through a linear modulation of partial charges, the CPHMD formalism can be combined, in a similar way, with other implicit solvent models such as Poisson–Boltzmann. However, since the Poisson–Boltzmann approach requires numerical solutions, the PMF for model compound titration, $U^{model}(\theta_i)$, is no longer analytic and has to be approximated by a continuous function, through, for example, cubic spline fitting to the grid data. The CPHMD formalism can also be extended to explicit solvent simulations although, with the current CPU and sampling capability, the convergence properties are expected to be poor.

The first series of applications of CPHMD to pH-dependent conformational processes has provided novel insights, inaccessible by traditional molecular simulations and current experiments, into the molecular mechanisms of pH-modulated peptide folding, formation of intermediate, and the aggregation preference of amyloidogenic peptides. Ongoing studies in our lab (JK and CLB, unpublished data) include the study of pH-dependent peptide folding and insertion at the membrane, as well as the exploration of backbone hydrogen bond registry shift in amyloid fibrils. The development and applications of CPHMD simulations have demonstrated that we are entering an era where *in silico* experimentation is becoming an indispensable complement to wet lab experiments in the investigation of many important and fascinating processes of biological and chemical nature.

ACKNOWLEDGEMENT

Financial support from the National Institute of Health (GM57513 and GM48807) is greatly appreciated.

REFERENCES

1. Belevich, I., Verkhovsky, M.I., Wikström, M. Proton-coupled electron transfer drives the proton pump of cytochrome C oxidase, Nature 2006, 440, 829–32.
2. Rastogi, V.K., Girvin, M.E. Structural changes linked to proton translocation by subunit C of the ATP synthase, Nature 1999, 402, 263–8.
3. Hunte, C., Screpanti, E., Venturi, M., Rimon, A., Padan, E., Michel, H. Structure of a Na^+/H^+ antiporter and insights into mechanism of action and regulation by pH, Nature 2005, 435, 1197–202.
4. Kelly, J.W. The environmental dependency of protein folding best explains prion and amyloid diseases, Proc. Natl. Acad. Sci. USA 1998, 95, 930–2.
5. Chen, Y.-Q., Kraut, J., Callender, R. pH-dependent conformational changes in *Escherichia coli* dihydrofolate reductase revealed by Raman difference spectroscopy, Biophys. J. 1997, 72, 936–41.
6. Matthew, J.B., Gurd, F.R., Garcia-Moreno, E.B., Flanagan, M.A., March, K.L., Shire, S.J. pH-dependent processes in proteins, CRC Crit. Rev. Biochem. 1985, 18, 91–197.
7. Harris, T.K., Turner, G.J. Structural basis of perturbed pK_a values of catalytic groups in enzyme active sites, IUBMB Life 2002, 53, 85–98.
8. Bashford, D. Macroscopic electrostatic models for protonation states in proteins, Front. Bioscience 2004, 9, 1082–99.
9. Mongan, J., Case, D.A. Biomolecular simulations at constant pH, Curr. Opin. Struct. Biol. 2005, 15, 157–63.
10. Ripoll, D.R., Vorobjev, Y.N., Liwo, A., Vila, J.A., Scheraga, H.A. Coupling between folding and ionization equilibria: effects of pH on the conformational preferences of polypeptides, J. Mol. Biol. 1996, 264, 770–83.
11. Dlugosz, M., Antosiewicz, J.M. Effects of solute-solvent proton exchange on polypeptide chain dynamics: a constant-pH molecular dynamics study, J. Phys. Chem. B 2005, 109, 13777–84.
12. Lee, M.S., Salsbury, F.R. Jr., Brooks, C.L. III. Constant-pH molecular dynamics using continuous titration coordinates, Proteins 2004, 56, 738–52.
13. Khandogin, J., Brooks, C.L. III. Constant pH molecular dynamics with proton tautomerism, Biophys. J. 2005, 89, 141–57.
14. Khandogin, J., Brooks, C.L. III. Toward the accurate first-principles prediction of ionization equilibria in proteins, Biochemistry 2006, 45, 9363–73.
15. Khandogin, J., Chen, J., Brooks, C.L. III. Exploring atomistic details of pH-dependent peptide folding, Proc. Natl. Acad. Sci. USA 2006, 103, 18546–50.
16. Bashford, D., Karplus, M. pK_as of ionizable groups in proteins: atomic detail from a continuum electrostatic model, Biochemistry 1990, 29, 10219–25.
17. Yang, A.-S., Gunner, M.R., Sampogna, R., Sharp, K., Honig, B. On the calculation of pK_a's in proteins, Proteins 1993, 15, 252–65.
18. Antosiewicz, J., McCammon, J.A., Gilson, M.K. Prediction of pH-dependent properties of proteins, J. Mol. Biol. 1994, 238, 415–36.
19. Simonson, T., Brooks, C.L. III. Charge screening and the dielectric constant of proteins: insights from molecular dynamics, J. Am. Chem. Soc. 1996, 118, 8452–8.
20. Nielsen, J.E., Vriend, G. Optimizing the hydrogen-bond network in Poisson–Boltzmann equation-based pK_a calculations, Proteins 2001, 43, 403–12.
21. You, T.J., Bashford, D. Conformation and hydrogen ion titration of proteins: a continuum electrostatic model with conformational flexibility, Biophys. J. 1995, 69, 1721–33.
22. Georgescu, R.E., Alexov, E.G., Gunner, M.R. Combining conformational flexibility and continuum electrostatics for calculating pK_as in proteins, Biophys. J. 2002, 83, 1731–48.

23. van Vlijmen, H.W., Schaefer, M., Karplus, M. Improving the accuracy of protein pK_a calculations: conformational averaging versus the average structure, Proteins 1998, 33, 145–58.
24. Mehler, E.L., Fuxreiter, M., Simon, I., Garcia-Moreno, B. The Role of hydrophobic microenvironments in modulating pK_a shifts in proteins, Proteins 2002, 48, 283–92.
25. Li, H., Robertson, A.D., Jensen, J.H. Very fast empirical prediction and rationalization of protein pK_a values, Proteins 2005, 61, 704–21.
26. Krieger, E., Nielsen, J.E., Spronk, C.A., Vriend, G. Fast empirical pK_a prediction by Ewald summation, J. Mol. Graph. Model. 2006, 25, 481–6.
27. Mertz, J.E., Pettitt, B.M. Molecular dynamics at a constant pH, Int. J. Supercomput. Appl. High Perform. Comput. 1994, 8, 47–53.
28. Sham, Y.Y., Chu, Z.T., Warshel, A. Consistent calculations of pK_as of ionizable residues in proteins: semi-microscopic and microscopic approaches, J. Phys. Chem. B 1997, 101, 4458–72.
29. Baptista, A.M., Teixeira, V.H., Soares, C.M. Constant-pH molecular dynamics using stochastic titration, J. Chem. Phys. 2002, 117, 4184–200.
30. Bürgi, R., Kollman, P.A., van Gunsteren, W.F. Simulating proteins at constant pH: an approach combining molecular dynamics and Monte Carlo simulation, Proteins 2002, 47, 469–80.
31. Mongan, J., Case, D.A., McCammon, J.A. Constant pH molecular dynamics in generalized Born implicit solvent, J. Comput. Chem. 2004, 25, 2038–48.
32. Dlugosz, M., Antosiewicz, J.M. Constant-pH molecular dynamics simulations: a test case of succinic acid, Chem. Phys. 2004, 302, 161–70.
33. Börjesson, U., Hünenberger, P.H. Explicit-solvent molecular dynamics simulation at constant pH: Methodology and application to small amines, J. Chem. Phys. 2001, 114, 9706–19.
34. Berendsen, H.J.C., Postma, J.P.M., van Gunsteren, W.F., Dinola, A., Haak, J.R. Molecular dynamics with coupling to an external bath, J. Chem. Phys. 1984, 81, 3684–90.
35. Börjesson, U., Hünenberger, P.H. pH-dependent stability of a decalysine α-helix studied by explicit-solvent molecular dynamics simulations at constant pH, J. Phys. Chem. B 2004, 108, 13551–9.
36. Kong, X., Brooks, C.L. III. λ-dynamics: A new approach to free energy calculations, J. Chem. Phys. 1996, 105, 2414–23.
37. Grey, M.J., Tang, Y., Alexov, E., McKnight, C.J., Raleigh, D.P., Palmer, A.G. III. Characterizing a partially folded intermediate of the villin headpiece domain under non-denaturing conditions: contribution of His41 to the pH-dependent stability of the N-terminal subdomain, J. Mol. Biol. 2006, 355, 1078–94.
38. Oda, Y., Yoshida, M., Kanaya, S. Role of histidine 124 in the catalytic function of ribonuclease HI from *Escherichia coli*, J. Biol. Chem. 1993, 268, 88–92.
39. Mehler, E.L., Guarnieri, F. A self-consistent, microenvironment modulated screened coulomb potential approximation to calculate pH-dependent electrostatic effects in proteins, Biophys. J. 1999, 75, 3–22.
40. Zagrovic, B., Sorin, E.J., Pande, V. β-hairpin folding simulations in atomistic detail using an implicit solvent model, J. Mol. Biol. 2001, 313, 151–69.
41. Simmerling, C., Strockbine, B., Roitberg, A.E. All-atom structure prediction and folding simulations of a stable protein, J. Am. Chem. Soc. 2002, 124, 11258–9.
42. Pitera, J.W., Swope, W. Understanding folding and design: replica-exchange simulations of Trp-cage miniproteins, Proc. Natl. Acad. Sci. USA 2003, 100, 7587–92.
43. Chen, J., Im, W., Brooks, C.L. III. Balancing solvation and intramolecular interactions: toward a consistent generalized Born force field, J. Am. Chem. Soc. 2006, 128, 3728–36.
44. Sugita, Y., Okamoto, Y. Replica-exchange molecular dynamics method for protein folding, Chem. Phys. Lett. 1999, 314, 141–51.
45. Im, W., Chen, J., Brooks, C.L. III. Peptide and protein folding and conformational equilibria: theoretical treatment of electrostatics and hydrogen bonding with implicit solvent models, Adv. Protein Chem. 2006, 72, 173–98.
46. Srinivasan, J., Trevathan, M.W., Beroza, P., Case, D.A. Application of a pairwise generalized Born model to proteins and nucleic acids: inclusion of salt effects, Theor. Chem. Acc. 1999, 101, 426–34.
47. Forsyth, W.R., Antosiewicz, J.M., Robertson, A.D. Empirical relationships between protein structure and carboxyl pK_a values in proteins, Proteins 2002, 48, 388–403.
48. Edgcomb, S.P., Murphy, K.P. Variability in the pK_a of histidine side-chains correlates with burial within proteins, Proteins 2002, 49, 1–6.

49. Creighton, T.E. Chemical properties of polypeptides, in Proteins: Structure and Molecular Properties. New York: W.H. Freeman and Company; 1993, p. 1–48.
50. Bierzynski, A., Kim, P.S., Baldwin, R.L. A salt bridge stabilizes the helix formed by isolated C-peptide of RNase A, Proc. Natl. Acad. Sci. USA 1982, 79, 2470–4.
51. Osterhout, J.J. Jr., Baldwin, R.L., York, E.J., Stewart, J.M., Dyson, H.J., Wright, P.E. ^{1}H NMR studies of the solution conformations of an analogue of the C-peptide of ribonuclease A, Biochemistry 1989, 28, 7059–64.
52. Shoemaker, K.R., Fairman, R., Schultz, D.A., Robertson, A.D., York, E.J., Stewart, J.M., Baldwin, R.L. Side-chain interactions in the C-peptide helix: Phe8-His12$^+$, Biopolymers 1990, 29, 1–11.
53. Khandogin, J., Raleigh, D.P., Brooks, C.L. III. Folding intermediate in the villin headpiece domain arises from disruption of a N-terminal hydrogen-bonded network, J. Am. Chem. Soc. 2007, 129, 3056–7.
54. Barrow, C.J., Yasuda, A., Kenny, P.T.M., Zagorski, M.G. Solution conformations and aggregational properties of synthetic amyloid β-peptides of Alzheimer's disease, J. Mol. Biol. 1992, 225, 1075–93.
55. Lacroix, E., Viguera, A.R., Serrano, L. Elucidating the folding problem of α-helices: local motifs, long-range electrostatics, ionic-strength dependence and prediction of NMR parameters, J. Mol. Biol. 1998, 284, 173–91.
56. Lazo, N.D., Grant, M.A., Condron, M.C., Rigby, A.C., Teplow, D.B. On the nucleation of amyloid β-protein monomer folding, Protein Sci. 2005, 14, 1581–96.
57. Petkova, A.T., Ishii, Y., Balbach, J.J., Antzutkin, O.N., Leapman, R.D., Delaglio, F., Tycko, R. A structural model for Alzheimer's β-amyloid fibrils based on experimental constraints from solid state NMR, Proc. Natl. Acad. Sci. USA 2002, 99, 16742–7.
58. Fraser, P.E., Nguyen, J.T., Surewicz, W.K., Kirschner, D.A. pH-dependent structural transitions of Alzheimer's amyloid peptides, Biophys. J. 1991, 60, 1190–201.
59. Kirkitadze, M.D., Condron, M.M., Teplow, D.B. Identification and characterization of key kinetic intermediates in amyloid β-protein fibrillogenesis, J. Mol. Biol. 2001, 312, 1103–19.
60. Rick, S.W., Lynch, D.L., Doll, J.D. A variational Monte Carlo study of argon, neon, and helium clusters, J. Chem. Phys. 1991, 95, 3506–20.
61. Patel, S., Brooks, C.L. III. CHARMM fluctuating charge force field for proteins: I parameterization and application to bulk organic liquid simulations, J. Comput. Chem. 2003, 25, 1–16.
62. Patel, S., Mackerell, A.D. Jr., Brooks, C.L. III. CHARMM fluctuating charge force field for proteins. II. Protein/solvent properties from molecular dynamics simulations using a nonadditive electrostatic model, J. Comput. Chem. 2004, 25, 1504–14.

CHAPTER 2

Extending Atomistic Time Scale Simulations by Optimization of the Action

A.S. Clarke,* S.M. Hamm,* and A.E. Cárdenas*

Contents		
1. Introduction		15
1.1 Boundary value formulation in length		17
2. Applications		22
3. Conclusions		25
Acknowledgements		26
References		27

1. INTRODUCTION

In computational chemistry, molecular dynamics (MD) is the most widely used methodology to study the kinetic and thermodynamic properties of atomic and molecular systems [1–3]. These properties are obtained by solving the microscopic equations of motion for the system under consideration. Due to the short time step that is needed to keep numerical stability, the time scales that can be reached in these simulations are not long, typically on the order of nanosecond to microsecond depending on the complexity of the system. For many systems, like biomolecules, this simulation time is not enough to sample conformational space or to study rare but important events.

Due to this limitation of straightforward MD algorithms, numerous approaches have been developed. Some of them have aimed to increase the time step size [4–15]. Typical improvements in this multiple steps algorithms are rather modest (one order of magnitude) so they do not provide a satisfactory solution to the timescale problem.

A different set of methodologies attempt to compute trajectories connecting conformations from the reactant state to conformations of the product state. Tran-

---
* Department of Chemistry, University of South Florida, Tampa, Fl 33620, USA

sition path sampling, MaxFlux, discrete path sampling, string methods and optimization of actions are examples of methodologies searching for these transition paths.

Transition path sampling (TPS) is a methodology that can be used to study slow activated processes. This technique first developed by the Chandler group [16,17] and further improved by Bolhuis et al. [18–21] is based on a polymer-like representation of the complete trajectory. It uses exact short time trajectories (a few picoseconds) to sample and calculate rates of rare events. These events are typically associated with a transition over a single significant barrier. The calculations of the rate constants are based on the decay of a correlation function. A Monte Carlo sampling of the short trajectories ensures the ergodicity in trajectory space. TPS ignores the waiting periods that the system spends await from the transition process. The methodology has been successfully applied to many processes, such as reactions and conformational changes [19,22–27]. However application of this algorithm to complex systems with rugged energy surfaces entails the identification of basin states separated by several barriers with different heights. For these systems, the assumption of separation of time scales (separating the transition time from the incubation time) is not easy to justify. The definition of the reaction coordinate or physical descriptor that allows the identification of the different transition states present during transitions of complex molecules can be a cumbersome task [28].

MaxFlux is a different algorithm that finds the path that maximizes the diffusive flux between two configurations. This maximization of the reactive flux method has been applied by the group of John Straub to study conformational transitions in peptides and aggregate formation [29,30]. This approach can be used to describe slow processes controlled by diffusion. A difficulty in this description is the necessity to specify the phenomenological friction constant. The value of the friction constant strongly influences calculations of rates and affect the transition pathways. A temperature-dependent nudged-elastic-band (NEB) algorithm based also on the maximization of the flux was recently proposed [31,32]. In this algorithm a discretized path is constructed with the different neighboring structures maintained equally spaced by the use of spring forces. Then, the path is minimized using a modified Verlet algorithm. This methodology has the same limitations as the MaxFlux approach.

A methodology that samples paths in the potential energy surface (instead of the Gibbs free energy as in TPS) is discrete path sampling (DPS) [33–36]. In this method, the fastest paths connecting local minima and transition state conformations are constructed. Phenomenological rate constants can be extracted using kinetic Monte Carlo or graph transformations and transition state theory. The algorithm has been applied to a small pentapeptide [33] and the GB1 hairpin [34]. Reliance on statistical rate theory is one of the drawbacks of this methodology. A satisfactory sampling of stationary points of the potential energy for more complex systems can be difficult as well.

In the string method [37–41] and the associated transition path theory (TPT) [42,43], a collection of reactive trajectories is obtained by performing a sampling of the equilibrium distribution of the system in a collection of hyperplanes parameterized by a string connecting two metastable states. This distribution density defines a transition region in configuration space in which the transition occurs with high probability. This is a more sophisticated method compared to MaxFlux and NEB but due to its inherent complexity has been used only for very simple systems such as alanine dipeptide [39].

Another set of algorithms have been introduced in the past few years in the group of Ron Elber based on the optimization of actions [44,45]. In this methodology an initial guess for the trajectory is constructed and the least action formalism is used to compute a finite-temperature trajectory connecting two boundary states. The first formulation of this methodology is based on a discretized version of the action:

$$S[X(t')] = \int_0^t L \, dt',$$

generating the Onsager–Machlup object function [46–50]:

$$S_{SDET}(\{X_i\}_{i=1}^{N-1}) \equiv \sum_i \Delta t \left(M \frac{X_{i+1} + X_{i-1} - 2X_i}{\Delta t^2} + \frac{dU}{dX_i} \right)^2.$$

In these equations, X is the coordinate vector for the system, M is the mass matrix, U the potential energy, X_0 and X_N are the fixed boundary conformations in the trajectory. This algorithm, called stochastic difference equation in time (SDET) has been used to compute *approximate* long time trajectories connecting the boundary states using a large time step Δt. These paths are obtained by sampling trajectory space using molecular dynamics or Monte Carlo according to a Gaussian distribution of errors. Using similar time formalisms, Passerone and Parrinello [51,52] and Bai and Elber [53] have computed *exact* trajectories for relatively short but rare processes.

In this review a variant of the SDET algorithm is summarized. In this more recent formulation called SDEL (for stochastic difference equation in length) the trajectory is parameterized as a function of its arc length and a unique path is obtained connecting the two boundary conformations [45,54,55]. In the next section, we will describe the algorithm and details of its numerical implementation to obtain conformational changes of peptides and folding mechanisms of protein systems.

1.1 Boundary value formulation in length

As it was mentioned previously the use of straightforward MD simulations to study slow processes is limited by the size of the time step that is required to obtain a stable trajectory. The SDEL algorithm allows the computation of atomically detailed trajectories connecting two known conformations of the molecule over long time scales. In contrast to normal MD, step sizes can be easily increased by factors of thousands (or more) without significative changes in many properties of the trajectory. The trade-off is that trajectories obtained with such a large step size are approximate: molecular motions that occur on a scale shorter than the step size are filtered out from the trajectories, and initial and final configurations must be known. This means that the algorithm can not be used to predict the final conformation of a molecular system such as a protein. This limits the applicability of the algorithm to situations in which the initial and final configurations are known by experiment or modeling. This is not an unbearable limitation. For many situations, we are interested to determine how a system changes during an event connecting known structures of the molecule. For example, the algorithm can be used to describe folding mechanisms: how a protein folds to its native conformation starting from an unfolded structure. The native structure of many proteins has been determined by X-ray diffraction or NMR spectroscopy. The unfolded state is less precisely controlled in folding experiments and therefore a sample of unfolded conformations should be obtained.

The Onsager–Machlup action methodology has a disadvantage: the need to know the total time of the trajectory in advance. Also, low resolution trajectories do not approach a physical limit when the step size increases, in contrast to SDEL as we are going to see below.

Similarly to the Onsager–Machlup action method the SDEL algorithm is based on the classical action. However, in this case the starting point is the action S parameterized according to the length of the trajectory:

$$S = \int_{Y_u}^{Y_f} \sqrt{2(E - U(Y))} dl \tag{1}$$

where Y_u and Y_f (lower and upper limits of integration) are the mass weighted coordinates ($Y = \sqrt{M}X$) of the initial and final conformation of the system, respectively, E is the total energy and U is the potential energy of the system and dl is an infinitesimal mass-weighted length element for the path connecting Y_u and Y_f. Using the least-action principle of classical mechanics one obtains a classical trajectory connecting these two states of the system when a stationary solution for the action is computed (i.e., $\delta S/\delta Y = 0$). These trajectories are calculated differently from usual MD simulations. First, the trajectory is obtained using double boundary conditions: the initial and final coordinates of the system are required as input. In contrast, in the MD algorithm the initial coordinates and velocities are used. Second, the trajectory in Eq. (1) is parameterized as a function of length and not as a function of time. Finally, in the SDEL formulation the total energy of the trajectory is fixed. In contrast, in an MD trajectory the total time is fixed once the step size and the number of steps are constrained in the calculation.

Exact trajectories are more expensive to compute using Eq. (1) than in normal MD because to evaluate and optimize the action the entire trajectory is needed. However, if the goal is to obtain an approximate trajectory with large step size between successive structures optimization of Eq. (1) generates a more stable path than in a straightforward MD algorithm. Numerically, an approximate trajectory is computed from Eq. (1) by using a large step size $\Delta l \gg dl$. Specifically, when we perform the substitution $dl \rightarrow \Delta l$ a discrete version of the action in Eq. (2) is obtained:

$$S \cong \sum_{i=0,\ldots,N} \sqrt{2(E - U(Y_i))} \Delta l_{i,i+1} \tag{2}$$

where the action is now a function of the coordinates of the N intermediate structures in the path, $\{Y_i\}_{i=1}^N$ (with the coordinates of the structures $Y_0 \equiv Y_u$ and $Y_{N+1} \equiv Y_f$ held fixed) and $\Delta l_{i,i+1}$ is the mass-weighted distance separating consecutive structures in the trajectory ($\Delta l_{i,i+1} = |Y_i - Y_{i+1}|$). The trajectory that makes S stationary is determined by optimization. Thus, the optimized trajectory represents a sequence of structures connecting the initial and final state of the system. Explicitly after optimization we obtain [56]:

$$\frac{\Delta^2 Y_i}{\Delta l^2} - \frac{1}{2(E - U(Y_i))}(\nabla U - (\nabla U \cdot \hat{e}_i) \cdot \hat{e}_i) = 0 \qquad (3)$$

with $\dfrac{\Delta^2 Y_i}{\Delta l^2} = \dfrac{2Y_i - Y_{i+1} - Y_{i-1}}{\Delta l^2}$ and $\hat{e}_i \equiv \dfrac{Y_{i+1} - Y_{i-1}}{|Y_{i+1} - Y_{i-1}|}$,

where the length step Δl can be a constant and therefore independent of the index i. The first term of the Eq. (3) (equivalent to the acceleration term in the Newton's equation of motion) depends on the step size. A larger step sizes, this "acceleration" contribution becomes smaller. At the limit when this inertial term can be neglected, we get

$$\frac{\delta S}{\delta Y_i} \approx \frac{1}{2(E - U(Y_i))}(\nabla U(Y_i) - (\nabla U(Y_i) \cdot \hat{e}_i)\hat{e}_i) = 0$$
$$\rightarrow \nabla U(Y_i) - (\nabla U(Y_i) \cdot \hat{e}_i)\hat{e}_i = 0 \quad \forall i. \qquad (4)$$

Equation (4) generates a path in which the force is minimized in all directions but the direction of the path. This is one of the definitions of the Steepest Descent Path (SDP) [57]. This suggests that SDEL provides a physically meaningful result even at low resolution (large step sizes).

To obtain an approximate trajectory using SDEL the following methodology is used (Figure 2.1): First, an initial guess for the path connecting the initial and final coordinates of the system is generated using a minimum energy path algorithm, for example a self-penalty work (SPW) method [58]. Then, the total energy of the system E (an input parameter in the algorithm) can be estimated by averaging the total energy obtained using several MD simulations of the system at the temperature of interest. An alternative method is to identify the highest and lowest value of the potential energy for the SPW trajectory and then compute the average thermal energy at the top of the barrier (i.e., $E = U_{high} - U_{low} + ((3L - 6)/2)k_B T$, where L is the number of atoms in the system). Next, a stationary solution for the action S is obtained by minimizing the square of the action gradient $\Theta = \sum (\partial S / \partial Y_i)^2$ using a simulated annealing protocol (Figure 2.1). The evaluation of the potential energy, forces and Hessian is performed using the Amber/OPLS force field parameters included in the molecular simulation package MOIL [59]. Finally the optimized trajectory is examined and its accuracy is estimated by the step size Δl. This step size should be small enough to provide a smooth representation of the path. The cut-off value Δl_c depends on the particularities of the system. If Δl is larger than this cut-off then the trajectory is not accepted and more intermediate structures are added to the path and re-optimized following the procedure just described.

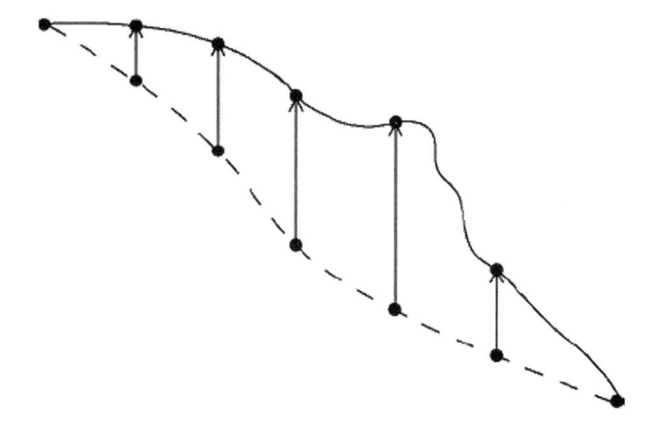

FIGURE 2.1 Simplified view of the optimization of a trajectory. The circles denote configuration snapshots taken along the initial guess for the trajectory (dotted line). The first and last circles are the fixed boundaries of the path. The solid line is the resulting trajectory after optimization of the target function Θ. The arrows indicate the direction of the gradient of the target function for that particular coordinate slice.

The SDEL algorithm has been efficiently parallelized using MPI libraries. In the parallelization scheme each node of a cluster of computers calculates the energy and derivatives for a particular segment of the path [60]. Inter-node communication is not heavy and the computation scales favorable with cluster size.

There are several advantages of SDEL compared to the methods mentioned previously:

(1) *The trajectories can be computed at room temperature or any other temperature of interest.*
(2) *Both the boundary conditions and the length parameterization enable to study very slow processes.* This is demonstrated in the next section.
(3) *The algorithm is easy to run in parallel with no costly communication between processors.* Ordinary PC clusters can be used.
(4) *All the trajectories are reactive.* This enhances the efficient use of computational resources. This is in contrast to initial value MD, in which many trajectories do not end at the desired state.
(5) *The SDEL formulation is very general.* It is not limited to processes with large energy barriers, single barriers or with exponential kinetics.
(6) *The algorithm produces an interpolation between the steepest descent path (SDP) and a true classical trajectory.* Hence, even trajectories with low-resolution can be useful in reaction path studies.

But the algorithm has several disadvantages:

(1) *The trajectories are approximate.* High frequency motions are not probed. These motions can be important in certain dynamical events.
(2) *The calculations are expensive.* With current processor speeds, computations of trajectories for systems with more than 1000 atoms require a parallel resource

of near 50 CPU-s. However, clusters of computers of this size are becoming common in computational chemistry groups.
(3) *The length formulation makes it difficult to estimate the timescale of the process.* Because of this limitation SDEL can provide information about the relative sequence of events but not absolute times.
(4) *Thermodynamic properties are approximate.* The filtering out of high frequency modes affects calculations of thermodynamic properties. However, we have found that the enthalpic thermodynamics properties of slow variables are affected only slightly.
(5) *The final solution depends on the initial guess for the trajectory.* For a large system, no optimization protocol will generate the true global minimum for the target function Θ in an acceptable time. In the applications of SDEL, the initial guess is an approximate minimum energy path obtained with a self-penalty walk algorithm and most of the solutions are local minimums near this initial guess. Implementation of unbiased procedures, sampling trajectories connecting structures in configuration space are currently pursued in our laboratory.
(6) *The current implementation of the algorithm uses implicit solvent model.* This is not a fundamental limitation of SDEL and computations of trajectories with an explicit description of the aqueous environment are possible. To include explicit solvent in the simulations, a time separation between the relaxation of the water molecules to equilibrium and the solvated molecule (for example, a protein) can be assumed. Under this assumption, the water configuration can be determined using thermal distribution for a fixed configuration of the molecule. A short MD simulation can be used to this end. This procedure was used before to include explicit water dynamics in the Onsager–Machlup action [48, 60]. However, inclusion of explicit solvent using this adiabatic approximation makes the computations slower for large systems.

A simpler way to include the effect of water interactions entails the extraction of configurations snapshots from SDEL trajectories and to immerse each one of these structures into a box with explicit water molecules. Then MD simulations can be performed until equilibration is reached [61]. A variant of this procedure is to use constant pressure and temperature MD to extract volume and enthalpy changes during the molecular event. These properties are probed by photothermal techniques. Space and Larsen demonstrated the application of this algorithm for a small β sheet peptide [62].

It is evident that the SDEL algorithm has very appealing advantages when it is applied to long time dynamics events. Meaningful trajectories can be obtained for processes that are difficult or impractical to study by other means. However, the lack of kinetic information makes SDEL an alternative to other algorithms that provide this information, albeit with limited time scale, like TPS and DPS. A non-Markovian hopping mechanism has been proposed recently to overcome this difficulty [63]. However, use of this scheme for large conformational changes of relatively large molecular systems (for example, protein folding) seems impractical at this time.

2. APPLICATIONS

The SDEL algorithm has been used successfully to study the folding dynamics of several peptides and protein systems. In these applications the solvent environment has been treated implicitly using the Generalized Born model [64,65]. The algorithm was first applied to study the folding of the B domain fragment of the *Staphylococcal* protein A [55]. This 60-residue three-helical protein has been studied by many groups using different computational strategies. A recent experimental assessment of the transition state for this folding process suggests the difficulties of atomic simulations in capturing all the features observed in the experiment [66]. The results from SDEL were similar to the results obtained using high-temperature MD simulations showing early formation of the most stable helix. The experiment indicates that the other two helices are more involved in the early formation pointing to the existence of some energetic frustration during the folding of this protein [67].

SDEL was also used to study the coil–helix transition of an alanine-rich peptide [68], the conformational transition of sugar puckering in deoxyadenosine [69], polymerase P [70], and the B–Z DNA transition [71,72]. The coil to helix study [68] demonstrated several properties of SDEL trajectories, like the filtering of high frequency modes and the preservation of thermodynamic properties from slow degrees of freedom when the trajectory resolution is decreased.

An interesting application of SDEL examined the folding mechanism of cytochrome c [73]. The folding kinetics of cytochrome c has been extensively studied experimentally by a variety of techniques [74–77]. Akiyama et al. [78] constructed a two dimensional view of the folding landscape by performing a simultaneous analysis of X-ray scattering and circular dichroism (CD) measurements. A mechanism was proposed in which chain collapse occurs before significant formation of secondary structure. SDEL was used to compute 26 folding trajectories for cytochrome c starting from coil conformations that fold to the native structure.

The analysis of these folding trajectories was in agreement with several experimental observations. For example, in Figure 2.2a we show the number of hydrogen bonds for each of the three helices as a function of the path length (starting from the unfolded conformations). The plot shows that the N and C helices form first and the 60's helix forms last. This result agrees with numerous experiments, such as hydrogen exchange [74,79]. Figure 2.2b shows a two-dimensional probability distribution of the progress of cytochrome c folding as a function of the radius of gyration and the number of helical residues (the experimental results of Akiyama et al. [78] are presented as four X in the figure). The general picture that emerges from these simulations is summarized in the folding pathway shown in Figure 2.2c. Initially the conformation is coiled-like without secondary structure modeling the strongly denatured states used in the experiments (upper left side Figure 2.2c). At the beginning of the folding event, there is a hydrophobic collapse without formation of secondary structure (upper right side Figure 2.2c). Then, the chain collapse continues, but this time with formation of the helices with the terminal helices forming first (lower right side Figure 2.2c). This structure represents a typical molten globule conformation. The structural features of this molten globule

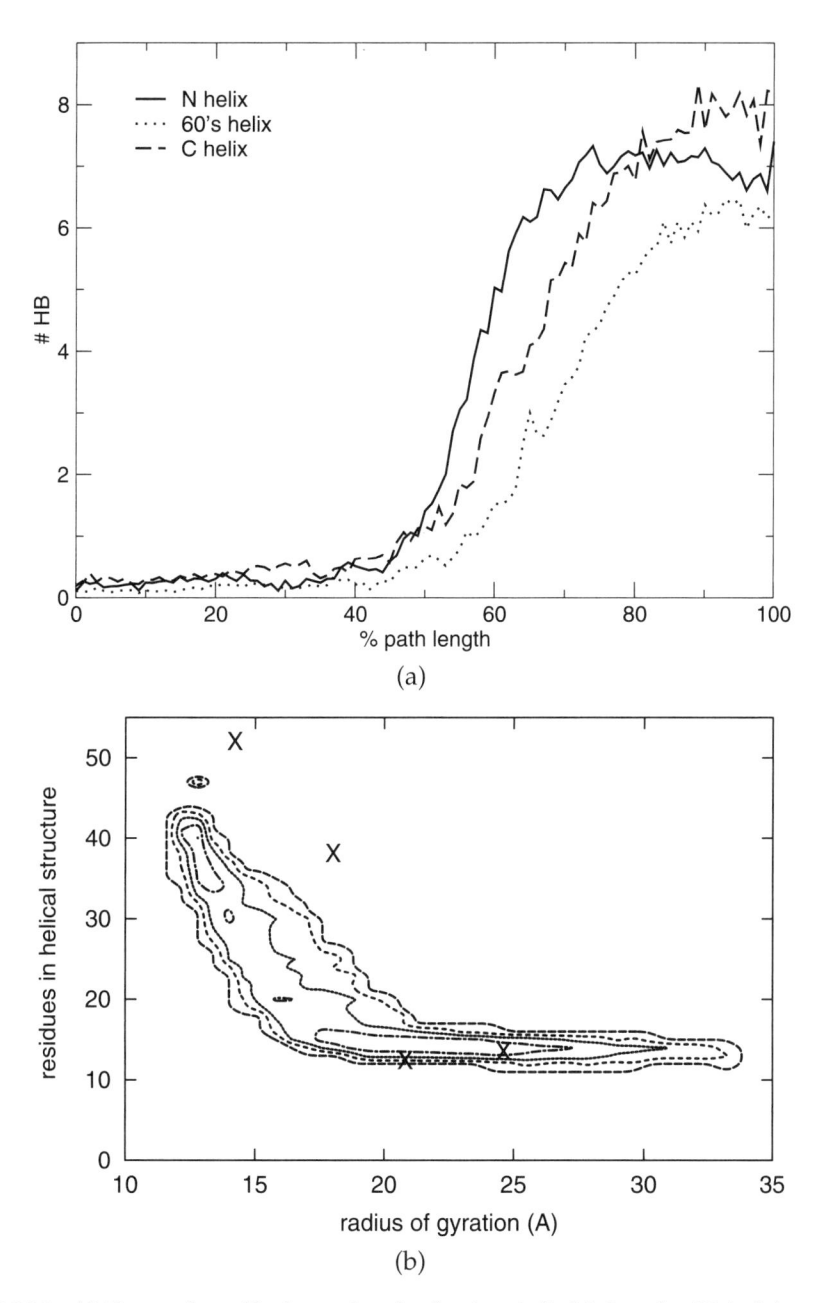

FIGURE 2.2 (a) The number of hydrogen bonds of a given helix (N, C or the 60's helix) as a function of the path length. Average over 26 trajectories is reported. (b) A contour plot summarizing the folding progress along two reactions coordinates: The radius of gyration and the number of residues in helical conformations. (c) Ribbon view of cytochrome C at four different positions along one of the folding trajectory.

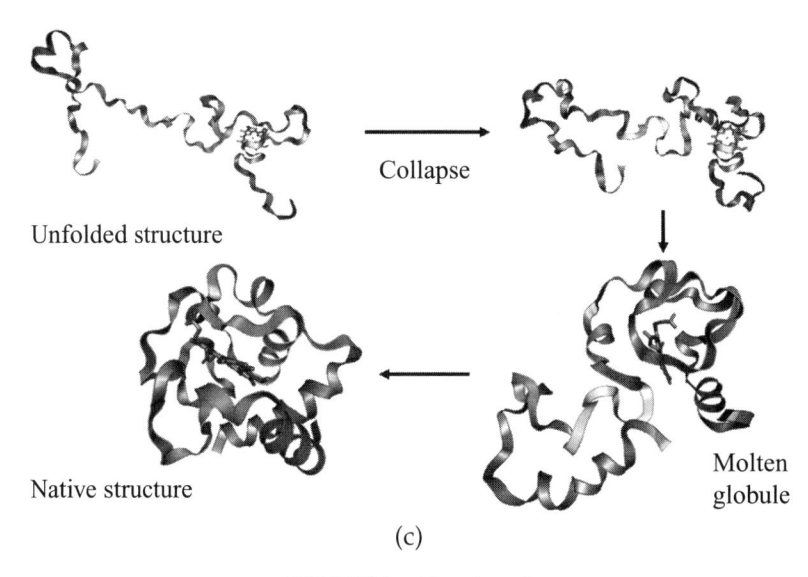

Unfolded structure → Collapse → Molten globule → Native structure

(c)

FIGURE 2.2 (*Continued*)

conformation are in agreement with recent fluorescence energy transfer experiments [80]. Further rearrangements are needed in order for the molten globule state to fold into the final native conformation (lower left side Figure 2.2c).

This methodology was similarly applied to cold shock proteins (CSP) [81]. They are a family of small single domain proteins, with a highly conserved sequence identity [82,83]. The known structures of CSPs consist of a secondary structure of two amphipathic β-sheets. The first β-sheet is formed by three anti-parallel β-strands and the second contains two anti-parallel β-strands. These two β-sheets form a hydrophobic core and a predominantly hydrophilic surface [84]. Many mutational forms of these proteins have been studied experimentally and hence they can be used as a model to study the effects of mutations on folding stability and mechanisms.

Two cold shock proteins where studied, 1C9O (from B. caldolyticus) and 1CSP (from B. subtilis). They differ by 12 amino acids and 1CSP has lower thermostability [85]. Figure 2.3 is a contour plot of the radius of gyrations versus native contacts sampled by room temperature folding trajectories for 1CSP. The collapse of the protein, showing no intermediate structures, agrees with experiment [86]. The conformational distribution in the range between 10 native contacts and 30 native contacts is much broader than the corresponding plot of 1C9O (data not shown). This difference correlates well with the tendency of 1CSP to unfold sooner than 1C9O.

The SDEL algorithm has also been applied to more complicated systems, such as the wild type human Cu, Zn superoxide dismutase (SOD) dimer. SOD is a 153-residue, homodimeric, anti-oxidant enzyme that dismutates superoxide ion to hydrogen peroxide and oxygen [87]. It is an eight-strand, flattened, beta-barrel protein with one copper and one zinc ion per monomer [88]. There are over 100

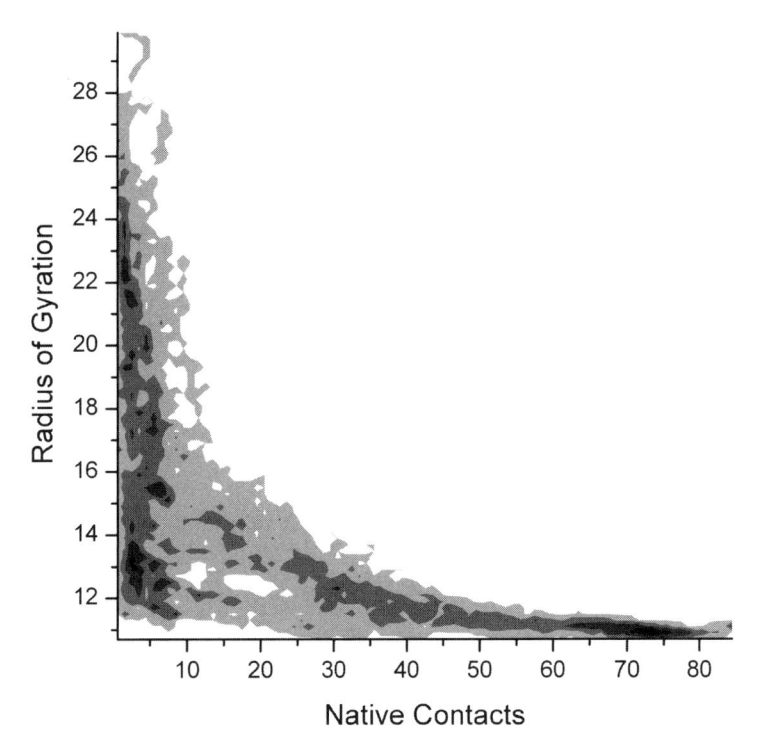

FIGURE 2.3 Radius of gyration as a function of native contacts. This probability density is the result of 26 folding trajectories for 1CSP. It shows a collapse of the protein followed by formation of native contacts.

known point mutations interspersed throughout the protein's sequence and structure that lead to Familial Amyotrophic Lateral Sclerosis (FALS). A mutation study of the roles of these residues is well suited to the mechanistic information the SDEL algorithm can provide.

A 1.8 Å resolution Apo-SOD crystal structure (PDB 1HL4 [89]) was used to generate SDEL trajectories of monomer folding and dimerization (Figure 2.4). The disulfide bridge between cysteine residues 57 and 146 was kept intact for the unfolding and refolding simulations. Initial analysis of a pair of trajectories showed a small population of folded but separated monomers. Interestingly, approximately 15–20% of each monomer's intra-subunit native contacts form when the subunit centers-of-mass are within a few angstroms of equilibrium position, with the rest of the native contacts forming when the monomers are farther away.

3. CONCLUSIONS

These applications demonstrate the potential of the SDEL algorithm as a tool to study conformational dynamics of large molecular systems such as peptides and proteins. This is the only algorithm from the methods discussed in the introduc-

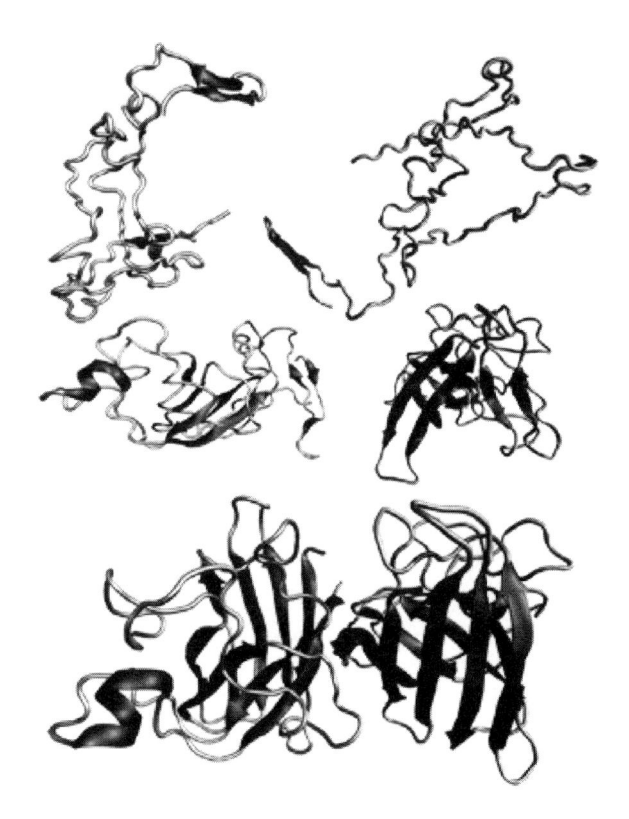

FIGURE 2.4 Snapshots of SOD trajectory. The top structure is the initial unfolded conformation. The second structure is an intermediate structure in the trajectory with partial secondary structure formation. The bottom structure is the folded dimer. The monomer size and separation are not shown to scale to enhance clarity. The center of mass distances in the first two snapshots [90] are 280 Å and 50 Å greater than the equilibrium dimer distance, respectively.

tion that can be used to compute trajectories for such complex processes that take milliseconds or longer, such as the folding of cytochrome c. Although the trajectories are approximate they still provide the structural insight that is needed to explain many coarse-grained experimental observations. If a more detailed and accurate description is required, snapshots taken from these SDEL trajectories can be used to extract thermodynamic information using MD, umbrella sampling or replica exchange methodologies.

ACKNOWLEDGEMENTS

This material is based in part upon work supported by the National Science Foundation under Grant No. 0447294. The authors would like to acknowledge the use of the services provided by Research Computing, University of South Florida.

REFERENCES

1. Hansson, T., Oostenbrink, C., van Gunsteren, W.F. Molecular dynamics simulations, Curr. Opin. Struct. Biol. 2002, 12, 190–6.
2. Karplus, M. Molecular dynamics simulations of biomolecules, Acc. Chem. Res. 2002, 35, 321–3.
3. Karplus, M., Kuriyan, J. Molecular dynamics and protein function, Proc. Natl. Acad. Sci. U.S.A. 2005, 102, 6679–85.
4. Minary, P., Tuckerman, M.E., Martyna, G.J. Long time molecular dynamics for enhanced conformational sampling in biomolecular systems, Phys. Rev. Lett. 2004, 93, 150201.
5. Phillips, J.C., Braun, R., Wang, W., Gumbart, J., Tajkhorshid, E., Villa, E., Chipot, C., Skeel, R.D., Kale, L., Schulten, K. Scalable molecular dynamics with NAMD, J. Comput. Chem. 2005, 26, 1781–802.
6. Franklin, J., Doniach, S. Adaptive time stepping in biomolecular dynamics, J. Chem. Phys. 2005, 123, 124909.
7. Janezic, D., Praprotnik, M., Merzel, F. Molecular dynamics integration and molecular vibrational theory. I. New symplectic integrators, J. Chem. Phys. 2005, 122, 174101.
8. Barash, D., Yang, L.J., Qian, X.L., Schlick, T. Inherent speedup limitations in multiple time step/particle mesh Ewald algorithms, J. Comput. Chem. 2003, 24, 77–88.
9. Skeel, R.D., Tezcan, I., Hardy, D.J. Multiple grid methods for classical molecular dynamics, J. Comput. Chem. 2002, 23, 673–84.
10. Qian, X.L., Schlick, T. Efficient multiple-time-step integrators with distance-based force splitting for particle-mesh-Ewald molecular dynamics simulations, J. Chem. Phys. 2002, 116, 5971–83.
11. Batcho, P.F., Case, D.A., Schlick, T. Optimized particle-mesh Ewald/multiple-time step integration for molecular dynamics simulations, J. Chem. Phys. 2001, 115, 4003–18.
12. Izaguirre, J.A., Catarello, D.P., Wozniak, J.M., Skeel, R.D. Langevin stabilization of molecular dynamics, J. Chem. Phys. 2001, 114, 2090–8.
13. Izaguirre, J.A., Reich, S., Skeel, R.D. Longer time steps for molecular dynamics, J. Chem. Phys. 1999, 110, 9853–64.
14. Tuckerman, M., Berne, B.J., Martyna, G.J. Reversible multiple time scale molecular-dynamics, J. Chem. Phys. 1992, 97, 1990–2001.
15. Ma, Q., Izaguirre, J.A., Skeel, R.D. Verlet-I/r-RESPA/Impulse is limited by nonlinear instabilities, Siam J. Sci. Comput. 2003, 24, 1951–73.
16. Bolhuis, P.G., Chandler, D., Dellago, C., Geissler, P.L. Transition path sampling: Throwing ropes over rough mountain passes, in the dark, Annu. Rev. Phys. Chem. 2002, 53, 291–318.
17. Bolhuis, P.G., Dellago, C., Chandler, D. Sampling ensembles of deterministic transition pathways, Faraday Discussions 1998, 110, 421–36.
18. van Erp, T.S., Moroni, D., Bolhuis, P.G. A novel path sampling method for the calculation of rate constants, J. Chem. Phys. 2003, 118, 7762–74.
19. Bolhuis, P.G. Transition-path sampling of beta-hairpin folding, Proc. Natl. Acad. Sci. U.S.A. 2003, 100, 12129–34.
20. van Erp, T.S., Bolhuis, P.G. Elaborating transition interface sampling methods, J. Comput. Phys. 2005, 205, 157–81.
21. Bolhuis, P.G. Transition path sampling on diffusive barriers, J. Phys.: Condens. Matter 2003, 15, S113–20.
22. Juraszek, J., Bolhuis, P.G. Sampling the multiple folding mechanisms of Trp-cage in explicit solvent, Proc. Natl. Acad. Sci. U.S.A. 2006, 103, 15859–64.
23. Bolhuis, P.G. Kinetic pathways of beta-hairpin (Un)folding in explicit solvent, Biophys. J. 2005, 88, 50–61.
24. Bolhuis, P.G., Chandler, D. Transition path sampling of cavitation between molecular scale solvophobic surfaces, J. Chem. Phys. 2000, 113, 8154–60.
25. Dellago, C., Bolhuis, P.G., Chandler, D. Efficient transition path sampling: Application to Lennard-Jones cluster rearrangements, J. Chem. Phys. 1998, 108, 9236–45.
26. Hagan, M.F., Dinner, A.R., Chandler, D., Chakraborty, A.K. Atomistic understanding of kinetic pathways for single base-pair binding and unbinding in DNA, Proc. Natl. Acad. Sci. U.S.A. 2003, 100, 13922–7.

27. Geissler, P.L., Dellago, C., Chandler, D. Chemical dynamics of the protonated water trimer analyzed by transition path sampling, Phys. Chem. Chem. Phys. 1999, 1, 1317–22.
28. Radhakrishnan, R., Schlick, T. Orchestration of cooperative events in DNA synthesis and repair mechanism unraveled by transition path sampling of DNA polymerase beta's closing, Proc. Natl. Acad. Sci. U.S.A. 2004, 101, 5970–5.
29. Huo, S.H., Straub, J.E. The MaxFlux algorithm for calculating variationally optimized reaction paths for conformational transitions in many body systems at finite temperature, J. Chem. Phys. 1997, 107, 5000–6.
30. Straub, J.E., Guevara, J., Huo, S.H., Lee, J.P. Long time dynamic simulations: Exploring the folding pathways of an Alzheimer's amyloid A beta-peptide, Acc. Chem. Res. 2002, 35, 473–81.
31. Crehuet, R., Field, M.J. A temperature-dependent nudged-elastic-band algorithm, J. Chem. Phys. 2003, 118, 9563–71.
32. Crehuet, R., Thomas, A., Field, M.J. An implementation of the nudged elastic band algorithm and application to the reaction mechanism of HGXPRTase from Plasmodium falciparum, J. Mol. Graph. Model. 2005, 24, 102–10.
33. Evans, D.A., Wales, D.J. The free energy landscape and dynamics of met-enkephalin, J. Chem. Phys. 2003, 119, 9947–55.
34. Evans, D.A., Wales, D.J. Folding of the GB1 hairpin peptide from discrete path sampling, J. Chem. Phys. 2004, 121, 1080–90.
35. Wales, D.J. Energy landscapes: calculating pathways and rates, Int. Rev. Phys. Chem. 2006, 25, 237–82.
36. Wales, D.J., Bogdan, T.V. Potential energy and free energy landscapes, J. Phys. Chem. B 2006, 110, 20765–76.
37. Weinan, E., Ren, W.Q., Vanden-Eijnden, E. Finite temperature string method for the study of rare events, J. Phys. Chem. B 2005, 109, 6688–93.
38. Weinan, E., Ren, W.Q., Vanden-Eijnden, E. Transition pathways in complex systems: Reaction coordinates, isocommittor surfaces, and transition tubes, Chem. Phys. Lett. 2005, 413, 242–7.
39. Ren, W., Vanden-Eijnden, E., Maragakis, P., E, W.N. Transition pathways in complex systems: Application of the finite-temperature string method to the alanine dipeptide, J. Chem. Phys. 2005, 123, 134109.
40. Maragliano, L., Fischer, A., Vanden-Eijnden, E., Ciccotti, G. String method in collective variables: Minimum free energy paths and isocommittor surfaces, J. Chem. Phys. 2006, 125, 024106.
41. Maragliano, L., Vanden-Eijnden, E. A temperature accelerated method for sampling free energy and determining reaction pathways in rare events simulations, Chem. Phys. Lett. 2006, 426, 168–75.
42. Vanden-Eijnden, E., E, W. Towards a theory of transition paths, J. Stat. Phys. 2006, 123, 503–23.
43. Metzner, P., Schutte, C., Vanden-Eijnden, E. Illustration of transition path theory on a collection of simple examples, J. Chem. Phys. 2006, 125, 084110.
44. Elber, R., Ghosh, A., Cardenas, A. Stochastic difference equation as a tool to compute long time dynamics, in Nielaba, P., Mareschal, M., Ciccotti, G., editors. Bridging the Time Scale Gap. Berlin: Springer Verlag; 2002.
45. Elber, R., Cardenas, A., Ghosh, A., Stern, H.A. Bridging the gap between long time trajectories and reaction pathways, Adv. Chem. Phys. 2003, 126, 93–129.
46. Machlup, S., Onsager, L. Fluctuations and irreversible process. 2. Systems with kinetic energy, Phys. Rev. 1953, 91, 1512–5.
47. Onsager, L., Machlup, S. Fluctuations and irreversible processes, Phys. Rev. 1953, 91, 1505–12.
48. Elber, R., Meller, J., Olender, R. Stochastic path approach to compute atomically detailed trajectories: Application to the folding of C peptide, J. Phys. Chem. B 1999, 103, 899–911.
49. Olender, R., Elber, R. Calculation of classical trajectories with a very large time step: Formalism and numerical examples, J. Chem. Phys. 1996, 105, 9299–315.
50. Siva, K., Elber, R. Ion permeation through the gramicidin channel: Atomically detailed modeling by the stochastic difference equation, Proteins: Struct., Funct., Genet. 2003, 50, 63–80.
51. Passerone, D., Parrinello, M. Action-derived molecular dynamics in the study of rare events, Phys. Rev. Lett. 2001, 8710, 108302.
52. Passerone, D., Ceccarelli, M., Parrinello, M. A concerted variational strategy for investigating rare events, J. Chem. Phys. 2003, 118, 2025–32.

53. Bai, D., Elber, R. Calculation of point-to-point short-time and rare trajectories with boundary value formulation, J. Chem. Theory Comput. 2006, 2, 484–94.
54. Elber, R., Ghosh, A., Cardenas, A. Long time dynamics of complex systems, Acc. Chem. Res. 2002, 35, 396–403.
55. Ghosh, A., Elber, R., Scheraga, H.A. An atomically detailed study of the folding pathways of protein A with the stochastic difference equation, Proc. Natl. Acad. Sci. U.S.A. 2002, 99, 10394–8.
56. Landau, L.D., Lifshitz, E.M. Mechanics, Oxford: Butterworth-Heinenann; 1976.
57. Ulitsky, A., Elber, R. A new technique to calculate steepest descent paths in flexible polyatomic systems, J. Chem. Phys. 1990, 92, 1510–1.
58. Czerminski, R., Elber, R. Self-avoiding walk between 2 fixed-points as a tool to calculate reaction paths in large molecular-systems, Int. J. Quantum Chem. 1990, 167–186.
59. Elber, R., Roitberg, A., Simmerling, C., Goldstein, R., Li, H.Y., Verkhivker, G., Keasar, C., Zhang, J., Ulitsky, A. Moil—a program for simulations of macromolecules, Comput. Phys. Commun. 1995, 91, 159–89.
60. Zaloj, V., Elber, R. Parallel computations of molecular dynamics trajectories using the stochastic path approach, Comput. Phys. Commun. 2000, 128, 118–27.
61. Pitici, F., Elber, R. Computer simulations of folding of cytochrome c variants, in Biophysical Society 50th Annual Meeting. Utah: Salt Lake City; 2006.
62. DeVane, R., Ridley, C., Larsen, R.W., Space, B., Moore, P.B., Chan, S.I. A molecular dynamics method for calculating molecular volume changes appropriate for biomolecular simulation, Biophys. J. 2003, 85, 2801–7.
63. Faradjian, A.K., Elber, R. Computing time scales from reaction coordinates by milestoning, J. Chem. Phys. 2004, 120, 10880–9.
64. Tsui, V., Case, D.A. Theory and applications of the generalized Born solvation model in macromolecular simulations, Biopolymers 2000, 56, 275–91.
65. Hawkins, G.D., Cramer, C.J., Truhlar, D.G. Pairwise solute descreening of solute charges from a dielectric medium, Chem. Phys. Lett. 1995, 246, 122–9.
66. Sato, S., Religa, T.L., Daggett, V., Fersht, A.R. Testing protein-folding simulations by experiment: B domain of protein A, Proc. Natl. Acad. Sci. U.S.A. 2004, 101, 6952–6.
67. Wolynes, P.G. Latest folding game results: Protein A barely frustrates, Proc. Natl. Acad. Sci. U.S.A. 2004, 101, 6837–8.
68. Cardenas, A.E., Elber, R. Atomically detailed simulations of helix formation with the stochastic difference equation, Biophys. J. 2003, 85, 2919–39.
69. Arora, K., Schlick, T. Deoxyadenosine sugar puckering pathway simulated by the stochastic difference equation algorithm, Chem. Phys. Lett. 2003, 378, 1–8.
70. Arora, K., Schlick, T. Conformational transition pathway of polymerase ss/DNA upon binding correct incoming substrate, J. Phys. Chem. B 2005, 109, 5358–67.
71. Lim, W., Feng, Y.P. Applying the stochastic difference equation to DNA conformational transitions: A study of B–Z and B–A DNA transitions, Biopolymers 2005, 78, 107–20.
72. Lim, W., Feng, Y.P. The stretched intermediate model of B–Z DNA transition, Biophys. J. 2005, 88, 1593–607.
73. Cardenas, A.E., Elber, R. Kinetics of cytochrome C folding: Atomically detailed simulations, Proteins: Struct., Funct., Genet. 2003, 51, 245–57.
74. Roder, H., Elove, G.A., Englander, S.W. Structural characterization of folding intermediates in cytochrome-c by H-exchange labeling and proton NMR, Nature 1988, 335, 700–4.
75. Elove, G.A., Chaffotte, A.F., Roder, H., Goldberg, M.E. Early steps in cytochrome-c folding probed by time-resolved circular-dichroism and fluorescence spectroscopy, Biochemistry 1992, 31, 6876–83.
76. Pollack, L., Tate, M.W., Darnton, N.C., Knight, J.B., Gruner, S.M., Eaton, W.A., Austin, R.H. Compactness of the denatured state of a fast-folding protein measured by submillisecond small-angle X-ray scattering, Proc. Natl. Acad. Sci. U.S.A. 1999, 96, 10115–7.
77. Hagen, S.J., Eaton, W.A. Two-state expansion and collapse of a polypeptide, J. Mol. Biol. 2000, 297, 781–9.
78. Akiyama, S., Takahashi, S., Kimura, T., Ishimori, K., Morishima, I., Nishikawa, Y., Fujisawa, T. Conformational landscape of cytochrome c folding studied by microsecond-resolved small-angle X-ray scattering, Proc. Natl. Acad. Sci. U.S.A. 2002, 99, 1329–34.

79. Hoang, L., Bedard, S., Krishna, M.M.G., Lin, Y., Englander, S.W. Cytochrome c folding pathway: Kinetic native-state hydrogen exchange, Proc. Natl. Acad. Sci. U.S.A. 2002, 99, 12173–8.
80. Lyubovitsky, J.G., Gray, H.B., Winkler, J.R. Structural features of the cytochrome c molten globule revealed by fluorescence energy transfer kinetics, J. Am. Chem. Soc. 2002, 124, 14840–1.
81. Cardenas, A.E., Clarke, A., Urahata, S. Effects of mutations on the folding pathways of a cold shock protein, Abstracts of Papers of the American Chemical Society 2006, 231.
82. Mueller, U., Perl, D., Schmid, F.X., Heinemann, U. Thermal stability and atomic-resolution crystal structure of the Bacillus caldolyticus cold shock protein, J. Mol. Biol. 2000, 297, 975–88.
83. Schindelin, H., Marahiel, M.A., Heinemann, U. Universal nucleic acid-binding domain revealed by crystal-structure of the Bacillus-subtilis major cold-shock protein, Nature 1993, 364, 164–8.
84. Newkirk, K., Feng, W.Q., Jiang, W.N., Tejero, R., Emerson, S.D., Inouye, M., Montelione, G.T. Solution NMR structure of the major cold shock protein (CSPA) from Escherichia-coli—identification of a binding epitope for DNA, Proc. Natl. Acad. Sci. U.S.A. 1994, 91, 5114–8.
85. Perl, D., Mueller, U., Heinemann, U., Schmid, F.X. Two exposed amino acid residues confer thermostability on a cold shock protein, Nat. Struct. Biol. 2000, 7, 380–3.
86. Perl, D., Welker, C., Schindler, T., Schroder, K., Marahiel, M.A., Jaenicke, R., Schmid, F.X. Conservation of rapid two-state folding in mesophilic, thermophilic and hyperthermophilic cold shock proteins, Nat. Struct. Biol. 1998, 5, 229–35.
87. Valentine, J.S., Hart, P.J. Misfolded CuZnSOD and amyotrophic lateral sclerosis, Proc. Natl. Acad. Sci. U.S.A. 2003, 100, 3617–22.
88. Strange, R.W., Antonyuk, S.V., Hough, M.A., Doucette, P.A., Valentine, J.S., Hasnain, S.S. Variable metallation of human superoxide dismutase: Atomic resolution crystal structures of Cu-Zn, Zn-Zn and as-isolated wild-type enzymes, J. Mol. Biol. 2006, 356, 1152–62.
89. Strange, R.W., Antonyuk, S., Hough, M.A., Doucette, P.A., Rodriguez, J.A., Hart, P.J., Hayward, L.J., Valentine, J.S., Hasnain, S.S. The structure of holo and metal-deficient wild-type human Cu, Zn superoxide dismutase and its relevance to familial amyotrophic lateral sclerosis, J. Mol. Biol. 2003, 328, 877–91.
90. Humphrey, W., Dalke, A., Schulten, K. VMD: visual molecular dynamics, J. Mol. Graphics 1996, 14, 33–8.

CHAPTER 3

Fishing for Functional Motions with Elastic Network Models

A.J. Rader*

Contents	1. Introduction	31
	2. Background	32
	3. Identification of Functional Motions	33
	4. Detailed EN Models	34
	5. Merging EN Models and MD Simulations	36
	6. Summary & Future Prospects	36
	References	37

1. INTRODUCTION

Computational investigations of biomolecular dynamics have benefited from a revival of normal mode analysis (NMA) inspired methods. Simulations of large macromolecules and long time-scale motions that are inaccessible to conventional molecular dynamics (MD) simulation procedures are now routinely performed using NMA of elastic network (EN) models. Consequently, this approach is gaining broad acceptance as evidenced by a dramatic increase in the number of publications utilizing it over the past five years. Unfortunately, the relative low computational cost and conceptually intuitive theory almost invite misuse and over interpretation. One must therefore exercise self-control in interpreting the results from a model that predicts harmonic motion about an assumed structural minimum. Central to accurate implementation of EN models is determining when such models can be used to deduce functional motions and when they cannot. In other words, which predicted modes of motion from EN models are functionally relevant? Since the predictive ability of EN models, especially for large structures has been recently reviewed [1–3], this article will focus on this question of how to connect the EN modes to physically relevant, functional biological motions.

* Department of Physics, Indiana University-Purdue University Indianapolis, Indianapolis, IN 46202, USA
E-mail: ajrader@iupui.edu

2. BACKGROUND

Due to their computational efficiency and theoretical elegance, elastic network (EN) models have increasingly become an ubiquitous computational tool to quickly identify and predict the large-scale motions for known structures of proteins and other biomolecules. Even as accessibility to greater computational power has increased dramatically, analysis utilizing such coarse-grained EN models has gained popularity. A key reason for the proliferation of these models is the observation that unlike standard normal mode analysis (NMA) that requires a computationally intensive energy minimization step, a similar NMA of EN models can be accomplished directly by assuming the EN model represents a structural minimum [4]. Interested readers are directed to a recent book on how EN models function as a coarse-grained NMA and the resulting range of applications [5]. Briefly, the general protocol is to map an experimentally determined molecular structure onto an EN model comprised of beads and springs. These beads are most often associated with the C^α-atom positions while the springs represent through space contacts [6,7]. The computational efficiency of EN models comes from the aforementioned elimination of an energy minimization stage and the approximation of a complicated, all-atom potential energy function by a harmonic function given by:

$$E = \frac{k}{2} \sum_{i,j}^{N} (R_{ij} - R_{ij}^0)^2 f(R_{ij}^0) \qquad (1)$$

where R_{ij}^0 and R_{ij} are the native and instantaneous distances between residues i and j; k is a uniform spring constant; and the $f(R_{ij}^0)$ is the Heaviside function (-1 if $R_{ij}^0 \leq R_c$, an interaction cutoff distance, and zero otherwise). Using this potential, the EN dynamics are then characterized by eigenvalue decomposition into a set of orthogonal modes of motion (eigenvectors) each with a relative weight (eigenfrequency).

The attractiveness of EN models comes from their ability to distill the complex, global cooperative motions of biomolecules into a few descriptive modes while using only a single adjustable parameter. It has been demonstrated that a set of low-frequency modes tend to identify the directions of specific conformational changes in protein structures [8,9]. Thermally averaged properties such as experimental B-factors can also be reliably predicted with an average correlation of 0.65 [10]. This agreement may be significantly higher for an individual protein as can be seen at the *i*GNM database [11]. Several recent reviews have delineated the range of applications and limitations [1,2].

Many successful applications of EN models have shown that the molecular shape plays a large role in determining the modes of motion, especially in the context of very large structures [2,3]. Thus although the standard EN protocol assigns nodes to the coordinate positions of actual amino acid residues, it has been demonstrated that grouping 2, 10, 20 or 40 residues into a single node produces essentially the same low-frequency modes as the traditional, single node per residue

EN model [12]. Embedding atomic-contacts and/or residues into a more coarsely grained EN model has also shown robust motions of very large structures under various monikers such as rotations-and-translations-of-blocks (RTB) [13], block normal mode (BNM) [14] and cluster normal mode analysis (cNMA) [15]. Abstracting the protein to a lattice representation has also been demonstrated to produce the same modes of motion [16]. Coarse-grained EN models constructed from the electron density of cryo-EM maps also produce results similar to EN models constructed from high-resolution X-ray structures indicating it is possible to determine molecular motion without knowing the detailed amino acid sequence or atomic structure [17,18]. Often these methods naively cluster residues along the backbone chain into a single node. More rational coarse-graining procedures have been proposed recently involving a self-consistent, Markov chain hierarchical procedure [19,20] or an independent calculation of rigid clusters [21].

The use of such renormalized or coarser-grained models of biomolecules tend to have the most success on very large structures (over 3000 residues). Recognition that exact atomic coordinates are unnecessary, has allowed the application of EN models to assist in the refinement of low-resolution experimental data from cryo-EM [18] and X-ray [22,23]. This refinement application involves perturbing a known or approximate structure along a set of low-frequency modes until it better matches the experimental data [24].

3. IDENTIFICATION OF FUNCTIONAL MOTIONS

The observable, functional motions of biomolecules are almost always described by a single or small set of low-frequency modes predicted by EN models since these modes of motion represent the pathway(s) of least resistance to motion. Identification of such modes should then allow an initial prediction about macromolecular motions in the absence of additional experimental data. Comparisons of calculated modes for sets of similar structures have been noticed that used to define consensus modes of motion as functional ones [25–27]. Thus the functionally relevant modes from EN models are the ones robust to changes in sequence and found in each such structure [28]. Analysis of five potassium channels with different sequences also found that the underlying low-frequency motions are very similar, presumably to perform the desired function of pore opening and closing [27]. Wider comparisons of EN modes for protein families [29,30] and super-families [31,32] have identified common dynamic features which transcend differences in sequence, pointing to an underlying evolutionary pressure to preserve specific, functional motions.

Remarkably, even non-equilibrium motions such as ligand-binding [33–35] and macromolecular assembly have been linked to these EN modes. Benchmarking against a set of additional, experimental distance constraints such as those determined from NMR or FRET has been shown to allow prediction of conformational changes [36]. Unfortunately how to distinguish the functionally relevant modes among all low-frequency modes without knowing the observable functional motions remains a daunting task. The comparison of modes from different EN models at different levels of resolution has identified the most robust modes, i.e. those mode that are consistently reproduced by multiple EN models, as the functional modes of motion [37]. In fact, the observed correlation between low-frequency modes from NMA and EN models [4,38,39] has been one of the reasons for the increased usage of EN models in general. More importantly, such analysis illustrates a way to get a handle on one of the question of how to determine which low-frequency modes say something meaningful about function.

Ultimately the goal is to use these efficiently calculated approximate motions in order to deduce actual, physically-realistic motions of proteins that currently elude other simulation methods. Many of these interesting motions involve decidedly non-equilibrium processes which necessarily require anharmonic motions. The resulting challenge becomes to how to incorporate the harmonic, normal modes from a EN model into an accurate descriptor of non-equilibrium processes including ligand binding and large-conformational changes. There are several unifying features among the various approaches have been suggested over the past few years in order to address this non-equilibrium concern. These approaches tend to employ one or more of the following features: embedding additional atomic details within a modified EN model, iterative or adaptive EN models, and coupling EN modes to molecular dynamics (MD) simulations.

4. DETAILED EN MODELS

The success of EN models using a single, universal spring constant regardless of residue type has inspired many efforts to introduce residue specific interactions into the EN model. Thus this single adjustable parameter is replaced by a set of residue or environment specific parameters to better mimic the biochemical properties which are critical for proper function and thus achieve an improved agreement with experiments. Often the standard EN model described by Eq. (1) has been altered in order to improve agreement with thermally averaged experimental data, typically B-factors. Introduction of a second bead per residue representing side-chains [40] produced minimal improvement in the B-factor correlation; however, introducing a stronger interaction for C^α residues that are sequence neighbors [41] that reflects a difference between covalent and non-covalent interactions yielded an improved average correlation from an initial value of 0.65 [10] to one of 0.75 [42]. The adoption of an all heavy atom EN model has been cited to improved the correlation with B-factors [43] but simply reflects a regression to the Tirion NMA model [4]. There are two potential dangers of describing the protein with sophisticated EN models containing many different terms. The addition of additional terms and details (i) eradicates the simple elegance of EN models containing a single adjustable parameter and (ii) increases the potential for over-fitting of available data. Often the good-intentioned desire to impart greater residue or chemical specificity can result in over-fitting to noisy experimental data.

It also remains to be seen whether or not improved fitting to an averaged property of the native state such as the B-factor will necessarily produce better agreement with the desired simulation state. This is particularly true when the desired simulation state involves a chemically active, dynamic set of structures. Take for example that the introduction of different springs to reflect chemical interactions (i.e. covalent, hydrogen, ionic, and van der Waals) tended to improve the correlation among individual low-frequency modes and large-scale conformational changes [44] but not B-factors [42]. EN models effectively coarse-grain the multidimensional molecular potential whereas including details to represent the atomic-level fluctuations necessarily *restricts* the range of accessible conformations. Deformations along low-frequency modes from EN models have been used to define pathways between two known conformations by using an iterative displacement and reassessment of EN contacts [9,45,46]. Recently such deformations have been shown to produce a plausible, *unknown* structure of the light-activated metarhodopsin II state that is consistent with experimental data [47]. Similar deformations using an all-atom model would push the structure out of its minimum invalidating the base assumption of the NMA. This subtle point illustrates the simple elegance of the single parameter EN model.

Further complicating the analysis from EN models is the requirement that one must keep in mind that an individual mode is actually expressed in the context of all other modes. Thus while a single mode may often indicate strong correlation between two residues or regions, it is crucial that one examine this correlation in the context of all other modes of commensurate frequency to avoid misinterpretation. In the specific case of an ion channel it has been demonstrated that although a single low-frequency mode corresponds to an important functional motion, the motional correlation between different regions observed in this single mode were not preserved when additional modes were considered [48]. This suggests that whenever possible, a number of alternative EN models or a set of similar structures be compared in order to establish the consensus correlated motions within a common low-frequency subspace of modes.

Applying such checks and balances, alternative EN models tuned for specific molecular properties may yield useful applications. The introduction of stronger springs to represent (pseudo)covalent bonds along the C^α trace improves correspondence with the all-atom density of states distribution [41] which can then be used for estimating allosteric effects [41,49]. A similar domain enhanced EN model that defines contacts within domains stronger than those between domains has been used to renormalize elasticity in order to identify conserved or swapped domains [50] that would be uncharacteristically flexible under standard EN models. Adding specific residue-residue potentials on top of such stronger intradomain contacts has the potential to monitor the effects of mutation on residue motion [51]. An iterative refitting of initial spring constants so as to best fit experimental B-factors using a self-consistent, perturbative adaptation of each spring constant generates a better agreement [52] in a way that reflects the local protein environment for each residue without imposing a large set of external adjustable parameters. Then with these specific set of spring constants in place it is possible to observe the effects of binding at the individual residue level. An EN model expressed in terms of contact-weighted vector orientations rather than atomic fluctuations was shown to yield improved agreement with NMR order parameters as well as information about correlated motions [53]. Building an EN model from internal pseudo-dihedral angles rather than Cartesian coordinates has the advantage of producing modes of motion within the space of pseudo-bond rotations rather than atomic fluctuations [54]. Such a method may alleviate the need to impose stronger springs along the C^α trace in order to better mimic physically realistic motions.

5. MERGING EN MODELS AND MD SIMULATIONS

Iterative interpolation between two known end states using only a subset of low-frequency EN normal modes provides a way to obtain more physically realistic large-scale motions in proteins [45]. Within such a framework, the energy landscape of the protein conformational change can be characterized by the intersection of low-frequency modes from each state [33,46,55]. Similar to targeted MD simulations, these iterative interpolations drive the conformational change along a coarse-grained energy landscape to a given end state. EN models have the ability to rapidly explore a larger conformational space than is accessible under comparable MD simulations. Thus the question of whether it is possible to use these more efficient calculations to accelerate and inspire MD simulations naturally arises.

The amplified collective modes (ACM) method biases the conformational search space in a standard molecular dynamics simulation to favor motions that correspond to the lowest k frequency modes [56,57]. Such biasing is accomplished by coupling these low-frequency modes to a higher temperature and interspersing MD simulation steps with NMA calculations of the EN model. An alternative, recent integration of EN models into MD simulations involves projecting the Cartesian coordinates onto the "rigid body" EN model basis vectors [58]. This allows the protein to evolve in time inspired by the low frequency, large amplitude motions without being bogged down in the high-frequency, local fluctuations.

6. SUMMARY & FUTURE PROSPECTS

Aside from the recent impact in refining experimental data, two applications of EN models that appear to have great potential for impact in computational biology are high-throughput analysis of multiple structures and coupling EN modes to MD simulations. Specifically, the identification of similarities among motions for protein (super)families [31,32] point to the underlying, evolutionarily conserved key features. The combination of EN models and MD simulations hold much promise because the former are able to explore conformational space efficiently and the latter contain atomic specificity. Simulations using amplified collective modes have characterized extremely large-scale, non-equilibrium folding motions [59] suggesting a wide range of possible applications. Other molecular dynamics procedures that rely upon a pathway between states such as milestoning [60] or transition path sampling [61] could benefit from widely-spaced pathway points generated by EN models. In conclusion, although a single EN model that universally explains protein structural dynamics might be preferable, it is unlikely that the proliferation of different, even competing EN models will diminish. The desire to simulate functional motions, residue specific biochemical processes, and mutational effects will continue to drive the introduction of innovative EN models that may differ in the network details.

REFERENCES

1. Bahar, I., Rader, A.J. Coarse-grained normal mode analysis in structural biology, Curr. Opin. Struct. Biol. 2005, 15, 586.
2. Ma, J. Usefulness and limitations of normal mode analysis in modeling dynamics of biomolecular complexes, Struct. (Camb.) 2005, 13, 373.
3. Tama, F., Brooks, C.L. Symmetry, form, and shape: guiding principles for robustness in macromolecular machines, Annu. Rev. Biophys. Biomol. Struct. 2006, 35, 115.
4. Tirion, M.M. Large amplitude elastic motions in proteins from a single-parameter, atomic analysis, Phys. Rev. Lett. 1996, 77, 1905.
5. Cui, Q., Bahar, I. Normal Mode Analysis: Theory and Applications to Biological and Chemical Systems, Boca Raton, FL: Chapman & Hall/CRC; 2006.
6. Haliloglu, T., Bahar, I., Erman, B. Gaussian dynamics of folded proteins, Phys. Rev. Lett. 1997, 79, 3090.
7. Bahar, I., Atilgan, A.R., Erman, B. Direct evaluation of thermal fluctuations in proteins using a single-parameter harmonic potential, Fold. Des. 1997, 2, 173.
8. Tama, F., Sanejouand, Y.H. Conformational change of proteins arising from normal mode calculations, Prot. Engin. 2001, 14, 1.
9. Petrone, P., Pande, V.S. Can conformational change be described by only a few normal modes?, Biophys. J. 2006, 90, 1583.
10. Kundu, S., Melton, J.S., Sorensen, D.C., Phillips, G.N. Jr. Dynamics of proteins in crystals: comparison of experiment with simple models, Biophys. J. 2002, 83, 723.
11. Yang, L.W., et al. iGNM: a database of protein functional motions based on Gaussian network model, Bioinformatics 2005, 21, 2978.
12. Doruker, P., Jernigan, R.L., Bahar, I. Dynamics of large proteins through hierarchical levels of coarse-grained structures, J. Comput. Chem. 2002, 23, 119.
13. Durand, P., Trinquier, G., Sanejouand, Y.H. A new approach for determining low-frequency normal modes in macromolecules, Biopolymers 1994, 34, 759.
14. Li, G., Cui, Q. A coarse-grained normal mode approach for macromolecules: an efficient implementation and application to Ca^{2+}– ATPase, Biophys. J. 2002, 83, 2457.
15. Schuyler, A.D., Chirikjian, G.S. Normal mode analysis of proteins: a comparison of rigid cluster modes with C_α coarse graining, Journal of Molecular Graphics and Modelling 2004, 22, 183.
16. Doruker, P., Jernigan, R.L. Functional motions can be extracted from on-lattice construction of protein structures, Proteins 2003, 53, 174.
17. Ming, D., Kong, Y., Lambert, M.A., Huang, Z., Ma, J. How to describe protein motion without amino acid sequence and atomic coordinates, Proc. Natl. Acad. Sci. 2002, 99, 8620.
18. Tama, F., Wriggers, W., Brooks III, C.L. Exploring global distortions of biological macromolecules and assemblies from low-resolution structural information and elastic network theory, J. Mol. Biol. 2002, 321, 297.
19. Chennubhotla, C., Rader, A.J., Yang, L.W., Bahar, I. Elastic network models for understanding biomolecular machinery: from enzymes to supramolecular assemblies, Phys. Biol. 2005, 2, S173.
20. Chennubhotla, C., Bahar, I. Markov propagation of allosteric effects in biomolecular systems: application to GroEL-GroES, Mol. Syst. Biol. 2006, 2, 36.
21. Gohlke, H., Thorpe, M.F. A natural coarse graining for simulating large biomolecular motion, Biophys. J. 2006, 91, 2115.
22. Delarue, M., Dumas, P. On the use of low-frequency normal modes to enforce collective movements in refining macromolecular structural models, Proc. Natl. Acad. Sci. 2004, 101, 6957.

23. Wu, Y., Ma, J. Refinement of F-actin model against fiber diffraction data by long-range normal modes, Biophys. J. 2004, 86, 116.
24. Tama, F., Miyashita, O., Brooks, I.I.I. Flexible multi-scale fitting of atomic structures into low-resolution electron density maps with elastic network normal mode analysis, J. Mol. Biol. 2004, 337, 985.
25. Zheng, W., Doniach, S. A comparative study of motor-protein motions by using a simple elastic-network model, Proc. Natl. Acad. Sci. 2003, 100, 13253.
26. Zheng, W., Brooks, B.R., Doniach, S., Thirumalai, D. Network of dynamically important residues in the open/closed transition in polymerases is strongly conserved, Structure 2005, 13, 565.
27. Shrivastava, I.H., Bahar, I. Common mechanism of pore opening shared by five different potassium channels, Biophys. J. 2006, 90, 3929.
28. Zheng, W., Brooks, B.R., Thirumalai, D. Low-frequency normal modes that describe allosteric transitions in biological nanomachines are robust to sequence variations, Proc. Natl. Acad. Sci. 2006, 103, 7664.
29. Maguid, S., Fernandez-Alberti, S., Ferrelli, L., Echave, J. Exploring the common dynamics of homologous proteins. Application to the globin family, Biophys. J. 2005, 89, 3.
30. Chen, S.C., Bahar, I. Mining frequent patterns in protein structures: a study of protease families, Bioinformatics 2004, 20, i77.
31. Leo-Macias, A., Lopez-Romero, P., Lupyan, D., Zerbino, D., Ortiz, A.R. An analysis of core deformations in protein superfamilies, Biophys. J. 2005, 88, 1291.
32. Carnevale, V., Raugei, S., Micheletti, C., Carloni, P. Convergent dynamics in the protease enzymatic superfamily, J. Am. Chem. Soc. 2006, 128, 9766.
33. Miyashita, O., Wolynes, P.G., Onuchic, J.N. Simple energy landscape model for the kinetics of functional transitions in proteins, J. Phys. Chem. B 2005, 109, 1959.
34. Xu, C., Tobi, D., Bahar, I. Allosteric changes in protein structure computed by a simple mechanical model: hemoglobin T ↔ R2 transition, J. Mol. Biol. 2003, 333, 153.
35. Tobi, D., Bahar, I. Structural changes involved in protein binding correlate with intrinsic motions of proteins in the unbound state, Proc. Natl. Acad. Sci. 2005, 102, 18908.
36. Zheng, W., Brooks, B.R. Normal-modes-based prediction of protein conformational changes guided by distance constraints, Biophys. J. 2005, 88, 3109.
37. Nicolay, S., Sanejouand, Y.H. Functional modes of proteins are among the most robust, Phys. Rev. Lett. 2006, 96, 078104.
38. Bahar, I., Atilgan, A.R., Demirel, M.C., Erman, B. Vibrational dynamics of folded proteins: significance of slow and fast motions in relation to function and stability, Phys. Rev. Lett. 1998, 80, 2733.
39. Van Wynsberghe, A., Li, G., Cui, Q. Normal-mode analysis suggests protein flexibility modulation throughout RNA polymerase's functional cycle, Biochem. 2004, 43, 13083.
40. Micheletti, C., Carloni, P., Maritan, A. Accurate and efficient description of protein vibrational dynamics: comparing molecular dynamics and Gaussian models, Proteins 2004, 55, 635.
41. Ming, D., Wall, M.E. Allostery in a coarse-grained model of protein dynamics, Phys. Rev. Lett. 2005, 95, 198103.
42. Kondrashov, D.A., Cui, Q., Phillips, G.N. Jr. Optimization and evaluation of a coarse-grained model of protein motion using X-ray crystal data, Biophys. J. 2006, 91, 2760.
43. Sen, T.Z., Feng, Y., Garcia, J.V., Kloczkowski, A., Jernigan, V.L. The extent of cooperativity of protein motions observed with elastic network models is similar for atomic and coarser-grained models, J. Chem. Theory Comput. 2006, 2, 696.
44. Jeong, J.I., Jang, Y., Kim, M.K. A connection rule for alpha-carbon coarse-grained elastic network models using chemical bond information, J. Mol. Graph. Model. 2006, 24, 296.
45. Kim, M.K., Chirikjian, G.S., Jernigan, R.L. Elastic models of conformational transitions in macromolecules, J. Mol. Graph. Model. 2002, 21, 151.
46. Miyashita, O., Onuchic, J.N., Wolynes, P.G. Nonlinear elasticity, proteinquakes, and the energy landscapes of functional transitions in proteins, Proc. Natl. Acad. Sci. 2003, 100, 12570.
47. Isin, B., Rader, A.J., Dhiman, H.K., Klein-Seetharaman, J., Bahar, I. Predisposition of the dark state of rhodopsin to functional changes in structure, Proteins 2006, 65, 970.
48. Van Wynsberghe, A.W., Cui, Q. Interpreting correlated motions using normal mode analysis, Structure 2006, 14, 1647.

49. Ming, D., Wall, M.E. Quantifying allosteric effects in proteins, Proteins 2005, 59, 697.
50. Song, G., Jernigan, R.L. An enhanced elastic network model to represent the motions of domain-swapped proteins, Proteins 2006, 63, 197.
51. Hamacher, K., McCammon, J.A. Computing the amino acid specificity of fluctuations in biomolecular systems, J. Chem. Theory Comput. 2006, 2, 873.
52. Erman, B. The Gaussian network model: precise prediction of residue fluctuations and application to binding problems, Biophys. J. 2006, 91, 3589.
53. Ming, D., Bruschweiler, R. Reorientational contact-weighted elastic network model for the prediction of protein dynamics: comparison with NMR relaxation, Biophys. J. 2006, 90, 3382.
54. Lu, M., Poon, B., Ma, J. A new method for coarse-grained elastic normal-mode analysis, J. Chem. Theory Comput. 2006, 2, 464.
55. Maragakis, P., Karplus, M. Large amplitude conformational change in proteins explored with a plastic network model: adenylate kinase, J. Mol. Biol. 2005, 352, 807.
56. He, J., Zhang, Z., Shi, Y., Liu, H. Efficiently explore the energy landscape of proteins in molecular dynamics simulations by amplifying collective motions, J. Chem. Phys. 2003, 119, 4005.
57. Zhang, Z., Shi, Y., Liu, H. Molecular dynamics simulations of peptides and proteins with amplified collective motions, Biophys. J. 2003, 84, 3583.
58. Essiz, S., Coalson, R.D. A rigid-body Newtonian propagation scheme based on instantaneous decomposition into rotation and translation blocks, J. Chem. Phys. 2006, 124, 144116.
59. Zhang, Z., Boyle, P.C., Lu, B.Y., Chang, J.Y., Wriggers, W. Entropic folding pathway of human epidermal growth factor explored by disulfide scrambling and amplified collective motion simulations, Biochem. 2006, 45, 15269.
60. Faradjian, A.K., Elber, R. Computing time scales from reaction coordinates by milestoning, J. Chem. Phys. 2004, 120, 10880.
61. Dellago, C., Bolhuis, P.G., Csajka, F.S., Chandler, D. Transition path sampling and the calculation of rate constants, J. Chem. Phys. 1998, 108, 1964.

CHAPTER 4

Alchemical Free Energy Calculations: Ready for Prime Time?

Michael R. Shirts[1*], **David L. Mobley**[**] and **John D. Chodera**[***]

Contents	1. Introduction	41
	2. Background	42
	3. Equilibrium Methods	43
	4. Nonequilibrium Methods	45
	5. Intermediate States	46
	6. Sampling	48
	6.1 Sampling using λ	48
	6.2 Short correlation time intermediates	49
	7. Applications	50
	7.1 Small molecule solvation	50
	7.2 Small molecule binding affinities	51
	8. Conclusion	51
	Acknowledgements	53
	References	53

1. INTRODUCTION

Two major goals of computational chemistry are to provide physical insight by modeling details not easily accessible to experiment and to make predictions in order to aid and guide experiment. Both of these goals frequently involve the calculation of free energy differences, since the free energy difference of a chemical process governs the balance of the different chemical species present and the amount of chemical work available.

The ability to rapidly and accurately calculate free energy changes in complex biochemical systems would make possible the computational design of new

---

[*] Department of Chemistry, Columbia University, New York, NY 10027, USA. E-mail: michael.shirts@columbia.edu
[**] Department of Pharmaceutical Chemistry, University of California, San Francisco, CA 94143, USA. E-mail: dmobley@gmail.com
[***] Department of Chemistry, Stanford University, Stanford, CA 94305, USA. E-mail: jchodera@gmail.com
[1] Corresponding author.

chemical entities, which has the potential to revolutionize a number of fields. Pharmaceutical chemistry would benefit through virtual high throughput screening, computational lead optimization, and virtual specificity screens, saving money and time on early product development [1]. Chemical biology would benefit from the creation of molecules to modulate the function of specific proteins in desired ways, or by the design of enzymes to catalyze particular reactions. Reliable methods for efficient free energy computation would be widely useful in many other fields, such as bioremediation and materials design.

2. BACKGROUND

When free energy methods were first applied to problems in drug discovery in the early 1990's, there was a great deal of excitement. This excitement cooled considerably when it became clear that free energies could not reliably be obtained for important applications such as ligand binding to receptors for drug design [2–4]. Inefficient early methods and limited computer power meant that converged free energies in complex heterogeneous systems were simply not achievable. Additionally, many subtle sources of errors in simulation methods were not initially understood, inhibiting comparison between research groups and hindering systematic improvement [5–10]. Many comprehensive reviews exist on the history and use of free energy calculations, both from earlier decades [11–13], and more recently [1,14–19]. Of special note is a recently released book with perhaps the most comprehensive treatment of free energy calculations yet [20].

As free energy methods have improved and computational power has continued to grow exponentially, this promise has begun in small part to be fulfilled. In the last five years, there has been a surge in research in free energy calculations. Although there remain formidable barriers to efficient property prediction and chemical design, recent developments in free energy methods have begun to yield converged, reproducible results for relatively complex phenomena such as ligand binding. In this review, we will survey recent literature on methodological improvements in free energy computation that have made this possible.

This review will specifically focus on current research using the most common rigorously exact method, that of "alchemical" transformation. In an alchemical transformation, a chemical species is transformed into another via a pathway of nonphysical (alchemical) states. Many physical processes, such as ligand binding or transfer of a molecule from gas to solvent, can be equivalently expressed as a composition of such alchemical transformations. Often, these alchemical processes are much more amenable to computational simulation than the physical process itself, especially in complex biochemical systems. A relative ligand binding affinity, for example, may be computed via a thermodynamic cycle by alchemically transforming one ligand to another both bound to a receptor and in solution (Figure 4.1). There are other successful non-alchemical approaches to the computation of free energy differences, such as phase equilibrium Monte Carlo methods for modeling multicomponent fluids [21] and potential of mean force methods [22].

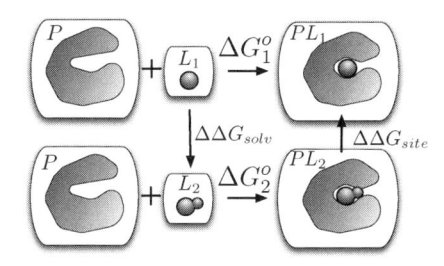

FIGURE 4.1 Thermodynamic cycle for relative binding free energies. Across the top and bottom are the absolute binding free energies (ΔG^o) for two compounds, L_1 and L_2, to a protein of interest. Alchemical free energy calculations can be used to calculate the free energy of turning L_1 into L_2 in water ($\Delta\Delta G_{solv}$) and in the binding site ($-\Delta\Delta G_{site}$). Then the relative binding free energy can be calculated using $\Delta G_1^o - \Delta G_2^o = \Delta\Delta G_{solv} + \Delta G_{site}$.

Alchemical approaches, however, generally allow for the larger range of conformational complexity typical of biochemical systems.

Our focus in this review is on the efficiency and convergence of free energy methods, rather than on accuracy, which is a function of the force field. Methods that lead to converged free energies are a prerequisite to validation or improvement of force fields. If the methods used are biased or incorrect, or the simulations are simply not converged, any conclusions about the underlying models will likely be wrong [5,6]. Within this relatively broad area of alchemical free energy calculations, we focus on two especially active areas of research: identifying the most efficient free energy methods and improving the sampling of free energy methods. We will also review selected applications highlighting the emerging role of alchemical free energy calculations in rational molecular design.

3. EQUILIBRIUM METHODS

In practice, an equilibrium alchemical free energy calculation is separated into multiple stages: (1) selection of alchemical intermediates (see Section 5); (2) generation of uncorrelated samples at each intermediate (see Section 6); and (3) estimation of the free energy difference between the states of interest using one of the analysis methods listed below. Each of these three stages present a number of different choices, and comparing among all possible options becomes exceedingly difficult. For this section, we restrict the comparison to methods that require equilibrium simulations along a single alchemical path.

The classical free energy at inverse temperature β between two states characterized by Hamiltonians H_0 and H_1 can be expressed as a ratio of their partition functions $Z = \int d\Gamma \, e^{-\beta H(\Gamma)}$:

$$\Delta F = -\beta^{-1} \ln \frac{Z_1}{Z_0}. \tag{1}$$

A relationship originally due to Zwanzig [23], which we will call *exponential averaging* (EXP), shows that the free energy difference could be computed directly by

$$\Delta F = -\beta^{-1} \ln \langle e^{-\beta \Delta E_{0\to 1}} \rangle_0 \qquad (2)$$

where $\Delta E_{0\to 1}$ is the energy difference $H_1(\Gamma) - H_0(\Gamma)$ of a point in phase space Γ, and is averaged over the equilibrium ensemble of one of the end states. This method is sometimes called *free energy perturbation* (FEP) in older literature, but this name has become disfavored as other methods for computing free energy changes by "perturbing" the potential have been developed.

However, EXP becomes inefficient when phase space overlap is low, that is, when the configurations sampled with one Hamiltonian are very improbable in the other, and visa versa. Because the statistical uncertainty of EXP grows exponentially as the overlap in phase space between the states decreases, this method requires an impractical amount of uncorrelated data to produce reasonable estimates when overlap is poor [24–26]. Introducing a number of alchemical intermediate states can allow much more efficient computation of free energy differences between Hamiltonians with little phase space overlap:

$$\Delta F = -\beta^{-1} \ln \frac{Z_N}{Z_0} = -\beta^{-1} \ln \frac{Z_1}{Z_0} \cdots \frac{Z_2}{Z_1} \cdots \frac{Z_N}{Z_{N-1}} = \sum_{n=1}^{N-1} \Delta F_{n \to n+1}. \qquad (3)$$

A superior estimator, generally called the *Bennett acceptance ratio* (BAR), was originally proposed by Bennett [27] and later rederived in alternate ways which illustrate important facets of its efficiency [24,28]. It makes use of expectations computed at both the initial and final states:

$$\Delta F = -\beta^{-1} \ln \frac{\langle f(\Delta E_{0\to 1}) \rangle_0}{\langle f(\Delta E_{1\to 0}) \exp[-\beta \Delta E_{0\to 1}] \rangle_1} \qquad (4)$$

where $f(\Delta E_{0\to 1}) = [1 + (n_0/n_1) \exp(\beta(\Delta E_{0\to 1} - \Delta F))]^{-1}$, and n_i is the number of configurations sampled from state i. By using information from both simulations, BAR usually allows many fewer intermediate states to be used for equivalent precision to EXP [24,29].

When multiple alchemical intermediates are used, the *weighted histogram analysis method* (WHAM) [30–32], can be applied to use information from all alchemical intermediates to obtain an optimal estimate of the free energy differences between all pairs:

$$\Delta F_k = -\beta^{-1} \ln \sum_{k=1}^{K} \left\langle \left[\sum_{k'=1}^{K} (n_{k'}/n_k) \exp[\beta(\Delta F_{k'} - (\Delta E_{k \to k'}))] \right]^{-1} \right\rangle_k. \qquad (5)$$

Maximum-likelihood [33] and Bayesian [34] extensions based on WHAM have been proposed, and WHAM has been shown to be equivalent to BAR in the case of two states [29,30].

An alternative approach, called *thermodynamic integration* (TI) [35] can be used by introducing a continuous series of intermediates in terms of a variable λ:

$$\Delta F = \int_0^1 d\lambda \left\langle \frac{\partial H}{\partial \lambda} \right\rangle_\lambda \qquad (6)$$

where the trapezoid rule or other quadrature methods [36,37] are used to approximate the integral. As with any numerical integration, this approach will lead to significant integration errors whenever the integrand $\langle \frac{\partial H}{\partial \lambda} \rangle$ is not well-behaved.

As in any experiment, computed free energies are meaningless without a measure of the expected uncertainty of the measurement. The variance of many estimators has been worked out for the case of uncorrelated samples [29], though block bootstrap methods can sometimes be used for more complex estimators or when it is undesirable to discard correlated data [38].

Because all of these methods involve the time-consuming equilibrium sampling of many uncorrelated configurations at a number of alchemical states, there has been much recent interest in comparing the efficiency and bias of the various estimators of free energy differences [24,29,39,40]. Efficiency can be measured as the inverse of the variance in the free energy given a fixed amount of data, so more efficient methods will require less data (and hence less computer time) to reach the same level of precision [29].

Theoretically, WHAM (and BAR for two states) is the statistically optimal way to compute free energies given potential energy difference between Hamiltonians [27,28]. Direct comparison between TI and other methods is complicated because it involves the derivative of the energy instead of an energy difference, and introduces integration error that is difficult to quantify.

In practice, however, it appears that TI can be as efficient as WHAM under conditions where the integrand is very smooth [29,39], such as charging or small changes in molecular size, but WHAM (or BAR) appears to be significantly more efficient than TI or EXP for free energies of larger molecular changes, sometimes by almost an order of magnitude [24,29,40].

4. NONEQUILIBRIUM METHODS

Free energy differences can also be computed from nonequilibrium simulations switching between two Hamiltonians, using measurements of the work $W_{0 \to 1}$ performed on the system during the switching process [41–46]. The Jarzynski relation [41]

$$\Delta F = -\beta^{-1} \ln \langle e^{-\beta W_{0 \to 1}} \rangle_0 \qquad (7)$$

and its subsequent generalization by Crooks [42]

$$\Delta F = -\beta^{-1} \ln \frac{\langle f(W_{0 \to 1}) \rangle_0}{\langle f(-W_{1 \to 0}) \rangle_1} \qquad (8)$$

hold for arbitrary $f(W)$ if the system is initially prepared in equilibrium and switching between the two Hamiltonians proceeds with the same schedule in all realizations.

In the case of an instantaneous transformation, where $W_{0 \to 1} = H_1 - H_0$, the Jarzynski relation can be seen as a generalization of EXP, and Crooks relation can be used to derive the BAR equation for nonequilibrium work [28,42]. A multi-state generalization of BAR has also been derived for nonequilibrium work measurements [47].

A number of straightforward applications of the Jarzynski relation (often termed "fast growth") soon followed [44,48–50]. It was quickly realized, however, that the poorly sampled tails of the work distribution in Eq. (7) contribute substantially to the free energy, posing difficulties in convergence similar to those that occur with EXP [45,48,51]. Indeed, convergence was found to require a number of simulations that increased exponentially in the typical dissipative work of the reverse process [45]. In practice, comparisons of fast growth methods and more standard equilibrium simulations seem to show that application of nonequilibrium measurements were often less efficient than equilibrium methods [39,52].

Recently, a new generation of algorithms have emerged that exploit path sampling methods [53,54] to speed convergence by enriching sampling of nonequilibrium trajectories that contribute the most to the expectation in Eq. (7). Sun transformed the problem of computing the expectation in Eq. (7) into thermodynamic integration of inverse temperature in path space, achieving reasonable statistical errors in cases where selecting initial configurations from equilibrium would have produced statistically unreliable results [55]. Ytreberg and Zuckerman employed a general formalism for importance-sampling of path functionals [56] and built on Sun's proposal to sample paths with dominant contributions to the expectation in Eq. (7) directly, achieving an increase in efficiency over the "fast growth" method by two orders of magnitude for a model potential [57].

Despite these methodological advances, it is still unclear whether nonequilibrium methods can be more efficient than their equilibrium counterparts. Careful theoretical and computational analysis of statistical error in the path importance-sampling approach revealed that the minimum statistical error was obtained for the fastest switching rates, which suggests that conventional umbrella sampling is more efficient than biased path sampling [58]. Additionally, a comparison of equilibrium and nonequilibrium methodologies for growing and charging a Lennard-Jones sphere in water demonstrated that the importance-sampled nonequilibrium approach can be more efficient than TI in cases where enthalpic barriers are large, suggesting the question of which method is most efficient may be dependent on the details of the system [39].

5. INTERMEDIATE STATES

To compute free energy differences between states with little overlap, it will usually be more efficient to compute the free energy along a pathway of intermediate states. Free energy calculations can be made significantly more efficient by optimizing the choice of intermediate states for increased phase space overlap [36, 59–61].

The simplest path to construct is linear in the two end point Hamiltonians:

$$H(\lambda) = \lambda H_1 + (1-\lambda)H_0. \tag{9}$$

However, there is strong consensus that when annihilating or creating particles, functional dependence of proportional to λ^n (where n is *any* integer), is insufficient, as singularities in the r^{-1} (Coulombic) or r^{-12} (Lennard-Jones) parts of the potential energy leads to results that converge very slowly [59,61,62]. In some cases, shrinking bonds while removing a molecule can eliminate the effect of these singularities, but this can lead to dynamical instabilities, and so is not recommended for molecular dynamics [62].

These problems can be avoided by removing the singularity as a function of λ, gradually rounding out the singularity at $r = 0$ for an r^{-n} potential to zero, called a "soft core" approach [59,63]. Further optimizations of the original soft core formulas have increased the efficiency noticeably [38,64–66]. At minimum, therefore, we suggest first turning off the charged interactions between the alchemically changing groups and their environment, and then altering the Lennard-Jones (or other highly repulsive core interactions) by means of a soft core pathway for highly reliable and relatively efficient calculations.

Although using a Lennard-Jones soft core has been proven to not be the optimal path, it has been shown to be relatively close to the global optimum in the space of λ pathways [61]. Some additional optimization may still be possible: Rodinger et al. proposed a pathway based on adding a fourth distance dimension [67]; as the particle is taken to infinity in this dimension, it is alchemically decoupled from the environment. This is similar to previous alchemical approaches where the extra dimension is treated as a dynamical variable [68,69], but in this case it is treated solely as an alchemical parameter. By transformation of the interval $r^2 \in [0, \infty)$ to $\lambda \in [0, 1]$, this can easily be shown to be a type of soft core, but with a different functional dependence on the alchemical parameter λ than previously proposed methods. The Beutler et al. soft core functional form is not particularly efficient for simultaneous removal of Coulombic and Lennard-Jones potentials; the 4D pathway appears to be more efficient, as do other proposed pathways [70], and comparisons between the methods might still lead to improvements in efficiency.

Despite impressive progress, there are other directions in which optimization of alchemical pathways can still be improved. For example, research in optimizing alchemical pathways has focused on minimizing the variance of TI, yet in most cases, WHAM/BAR is a more efficient method. While the uncertainties of various pathways using TI and WHAM/BAR are strongly related, they are not identical, so further research will be needed to clarify the best pathway for BAR. Additionally, some alchemical intermediates may suffer from long correlation times, and what might be the minimum variance pathway in theory may not be the minimum variance pathway in practice when the statistical efficiency is included.

Clever use of intermediates can provide other ways improve efficiency of free energy calculations. For example, Oostenbrink et al. used a simulation of a single unphysical soft core intermediate state to simultaneously find the free energy to a large number of different molecules [71–74], allowing for the computation of relative free energies between any two of the molecules from a single simulation.

6. SAMPLING

For both equilibrium (Section 3) and nonequilibrium (Section 4) alchemical free energy methods, it is necessary to generate a number of *uncorrelated* configurations at one or more Hamiltonians. In many cases, a large number of such configurations must be generated to obtain sufficiently precise results. This is the time-consuming step in most free energy calculations, as slow or infrequent conformational changes, like protein or ligand conformational rearrangement in binding free energy calculations, make it very costly to generate uncorrelated configurations.

While a number of methods to enhance conformational sampling have recently been explored, the most notable of which are the class of *generalized ensemble methods* (see [75] for a review), we focus on specifically on methods that take advantage of the choice of, or the coupling between, multiple alchemical intermediates.

6.1 Sampling using λ

Frequently, correlation times are much shorter for some alchemical intermediates than others, because there may be uncharged or uncoupled states that experience reduced energetic barriers. An obvious way to reduce correlation times throughout *all* intermediates is by permitting exchanges between them, so that individual trajectories can visit intermediate states with short correlation times, rapidly decorrelate, and return. One simple approach is to allow Metropolis Monte Carlo moves between alchemical intermediates [76], which unfortunately results in the system spending most of the time in the few alchemical states of lowest free energy. To ensure near-equal visitation of intermediates (and hence more rapid decorrelation), a biasing weight can be applied to each state, in analogy to simulated tempering [77]. Though these weights must be estimated iteratively, a number of algorithms have emerged to do this automatically and efficiently [78–80]. After the weights have been fixed, free energies can be estimated using WHAM [79] or even rapidly estimated from the history of Monte Carlo moves between weighted intermediates [78].

An alchemical analog of the popular replica-exchange algorithm [81,82] can be employed to avoid the need for biasing weights altogether. Exchanges can be made between only the λ values (called *Hamiltonian exchange*) [83–85], or among other parameters, such as temperature, as well [86]. A biasing potential that keeps replicas spread out can also be employed, though TI must be used to extract the free energy difference [87].

Alternatively, it is possible to treat λ as a dynamical variable, with fictitious mass [88,89]. Because this is unlikely to sample the entire allowed range of λ when the free energy varies more than $k_B T$, a biasing potential can be employed to ensure more uniform sampling, and WHAM can be used to extract the free energy difference at the endpoints [89].

Abrams et al. introduced a novel method for efficiently sampling in λ by heating the λ degrees of freedom to increase motion, but adiabatically decoupling them from the rest of the system by greatly increasing the mass to avoid perturbing the dynamics of the other degrees of freedom. In one example, this appeared to be twice as efficient as a similar calculation using BAR [64].

Of course, these methods are not a cure-all; in addition to standard cautions regarding the efficiency of replica-exchange and generalized ensemble simulations [90,91], it should be noted that degrees of freedom whose interactions are *not* attenuated in any intermediate will not experience enhanced sampling. Slow receptor conformational changes or ion motions [92] in a ligand binding free energy calculation may cause sampling problems. To improve mixing and reduce correlation times for each state, generalized ensemble methods can also be employed for each intermediate state separately to reduce correlation times [85,86].

6.2 Short correlation time intermediates

Above, we discussed how free energy calculations can be made more efficient by choosing intermediate states or pathways with high overlap between neighboring states. Judicious choices of intermediates states can help isolate and resolve sampling problems by reducing the correlation times at these intermediates.

For example, in absolute binding free energy calculations, at intermediate states where interactions between the ligand and receptor are weak, the ligand must explore a much larger volume to properly sample its configuration space. This can lead to long correlation times, as well as the an improper standard state for binding free energies [93–95]. To resolve this problem, several groups have used an alternate thermodynamic cycle, first restraining the ligand position [96] or additionally the orientation in the binding site prior to turning off interactions between the ligand and the receptor, and then releasing the restraints [94, 97–101], leading to increased efficiencies. This gain comes because the restraints effectively reduce the amount of configuration space that must be sampled at the non-interacting state and at intermediate alchemical states. This reduces correlation times and enhances sampling.

The use of restraints may provide a fairly general strategy for isolating sampling problems and reducing correlation times. Restraints have also been applied to ligand internal degrees of freedom using RMSD restraints [98,99], and could easily be applied to protein degrees of freedom as well, with some potential benefits [100]. Simulations of alchemical intermediates performed with restraints can be fairly short, as many degrees of freedom are no longer accessible, and sampling effort can be focused on obtaining converged values for free energies associated with turning on and off the restraints. Given that alchemical free energy calculations often require many intermediate states, the increase in efficiency from restraints may be large. This approach of isolating sampling problems using restraints may prove especially important, given that conformational degrees of freedom can present very severe sampling problems [5].

Recent work has also pointed out significant sampling problems associated with ligand orientational degrees of freedom in the context of ligand binding [97, 102] and suggested ways to combine contributions from multiple ligand orientations (generated by docking methods, for example) rather than the more traditional approach of simply running simulations long enough to sample all relevant orientations. Since time scales for reorientation of even very small ligands can be long [97] this appears to result in significant efficiency gains as well [97], and will likely be especially important whenever ligand bound orientations are not already known, or where there are multiple relevant orientations.

7. APPLICATIONS

In this review, we have attempted to demonstrate that the potential for precise and efficient alchemical free energy calculations has greatly increased in recent years. However, to what extent is it possible to say that free energy calculations have begun to fulfill this potential? What is the evidence in the last several years that free energy methods actually are "ready for prime time?"

7.1 Small molecule solvation

One area where free energy calculations are showing their utility is in the computation of solvation and transfer free energies of small molecules for the purpose of understanding and improving force field models. Free energies are ideal quantities for comparing to experiment, as they can be measured much more directly than most structural parameters observed by spectroscopic means. Computing the solvation energy of small solutes has been possible for twenty years [103], but early studies were restricted to relative free energies, and were costly enough to perform at high precision that quantitative force-field improvement via evaluation of many solvation free energies was out of the question.

In the last several years, improved alchemical methodologies and increased computer power have made it possible to compute the free energies of large sets of small molecules with precision of less than 0.1 kcal/mol, sufficient to verify the validity of force-field parameters for small molecules [38,64,104–110]. Similar gains in efficiency were obtained for small molecule solvation through variations on Monte Carlo sampling of the Gibbs ensemble [111–114].

Advances in free energy methods have lead to a number of recent discoveries about biomolecular force fields from small molecule solvation simulations that were not previously possible. Oostenbrink et al. demonstrated that it was impossible to simultaneously fit bulk liquid properties, free energies of hydration, and free energies of solvation in cyclohexane to experiment with the same fixed charge models [109]. Mobley et al. demonstrated that using higher levels of quantum mechanical calculations for calculating atomic partial charges did not improve hydration free energies, suggesting at least part of the problem lies in other parts of the force field [38]. Hess et al. obtained both enthalpy and entropy estimates for free energies of solvation and found that contemporary force fields may produce relatively accurate free energies and other properties even though the enthalpy and entropies may have systematic errors [108]. The ability to explore these issues presents many more opportunities for the rational design of molecular force fields than existed previously.

7.2 Small molecule binding affinities

A "holy grail" of computational chemistry has been the estimation of ligand binding affinities or free energies to receptors. To this end, alchemical free energy calculations have often been applied to calculate relative binding free energies of structurally similar ligands. This has been an area of active research for nearly 20 years, and the literature, even from the last several years, is too abundant for an exhaustive review. But recent work suggests that binding free energies may be another area where free energy calculations are ready for broader use.

In particular, several recent studies have highlighted the advantages of alchemical relative free energy calculations compared to approximate methods like MM-GBSA and docking [115–118]. Alchemical methods have also been successfully applied in a lead-optimization context (Ref. [17] and references therein). It appears that in this area, free energy calculations are already useful, but with large remaining hurdles to their more widespread use such as the computational cost and the difficulty of setting them up [117,119–121]. Several interesting applications have been to the estrogen receptor [71,74], neutrophil elastase [118], enantioselectivity in lipase B [122], and the src kinase SH2 domain [119].

In principle, if absolute binding free energies can be calculated, relative binding free energies follow as a by-product, without any of the difficulties of setting up complex dual state topologies [119,120] or identifying suitable intermediates [71, 72,123]. Absolute binding free energy calculations involve a number of complications not encountered in relative binding free energies, not the least of which is a much smaller conformational overlap between end states [93,94,124]. They have, however, been applied with some success to several systems just in the past several years [96,102,125]. RMS errors relative to experiment remain high in many of these studies, and RMS errors between different computational experiments are sometimes over 1 kcal/mol for larger ligands, even when the same force field parameters are used, so more work remains to be done. Nevertheless, they show great promise, outperforming docking methods and even successfully making quantitative binding affinity predictions [101]. These also provide a more stringent test of free energy methods than do relative free energy calculations, and have highlighted several potential challenges for binding free energy calculations generally [97,100,102]. An alternative approach to binding free energy calculations is via potentials of mean force [126,127], but these are outside the scope of this review.

Overall, the outlook in the context of protein–ligand binding is extremely promising, and these calculations are beginning to reach the point where computational costs can be affordable even for local computer clusters and absolute binding affinities.

8. CONCLUSION

There is still much work to do in determining optimal free energy methods, and even more in making them easy for the average practitioners to apply. With the proliferating combinations of methods, more systematic comparisons are mandatory. Most free energy methods work efficiently on low-dimensional toy problems that have been common for testing these methods, such as coupled harmonic oscillators or one-dimensional potentials, or small molecular changes like enlarging particle radii. But since most practical free energy calculations involve annihilating at least one particle, we suggest that all comparisons of new methods for free energy differences in biomolecular systems at least include the measurement of the free energy annihilation of water or a small solute in water, or the napthalene-to-napthalene null transformation [62], for which the answer is by definition zero.

As free energy methods see more widespread use, assessment of convergence becomes even more important. While much progress has been made in this area in recent years, a great deal remains to be done before automatic diagnostics are able to replace careful analysis. Reverse cumulative averaging [128] can automatically detect when the simulation has reached equilibrium, provided the simulation has sampled all relevant regions of configuration space. Measures of phase space overlap [129,130] can help determine when the choice of alchemical intermediates is inadequate. It is possible to efficiently estimate correlation times [131–135] to verify if each alchemical intermediate contain a sufficient number of uncorrelated samples for convergence, but these correlation time estimates will be unreliable if some conformational states have not yet been adequately sampled. A simple way to diagnose this problem is to initiate separate free energy calculations from very different starting structures to ensure that the same result is recovered to within statistical uncertainty; this idea was suggested by Gelman and Rubin many years ago in the Markov chain Monte Carlo literature [136], but is not commonly applied. Free energy estimates without associated uncertainty analysis and convergence tests are very often meaningless.

In addition to applying methods to new systems, it is useful to have standard, nontrivial test systems to compare new methods. Two such systems that have been used as benchmarks by multiple groups for ligand binding are the engineered cavity mutants of T4 lysozyme [94,97,99,101,124,137] and FKBP-12 binding protein [96,98,102,125,138,139]. The T4 lysozyme system has the advantage of binding relatively small molecules, removing some of the sampling issues, and being relatively small. It is also very well characterized experimentally [140–144]. FKBP ligands are much larger, with many rotatable bonds and complicated heterocycles, but the protein itself is even smaller (only 107 residues) and compact, has very little conformational change, and high-quality experimental studies with a variety of ligands over a very large range of binding affinities [145–147]. Although these model systems have little direct practical biological relevance, they provide a valuable resource for testing and comparing new methods. Until we can demonstrate that current methods work adequately on these simpler systems (for example, that calculations converge when started from independent structures) it seems unprofitable to focus on much larger systems where any sampling challenges are likely to be worse.

In an effort to make alchemical free energy calculations more accessible to non-experts, and help scientists avoid some of the numerous pitfalls possible in these calculations, we are developing a community-based web site, Alchemistry (http://www.alchemistry.org). This site is intended as a repository for background information, tutorials, input files and results for standard test cases, and best practices for alchemical free energy calculations, as well as tools for simulations and links to recent literature. We hope this will develop as a collaborative tool that will make alchemical methods more accessible to the community at large.

Free energy methods are rapidly becoming standard tools for computational chemists, as computational power grows and methods increase in efficiency. As methods continue to be standardized, and as understanding of error bounds and limitations continues to grow, free energy calculations will contribute more and more substantially to rational design in biological and other molecular systems.

ACKNOWLEDGEMENTS

The authors would like to thank Chris Oostenbrink, Jed Pitera, and M. Scott Shell for many useful comments on earlier drafts of this review.

REFERENCES

1. Jorgensen, W.L. The many roles of computation in drug discovery, Science 2004, 303(5665), 1813–8.
2. Chipot, C., Pearlman, D.A. Free energy calculations: the long and winding gilded road, Mol. Simulation 2002, 28(1–2), 1–12.
3. Chipot, C., Shell, M.S., Pohorille, A. Introduction, in Chipot, C., Pohorille, A., editors. Free Energy Calculations: Theory and Applications in Chemistry and Biology. Springer Series in Chemical Physics, vol. 86. Berlin and Heidelberg: Springer; 2007, p. 1–32.
4. Mark, A.E. Calculating free energy differences using perturbation theory, in Schleyer, P.V.R., Allinger, N.L., Clark, T., Gasteiger, J., Kollman, P.A., Schafer, H.F. III, Schreiner, P.R., editors. Encyclopedia of Computational Chemistry, vol. 2. Chichester: Wiley and Sons; 1998, p. 1070–83.
5. Leitgeb, M., Schröder, C., Boresch, S. Alchemical free energy calculations and multiple conformational substates, J. Chem. Phys. 2005, 122, 084109.
6. Lu, N., Adhikari, J., Kofke, D.A. Variational formula for the free energy based on incomplete sampling in a molecular simulation, Phys. Rev. E 2003, 68, 026122.
7. Berendsen, H.J.C. Incomplete equilibration: A source of error in free energy computations, in Renugopalakrishnan, V., Carey, P.R., Smith, I.C.P., Huang, S.G., Storer, A.C., editors. Proteins: Structure, Dynamics, and Design. Leiden: ESCOM; 1991, p. 384–92.
8. Steinbach, P.J., Brooks, B.R. New spherical-cutoff methods for long-range forces in macromolecular simulation, J. Comp. Chem. 1994, 15(7), 667–83.
9. Lisal, M., Kolafa, J., Nezbeda, I. An examination of the five-site potential (TIP5P) for water, J. Chem. Phys. 2002, 117(19), 8892–7.
10. Mark, P., Nilsson, L. Structure and dynamics of liquid water with different long-range interaction truncation and temperature control methods in molecular dynamics simulations, J. Comp. Chem. 2002, 23(13), 1211–9.
11. Kollman, P.A. Free energy calculations: Applications to chemical and biochemical phenomena, Chem. Rev. 1993, 7(93), 2395–417.
12. Beveridge, D.L., DiCapua, F.M. Free energy via molecular simulation: Applications to chemical and biomolecular systems, Annu. Rev. Biophys. Biophys. Chem. 1989, 18, 431–92.
13. Jorgensen, W.L. Free energy calculations: A breakthrough for modeling organic chemistry in solution, Acc. Chem. Res. 1989, 22(5), 184–9.
14. Rodinger, T., Pomès, R. Enhancing the accuracy, the efficiency and the scope of free energy simulations, Curr. Opin. Struct. Biol. 2005, 15, 164–70.

15. Kofke, D.A. Free energy methods in molecular simulation, Fluid Phase Equilibria 2005, 228–229, 41–8.
16. Raha, K., Kenneth, M., Merz, J. Calculating binding free energy in protein-ligand interaction, Annu. Rep. Comp. Chem. 2005, 1, 113–30.
17. Reddy, M.R., Erion, M.D., editors. Free Energy Calculations in Rational Drug Design. New York: Kluwer Academic; 2001.
18. Frenkel, D., Smit, B. Understanding Molecular Simulation: From Algorithms to Applications, San Diego, CA: Academic Press; 2002.
19. Brandsdal, B.O., Österberg, F., Almlöf, M., Feierberg, I., Luzhkov, V.B., Åqvist, J. Free energy calculations and ligand binding, Adv. Prot. Chem. 2003, 66, 123–58.
20. Chipot, C., Pohorille, C., editors. Free Energy Calculations: Theory and Applications in Chemistry and Biology, vol. 86. Berlin and Heidelberg: Springer; 2007.
21. Shell, M.S., Panagiotopoulos, A. Calculating free energy differences using perturbation theory, in Chipot, C., Pohorille, A., editors. Free Energy Calculations: Theory and Applications in Chemistry and Biology. Springer Series in Chemical Physics, vol. 68. Berlin and Heidelberg: Springer; 2007, p. 353–88.
22. Darve, E. Thermodynamic integration using constrained and unconstrained dynamics, in Chipot, C., Pohorille, A., editors. Free Energy Calculations: Theory and Applications in Chemistry and Biology. Springer Series in Chemical Physics, vol. 86. Berlin and Heidelberg: Springer; 2007, p. 119–70.
23. Zwanzig, R.W. High-temperature equation of state by a perturbation method. I. Nonpolar gases, J. Chem. Phys. 1954, 22(8), 1420–6.
24. Lu, N.D., Singh, J.K., Kofke, D.A. Appropriate methods to combine forward and reverse free-energy perturbation averages, J. Chem. Phys. 2003, 118(7), 2977–84.
25. Lu, N.D., Kofke, D.A. Accuracy of free-energy perturbation calculations in molecular simulation. I. Modeling, J. Chem. Phys. 2001, 114(17), 7303–11.
26. Lu, N.D., Kofke, D.A. Accuracy of free-energy perturbation calculations in molecular simulation. II. Heuristics, J. Chem. Phys. 2001, 115(15), 6866–75.
27. Bennett, C.H. Efficient estimation of free energy differences from Monte Carlo data, J. Comp. Phys. 1976, 22, 245–68.
28. Shirts, M.R., Bair, E., Hooker, G., Pande, V.S. Equilibrium free energies from nonequilibrium measurements using maximum-likelihood methods, Phys. Rev. Lett. 2003, 91(14), 140601.
29. Shirts, M.R., Pande, V.S. Comparison of efficiency and bias of free energies computed by exponential averaging, the Bennett acceptance ratio, and thermodynamic integration, J. Chem. Phys. 2005, 122, 144107.
30. Ferrenberg, A.M., Swendsen, R.H. Optimized Monte Carlo data analysis, Phys. Rev. Lett. 1989, 63(12), 1195–8.
31. Kumar, S., Bouzida, D., Swendsen, R.H., Kollman, P.A., Rosenberg, J.M. The weighted histogram analysis method for free-energy calculations on biomolecules. I. The method, J. Comp. Chem. 1992, 13(8), 1011–21.
32. Souaille, M., Roux, B. Extension to the weighted histogram analysis method: combining umbrella sampling with free energy calculations, Comp. Phys. Comm. 2001, 135(1), 40–57.
33. Bartels, C., Karplus, M. Multidimensional adaptive umbrella sampling: Applications to main chain and side chain peptide conformations, J. Comp. Chem. 1997, 18(12), 1450–62.
34. Gallicchio, E., Andrec, M., Felts, A.K., Levy, R.M. Temperature weighted histogram analysis method, replica exchange, and transition paths, J. Phys. Chem. B 2005, 109, 6722–31.
35. Kirkwood, J.G. Statistical mechanics of fluid mixtures, J. Chem. Phys. 1935, 3, 300–13.
36. Resat, H., Mezei, M. Studies on free energy calculations. I. Thermodynamic integration using a polynomial path, J. Chem. Phys. 1993, 99(8), 6052–61.
37. Guimaraes, C.R.W., de Alencastro, R.B. Thermodynamic analysis of thrombin inhibition by benzamidine and p-methylbenzamidine via free-energy perturbations: Inspection of intraperturbed-group contributions using the finite difference thermodynamic integration (FDTI) algorithm, J. Phys. Chem. B 2002, 106(2), 466–76.
38. Mobley, D.L., Dumont, È., Chodera, J.D., Dill, K.A. Comparison of charge models for fixed-charge force fields: Small-molecule hydration free energies in explicit solvent, J. Phys. Chem. B 2007, 111, 2242–54.

39. Ytreberg, F.M., Swendsen, R.H., Zuckerman, D.M. Comparison of free energy methods for molecular systems, J. Chem. Phys. 2006, 125, 184114.
40. Rick, S.W. Increasing the efficiency of free energy calculations using parallel tempering and histogram reweighting, J. Chem. Theo. Comput. 2006, 2, 939–46.
41. Jarzynski, C. Nonequilibrium equality for free energy differences, Phys. Rev. Lett. 1997, 78(14), 2690–3.
42. Crooks, G.E. Path-ensemble averages in systems driven far from equilibrium, Phys. Rev. E 2000, 61(3), 2361–6.
43. Jarzynski, C. Targeted free energy perturbation, Phys. Rev. E 2002, 65 (4/pt. 2A), 046122.
44. Hendrix, D.A., Jarzynski, C. A "fast growth" method of computing free energy differences, J. Chem. Phys. 2001, 114(14), 5961–74.
45. Jarzynski, C. Rare events and the convergence of exponentially averaged work values, Phys. Rev. E 2006, 73, 046105.
46. Roitberg, A.E. Non-equilibrium approaches to free energy calculations, Annu. Rep. Comp. Chem. 2005, 1, 103–11.
47. Maragakis, P., Spichty, M., Karplus, M. Optimal estimates of free energies from multistate nonequilibrium work data, Phys. Rev. Lett. 2006, 96, 100602.
48. Jarzynski, C. Equilibrium free-energy differences from nonequilibrium measurements: A master-equation approach, Phys. Rev. E 1997, 56 (5/pt. A), 5018–35.
49. Hummer, G. Fast-growth thermodynamic integration: Error and efficiency analysis, J. Chem. Phys. 2001, 114(17), 7330–7.
50. Hummer, G. Fast-growth thermodynamic integration: Results for sodium ion hydration, Mol. Simulation 2002, 28(1–2), 81–90.
51. Lua, R.C., Grosberg, A.Y. Practical applicability of the Jarzynski relation in statistical mechanics: A pedagogical example, J. Phys. Chem. B 2005, 109, 6805–11.
52. Oostenbrink, C., van Gunsteren, W.F. Calculating zeros: Non-equilibrium free energy calculations, Chem. Phys. 2006, 323, 102–8.
53. Bolhuis, P.G., Chandler, D., Dellago, C., Geissler, P.L. Transition path sampling: Throwing ropes over rough mountain passes, in the dark, Annu. Rev. Phys. Chem. 2002, 53, 291–318.
54. Dellago, C., Bolhuis, P.G., Geissler, P.L. Lecture notes for the International School of Solid State Physics—34th course: Computer Simulations in Condensed Matter: from Materials to Chemical Biology, Springer Lecture Notes in Physics, Erice, Sicily, 2006, Ch. Transition path sampling methods.
55. Sun, S.X. Equilibrium free energies from path sampling of nonequilibrium trajectories, J. Chem. Phys. 2003, 118(13), 5769–75.
56. Zuckerman, D.M., Woolf, T.B. Dynamic reaction paths and rates through importance-sampled stochastic dynamics, J. Chem. Phys. 1999, 111(21), 9475–84.
57. Ytreberg, F.M., Zuckerman, D.M. Single-ensemble nonequilibrium path-sampling estimates of free energy differences, J. Chem. Phys. 2004, 120(23), 10876–9.
58. Oberhofer, H., Dellago, C., Geissler, P.L. Biased sampling of nonequilibrium trajectories: Can fast switching simulations outperform conventional free energy calculation methods?, J. Phys. Chem. B 2005, 109, 6902–15.
59. Beutler, T.C., Mark, A.E., van Schaik, R.C., Gerber, P.R., van Gunsteren, W.F. Avoiding singularities and numerical instabilities in free energy calculations based on molecular simulations, Chem. Phys. Lett. 1994, 222, 529–39.
60. Gelman, A., Meng, X.L. Simulating normalizing constants: From importance sampling to bridge sampling to path sampling, Statist. Sci. 1998, 13(2), 163–85.
61. Blondel, A. Ensemble variance in free energy calculations by thermodynamic integration: theory, optimal alchemical path, and practical solutions, J. Comp. Chem. 2004, 25(7), 985–93.
62. Pitera, J.W., Van Gunsteren, W.F. A comparison of non-bonded scaling approaches for free energy calculations, Mol. Simulation 2002, 28(1–2), 45–65.
63. Zacharias, M., Straatsma, T.P., McCammon, J.A. Separation-shifted scaling, a new scaling method for Lennard-Jones interactions in thermodynamic integration, J. Phys. Chem. 1994, 100(12), 9025–31.
64. Shirts, M.R., Pande, V.S. Solvation free energies of amino acid side chains for common molecular mechanics water models, J. Chem. Phys. 2005, 122, 134508.

65. Shirts, M.R. Unpublished data.
66. Mobley, D.L. Unpublished data.
67. Rodinger, T., Howell, P.L., Pomès, R. Absolute free energy calculations by thermodynamic integration in four spatial dimensions, J. Chem. Phys. 2005, 123, 034104.
68. van Schaik, R.C., Berendsen, H.J.C., Torda, A.E., van Gunsteren, W.F. Ligand–receptor interactions, J. Mol. Biol. 1993, 234(3), 751–62.
69. Beutler, T.C., van Gunsteren, W.F. Molecular dynamics free energy calculation in four dimensions, J. Chem. Phys. 1994, 101(2), 1417–22.
70. Anwar, J., Heyes, D.M. Robust and accurate method for free-energy calculation of charged molecular systems, J. Chem. Phys. 2005, 122, 224117.
71. Oostenbrink, C., van Gunsteren, W.F. Free energies of binding of polychlorinated biphenyls to the estrogen receptor from a single simulation, Proteins 2004, 54(2), 237–46.
72. Oostenbrink, C., van Gunsteren, W.F. Single-step perturbations to calculate free energy differences from unphysical reference states: limits on size, flexibility, and character, J. Comp. Chem. 2003, 24(14), 1730–9.
73. Oostenbrink, C., van Gunseren, W.F. Efficient calculation of many stacking and pairing free energies in DNA from a few molecular dynamics simulations, Chem. Eur. J. 2005, 11, 4340–8.
74. Oostenbrink, C., van Gunsteren, W.F. Free energies of ligand binding for structurally diverse compounds, Proc. Natl. Acad. Sci. 2005, 102(19), 6750–4.
75. Okamoto, Y. Generalized-ensemble algorithms: Enhanced sampling techniques for Monte Carlo and molecular dynamics simulations, J. Mol. Graph. Model. 2004, 22, 425–39.
76. Pitera, J.W., Kollman, P.A. Exhaustive mutagenesis in silico: Multicoordinate free energy calculations in proteins and peptides, Proteins 2000, 41, 385–97.
77. Marinari, E., Parisi, G. Simulated tempering: a new Monte Carlo scheme, Europhys. Lett. 1992, 19(6), 451–8.
78. Park, S., Ensign, D.L., Pande, V.S. Bayesian update method for adaptive weighted sampling, Phys. Rev. E 2006, 74, 066703.
79. Mitsutake, A., Okamoto, Y. Replica-exchange simulated tempering method for simulations of frustrated systems, Chem. Phys. Lett. 2000, 332, 131–8.
80. Fasnacht, M., Swendsen, R.H., Rosenberg, J.M. Adaptive integration method for Monte Carlo simulations, Phys. Rev. E 2004, 69, 056704.
81. Hansmann, U.H.E. Parallel tempering algorithm for conformational studies of biological molecules, Chem. Phys. Lett. 1997, 281, 140–50.
82. Sugita, Y., Okamoto, Y. Replica-exchange molecular dynamics method for protein folding, Chem. Phys. Lett. 1999, 314, 141–51.
83. Fukunishi, H., Watanabe, O., Takada, S. On the Hamiltonian replica exchange method for efficient sampling of biomolecular systems: Application to protein structure prediction, J. Chem. Phys. 2002, 116(20), 9058–67.
84. Murata, K., Sugita, Y., Okamoto, Y. Free energy calculations for DNA base stacking by replica-exchange umbrella sampling, Chem. Phys. Lett. 2004, 385, 1–7.
85. Woods, C.J., Essex, J.W., King, M.A. Enhanced configurational sampling in binding free energy calculations, J. Phys. Chem. B 2003, 107, 13711–8.
86. Sugita, Y., Kitao, A., Okamoto, Y. Multidimensional replica-exchange method for free-energy calculations, J. Chem. Phys. 2000, 113(15), 6042–51.
87. Rodinger, T., Howell, P.L., Pomès, R. Distributed replica sampling, J. Chem. Theo. Comput. 2006, 2, 725–31.
88. Bitetti-Putzer, R., Yang, W., Karplus, M. Generalized ensembles serve to improve the convergence of free energy simulations, Chem. Phys. Lett. 2003, 377, 633–41.
89. Kong, X.J., Brooks, C.L. λ-dynamics: a new approach to free energy calculations, J. Chem. Phys. 1996, 105(6), 2414–23.
90. Zhang, W., Wu, C., Duan, Y. Convergence of replica exchange molecular dynamics, J. Chem. Phys. 2005, 123, 154105.
91. Zuckerman, D.M., Lyman, E. A second look at canonical sampling of biomolecules using replica exchange simulation, J. Chem. Theo. Comput. 2006, 2, 1200–2.
92. Donnini, S., Mark, A.E., Juffer, A.H., Villa, A. Molecular dynamics free energy calculation in four dimensions, J. Comp. Chem. 2005, 26(2), 115–22.

93. Gilson, M.K., Given, J.A., Bush, B.L., McCammon, J.A. A statistical-thermodynamic basis for computation of binding affinities: A critical review, Biophys. J. 1997, 72(3), 1047–69.
94. Boresch, S., Tettinger, F., Leitgeb, M., Karplus, M. Absolute binding free energies: A quantitative approach for their calculation, J. Phys. Chem. A 2003, 107(35), 9535–51.
95. Roux, B., Nina, M., Pomes, R., Smith, J.C. Thermodynamic stability of water molecules in the bacteriorhodopsin proton channel: a molecular dynamics free energy perturbation study, Biophys. J. 1996, 71(2), 670–81.
96. Shirts, M.R. Ph.D. dissertation, Stanford (2005).
97. Mobley, D.L., Chodera, J.D., Dill, K.A. On the use of orientational restraints and symmetry corrections in alchemical free energy calculations, J. Chem. Phys. 2006, 125, 084902.
98. Wang, J., Deng, Y., Roux, B. Absolute binding free energy calculations using molecular dynamics simulations with restraining potentials, Biophys. J. 2006, 91, 2798–814.
99. Deng, Y., Roux, B. Calculation of standard binding free energies: Aromatic molecules in the T4 lysozyme L99A mutant, J. Chem. Theo. Comput. 2006, 2, 1255–73.
100. Mobley, D.L., Chodera, J.D., Dill, K.A. Confine and release: Obtaining correct free energies in the presence of protein conformational change, J. of Chem. Theo. Comput., 2007 (in press, DOI: 10.1021/ct700032n).
101. Mobley, D.L., Graves, A.P., Chodera, J.D., McReynolds, A.C., Shoichet, B.K., Dill, K.A. Predicting absolute ligand binding free energies to a simple model site, J. Mol. Biol. 2007, in press.
102. Jayachandran, G., Shirts, M.R., Park, S., Pande, V.S. Parallelized-over-parts computation of absolute binding free energy with docking and molecular dynamics, J. Chem. Phys. 2006, 125, 084901.
103. Jorgensen, W.L., Ravimohan, C. Monte Carlo simulation of differences in free energies of hydration, J. Chem. Phys. 1985, 83(6), 3050–4.
104. Villa, A., Mark, A.E. Calculation of the free energy of solvation for neutral analogs of amino acid side chains, J. Comp. Chem. 2002, 23(5), 548–53.
105. Shirts, M.R., Pitera, J.W., Swope, W.C., Pande, V.S. Extremely precise free energy calculations of amino acid side chain analogs: Comparison of common molecular mechanics force fields for proteins, J. Chem. Phys. 2003, 119(11), 5740–61.
106. Deng, Y., Roux, B. Hydration of amino acid side chains: Nonpolar and electrostatic contributions calculated from staged molecular dynamics free energy simulations with explicit water molecules, J. Chem. Phys. 2004, 108, 16567–76.
107. Maccallum, J.L., Tieleman, D.P. Calculation of the water-cyclohexane transfer free energies of neutral amino acid side-chain analogs using the OPLS all-atom force field, J. Comp. Chem. 2003, 24(15), 1930–5.
108. Hess, B., van der Vegt, N.F.A. Hydration thermodynamic properties of amino acid analogues: A comparison of biomolecular force fields and water models, J. Phys. Chem. B 2006, 110, 17616–26.
109. Oostenbrink, C., Villa, A., Mark, A.E., Gunsteren, W.F.V. A biomolecular force field based on the free enthalpy of hydration and solvation: The GROMOS force-field parameter sets 53a5 and 53a6, J. Comp. Chem. 2004, 25(13), 1656–76.
110. Xu, Z., Luo, H.H., Tieleman, D.P. Modifying the OPLS-AA force field to improve hydration free energies for several amino acid side chains using new atomic charges and an off-plane charge model for aromatic residues, J. Comp. Chem. 2007, 28, 689–97.
111. Wick, C.D., Martin, M.G., Siepmann, J.I. Transferable potentials for phase equilibria. 4. United-atom description of linear and branched alkenes and alkylbenzenes, J. Phys. Chem. B 2000, 104(33), 8008–16.
112. Chen, B., Potoff, J.J., Siepmann, J.I. Monte Carlo calculations for alcohols and their mixtures with alkanes. Transferable potentials for phase equilibria. 5. United-atom description of primary, secondary, and tertiary alcohols, J. Phys. Chem. B 2001, 105(15), 3093–104.
113. Stubbs, J.M., Potoff, J.J., Siepmann, J.I. Transferable potentials for phase equilibria. 6. United-atom description for ethers, glycols, ketones, and aldehydes, J. Phys. Chem. B 2004, 108(45), 17596–605.
114. Wick, C.D., Stubbs, J.M., Rai, N., Siepmann, J.I. Transferable potentials for phase equilibria. 7. Primary, secondary, and tertiary amines, nitroalkanes and nitrobenzene, nitriles, amides, pyridine, and pyrimidine, J. Phys. Chem. B 2005, 109(40), 18974–82.
115. Pearlman, D.A., Charifson, P.S. Are free energy calculations useful in practice? A comparison with rapid scoring functions for the p38 MAP kinase protein system, J. Med. Chem. 2001, 44, 3417–23.

116. Pearlman, D.A. Evaluating the molecular mechanics Poisson–Boltzmann surface area free energy method using a congeneric series of ligands to P38 MAP kinase, J. Med. Chem. 2005, 48, 7796–807.
117. Michel, J., Verdonk, M.L., Essex, J.W. Protein–ligand binding free energy predictions by implicit solvent simulation: A tool for lead optimization?, J. Med. Chem. 2006, 49, 7427–39.
118. Steinbrecher, T., Case, D.A., Labahn, A. A multistep approach to structure-based drug design: Studying ligand binding at the human neutrophil elastase, J. Med. Chem. 2006, 49, 1837–44.
119. Chipot, C. Free energy calculations in biomolecular simulations: How useful are they in practice?, in Leimkuhler, B., Chipot, C., Elber, R., Laaksonen, A., Mark, A., Schlick, T., Schütte, C., Skeel, R., editors. Lecture Notes in Computational Science and Engineering: New Algorithms for Molecular Simulation, vol. 49. New York: Springer; 2005, p. 183–209.
120. Chipot, C., Rozanska, X., Dixit, S.B. Can free energy calculations be fast and accurate at the same time? Binding of low-affinity, non-peptide inhibitors to the SH2 domain of the src protein, J. Comp. Aid. Mol. Des. 2005, 19, 765–70.
121. de Graaf, C., Oostenbrink, C., Keizers, P.H.J., van Vugt-Lussenburg, B., Commandeur, J.N.M., Vermeulen, N.P.E. Free energies of binding of R- and S-propranolol to wild-type and F483A mutant cytochrome P450 2D6 from molecular dynamics simulations, Eur. Biophys. J. 2007, in press, DOI: 10.1007/s00249-006-0126-y.
122. Zhou, Y., Oostenbrink, C., van Maanen, E.M.T., Hagen, W.R., de Leeuw, S.W., Jongejan, J.A. Molecular modeling of the enantioselectivity of candida antarctica lipase b—free energy calculation, J. Comp. Chem, in press.
123. Christ, C., van Gunsteren, W. Enveloping distribution sampling: A method to calculate free energy differences from a single simulation, J. Chem. Phys. 2007, 126, 184110.
124. Hermans, J., Wang, L. Inclusion of the loss of translational and rotational freedom in theoretical estimates of free energies of binding: application to a complex of benzene and mutant T4 Lysozyme, J. Am. Chem. Soc. 1997, 119, 2707–14.
125. Fujitani, H., Tanida, Y., Ito, M., Shirts, M.R., Jayachandran, G., Snow, C.D., Pande, E.J.S.V.S. Direct calculation of the binding free energies of FKBP ligands, J. Chem. Phys. 2005, 123, 084108.
126. Woo, H.-J., Roux, B. Calculation of absolute protein-ligand binding free energy from computer simulation, Proc. Natl. Acad. Sci. 2005, 102, 6825–30.
127. Lee, M.S., Olsen, M.A. Calculation of absolute protein-ligand binding affinity using path and endpoint approaches, Biophys. J. 2006, 90, 864–77.
128. Yang, W., Bitetti-Putzer, R., Karplus, M. Free energy simulations: Use of reverse cumulative averaging to determine the equilibrated region of and the time required for convergence, J. Chem. Phys. 2004, 120(6), 2610–28.
129. Wu, D., Kofke, D.A. Phase-space overlap measures. I. Fail-safe bias detection in free energies calculated by molecular simulation, J. Chem. Phys. 2005, 123, 054103.
130. Wu, D., Kofke, D.A. Phase-space overlap measures. II. Design and implementation of staging methods for free-energy calculations, J. Chem. Phys. 2005, 123, 084109.
131. Müller-Krumbhaar, H., Binder, K. Dynamic properties of the Monte Carlo method in statistical mechanics, J. Stat. Phys. 1973, 8(1), 1–23.
132. Swope, W.C., Andersen, H.C., Berens, P.H., Wilson, K.R. A computer simulation method for the calculation of equilibrium constants for the formation of physical clusters of molecules: Application to small water clusters, J. Chem. Phys. 1982, 76(1), 637–49.
133. Flyvbjerg, H., Petersen, H.G. Error estimates on averages of correlated data, J. Chem. Phys. 1989, 91(1), 461–6.
134. Janke, W. Statistical analysis of simulations: Data correlations and error estimation, in Grotendorst, J., Marx, D., Murmatsu, A., editors. Quantum Simulations of Complex Many-Body Systems: From Theory to Algorithms, vol. 10. Jülich: John von Neumann Institute for Computing; 2002, p. 423–45.
135. Chodera, J.D., Swope, W.C., Pitera, J.W., Seok, C., Dill, K.A. Use of the weighted histogram analysis method for the analysis of simulated and parallel tempering simulations, J. Chem. Theo. Comput. 2007, 3(1), 26–41.
136. Gelman, A., Rubin, D.B. Inference from iterative simulation using multiple sequences, Statist. Sci. 1992, 7(4), 457–72.

137. Mann, G., Hermans, J. Modeling protein-small molecule interactions: Structure and thermodynamics of noble gases binding in a cavity in mutant phage T4 lysozyme L99A, J. Mol. Biol. 2000, 302, 979–89.
138. Pearlman, D.A., Connelly, P.R. Determination of the differential effects of hydrogen bonding and water release on the binding of FK506 to native and Tyr82 → Phe82 FKBP-12 proteins using free energy simulations, J. Mol. Biol. 1995, 248(3), 696–717.
139. Lamb, M.L., Jorgensen, W.L. Investigations of neurotrophic inhibitors of FK506 binding protein via Monte Carlo simulations, J. Med. Chem. 1998, 41(21), 3928–39.
140. Morton, A., Matthews, B.W. Specificity of ligand binding in a buried nonpolar cavity of T4 lysozyme: linkage of dynamics and structural plasticity, Biochemistry 1995, 34, 8576–88.
141. Morton, A., Baase, W.A., Matthews, B.W. Energetic origins of specificity of ligand binding in an interior nonpolar cavity of T4 lysozyme, Biochemistry 1995, 34, 8564–75.
142. Wei, B.Q., Baase, W.A., Weaver, L.H., Matthews, B.W., Shoichet, B.K. A model binding site for testing scoring functions in molecular docking, J. Mol. Biol. 2002, 322, 339–55.
143. Wei, B.Q., Weaver, L.H., Ferrari, A.M., Matthews, B.W., Shoichet, B.K. Testing a flexible-receptor docking algorithm in a model binding site, J. Mol. Biol. 2004, 337, 1161–82.
144. Graves, A.P., Brenk, R., Shoichet, B.K. Decoys for docking, J. Med. Chem. 2005, 48, 3714–28.
145. Holt, D.A., Luengo, J.I., Yamashita, D.S., Oh, H.J., Konialian, A.L., Yen, H.K., Rozamus, L.W., Brandt, M., Bossard, M.J., Levy, M.A., Eggleston, D.S., Liang, J., Schultz, L.W., Stout, T.J., Clardy, J. Design, synthesis, and kinetic evaluation of high-affinity FKBP ligands and the X-ray crystal structures of their complexes with FKBP12, J. Am. Chem. Soc. 1993, 115(22), 9925–38.
146. Sich, C., Improta, S., Cowley, D.J., Guenet, C., Merly, J.-P., Teufel, M., Saudek, V. Solution structure of a neurotrophic ligand bound to FKBP12 and its effects on protein dynamics, Europ. J. Biochem. 2000, 267, 5342–55.
147. Tatlock, J., Kalish, V., Parge, H.E., Knighton, D.R., Showalter, R., Lewis, C., French, J.V., Villafranca, J.E. High-affinity FKBP-12 ligands derived from (R)-$(-)$-carvone synthesis and evaluation of FK506 pyranose ring replacements, Bioorg. Med. Chem. 1995, 5(21), 2489–94.

Section 2
Biological and Biophysical Applications

Section Editor: Heather Carlson

University of Michigan
College of Pharmacy
428 Church Street
Ann Arbor
MI 48109-1065
USA

CHAPTER 5

Linear Quantitative Structure–Activity Relationships for the Interaction of Small Molecules with Human Cytochrome P450 Isoenzymes

Thomas Fox[1*] **and Jan M. Kriegl**[*]

Contents		
	1. Introduction	63
	2. The Cytochrome P450 Superfamily	64
	3. Methodological Overview	66
	3.1 Quantitative structure–activity relationships	66
	3.2 Molecular descriptors	66
	3.3 Statistical methods	67
	3.4 Assessment of model quality	67
	4. Applications	68
	4.1 Single-parameter relationships	68
	4.2 Models based on alignment-independent molecular descriptors	69
	4.3 3D-QSAR models	71
	5. Discussion and Outlook	73
	References	75

1. INTRODUCTION

Drug–drug interactions have become an important issue in human health care [1–4]. Many of the major pharmacokinetic interactions between drugs are due to hepatic cytochrome P450 (CYP) enzymes being inhibited by concomitant administration of other drugs. Additionally, genetic polymorphisms of important CYP

---
[*] Computational Chemistry, Department of Lead Discovery, Boehringer Ingelheim Pharma GmbH & Co. KG, 88397 Biberach, Germany. E-mail: thomas.fox@bc.boehringer-ingelheim.com
[1] Corresponding author.

isoenzymes are known to affect the biotransformation of drugs [5]. Therefore, during early stages of drug discovery, the inhibitory potency of a new molecule against a panel of CYP isoenzymes is often used as a first-tier screen to identify if it is likely to lead to complications of this type.

High-throughput *in-vitro* assays have been established over the last couple of years to assess the stability or inhibitory potency of compounds towards CYP mediated metabolism [6–8]. Still, their screening rate is far too low to cope with the requirements emerging from large compound collections. Computational tools to address these data sets or even larger virtual libraries are highly desirable. Moreover, they can be used to guide library design and to evaluate compounds prior to synthesis or acquisition.

In the past, understanding the interactions between CYPs and their ligands was hampered as structural information was only available on bacterial CYPs, from which homology models of mammalian CYPs were built (c.f. Refs. [9–12] and references therein). Only recently, X-ray structures of mammalian and human cytochromes have been solved (c.f. [10] and references therein, [13–16]). Therefore, many studies have tried to address this field with ligand-based techniques that may be used either to gain a qualitative classification or a quantitative prediction of the experimental outcome. Especially Quantitative Structure–Activity Relationships (QSARs) and pharmacophore modeling have been widely used to identify structural motifs which explain interactions between small molecules and the CYP active site. Due to the importance of these approaches in drug discovery, a number of reviews have appeared in recent years [10,11,17–26]. These studies often have focused on internal consistency of the resulting models and on biological plausibility in terms of key interactions between ligand and active site.

In this article we summarize many of the linear quantitative models which have been developed to predict interactions of small molecules with CYPs, including inhibition and metabolism. Special emphasis will be given to model quality assessment. Another important issue will be the level of abstraction which was incorporated to characterize the ligand properties. In the following we restrict ourselves to the human CYP1, CYP2, and CYP3 families due to their outmost importance in pharmaceutical research. Studies on closely related non-human isoforms will be included in some cases.

2. THE CYTOCHROME P450 SUPERFAMILY

The CYP superfamily of heme-containing isoenzymes plays an important role in the oxidative degradation (phase I metabolism) of a wide range of endogenous compounds, drugs, and other xenobiotics [27–30]. Various P450 enzyme gene polymorphisms, associated with a varying degree of metabolic activity along the CYP pathway, further complicate the assessment of CYP related adverse drug effects [5,31].

The mean expression levels of the most important human CYPs in human hepatic tissue and their contribution to drug metabolism are summarized in Table 5.1. These members of the CYP1, CYP2 and CYP3 families account for the oxidation more than 90% of all drugs currently on the market.

TABLE 5.1 Major human hepatic CYPs and drug oxidations [1,3,29,142–145]

CYP enzyme	%P450 complement	% drug oxidations	Selected drug substrates	Inhibitor(s)
1A2	13	8.2	paracetamol, caffeine, ondansetron, tamoxifen	furafylline
2A6	4	2.5	coumarin, nicotine	8-methoxypsoralen
2B6	1	3.4	benzphetamine, verapamil	orphenadrine
2C8/9	18	15.8	diclofenac, tolbutamide, (S)-warfarin	sulphaphenazole
2C19	1	8.3	diazepam, omeprazole, (R)-warfarin	fluconazole
2D6	2.5	18.8	codeine, desipramine, dextromethorphan, fluoxetine, haloperidol, imipramine	quinidine, ajmaline
2E1	7	4.1	paracetamol, caffeine, chlorzoxazone	pyridine
3A4	28	34	clarithromycin, codeine, diazepam, erythromycin, indinavir, midazolam, nifedipine, quinidine, terfenadine, verapamil	ketoconazole, itraconazole

The most abundant enzymes are CYP3A4, 2C8, 2C9, and 1A2. CYP1A2 plays a significant role in the metabolism of aromatic amines, estradiol, and other drugs. CYP2C9 is known to metabolize acidic substrates such as, for instance, antiinflammatory agents. Together with CYP2C19, it contributes to more than 20% of the phase I metabolism of clinically applied drugs [32]. With approximately 97% sequence identity, the polymorphic CYP2C19 is very closely related to CYP2C9. Selectivity between these isoenzymes is predominantly attributed to different access channels [33].

The CYP3 family constitutes the major portion of the human hepatic cytochrome P450. It is known to metabolize the majority of all commercially available drugs [34]. X-ray crystallographic structures of CYP3A4–ligand complexes have shown a considerable flexibility of the active site [15], and substrates of CYP3A4 cover a wide range of lipophilic drugs of different size and shape [35, 36]. In addition, CYP3A4 is known for its atypical interaction with its substrates *in vitro* [37,38]. It has been proposed that there are multiple binding modes possible [39], and there is evidence that CYP3A4 may accommodate more than one substrate simultaneously [40]. This complexity of the biological findings additionally contributes to the challenge of constructing predictive models.

The isoenzyme CYP2D6 represents only 1.5% of the human hepatic cytochrome P450. However, it participates in the metabolism of approximately 20% of clinically prescribed drugs. Due to gene multiplicity and polymorphisms, large inter-individual differences exist in CYP2D6 activity. The interest in this isoenzyme started early, and a relatively large number of small-molecule models have been derived [41–45].

Compared with other members of the CYP superfamily, relatively few clinically relevant pharmaceuticals are metabolized by CYP2A6. However, it is involved in the metabolism of a variety of toxic compounds, including nicotine or other tobacco-related xenobiotics such as nitrosamines [46]. CYP2B6 metabolizes several clinically important drugs, but only few data on CYP2B6 binding have been published so far. We therefore include also QSAR studies on CYP2B1 and CYP2B4. These closely related isoenzymes are found in rats and rabbits, respectively. With its relatively small and restricted active site, CYP2E1 is involved in the metabolism of low molecular weight solvents, aromatic hydrocarbons, and anesthetics [47,48].

3. METHODOLOGICAL OVERVIEW

3.1 Quantitative structure–activity relationships

The simplest approach to predicting the biological property of a compound is via QSAR, a well established methodology in computational chemistry and drug design [49,50]. In brief, this method attempts to find a quantitative mathematical relationship between the chemical structure of a set of compounds and their biological properties. The molecular features are encoded in measured or calculated properties, which are referred to as molecular descriptors.

3.2 Molecular descriptors

Numerous descriptors have been developed [51]. Depending on the information about the spatial distribution of atoms which is necessary to compute them, they are often grouped into 1D, 2D, and 3D descriptors. In the following we distinguish between descriptors which are invariant with respect to translation and rotation of the compounds in a given data set (alignment independent descriptors), and descriptors which require a prior alignment in a fixed coordinate system (alignment dependent descriptors).

Alignment independent descriptors can be derived from the molecular formula, from the 2D molecular graph, or from a 3D conformation. They cover global molecular properties, but also 3D chemical information at the molecular surface as provided, for instance, by the Molecular Surface Weighted Holistic Invariant Molecular descriptors (MS-WHIM) [52,53]. The class of alignment independent descriptors includes also properties which are obtained from QM calculations, such as dipole moment or orbital energies.

Describing the compounds with molecular interaction fields (MIFs) leads to the class of alignment dependent approaches, often termed 3D-QSAR [54,55]. Here an alignment rule to superimpose the compounds has to be found prior to the descriptor calculation. The most common 3D-QSAR approaches are CoMFA [56, 57], CoMSiA [58], or GRID/GOLPE [59,60]. 3D-QSAR techniques use relevant conformers of ligands to suggest functional groups, the geometry of structural features, and/or regions of electrostatic and steric interactions essential for activity. Later, several approaches were developed to transform the alignment-dependent information contained in MIFs into alignment-independent descriptors (e.g. GRIND [61]; Volsurf [62,63]).

3.3 Statistical methods

In this paper we focus on linear relationships between descriptors and biological properties, which are detected by statistical techniques such as multiple linear regression (MLR) or partial least squares analysis (PLS) [64]. A further development are hierarchical PLS models [65–67], which can be employed if the descriptors can be grouped into several subsets: After a PLS analysis for each subset, the results of these base-level PLS models are combined into a top-level PLS analysis.

In addition to these linear techniques, in the last few years a number of non-linear approaches originally developed for pattern recognition tasks have seen a considerable rise in attention, very often in the context of classification models [68–83]. However, these studies are beyond the scope of this review and have been discussed elsewhere [22,24,84].

3.4 Assessment of model quality

In the ideal case, a QSAR model should be developed in four stages: data preparation, model generation, model validation, and assessment of the applicability domain [85]. Data preparation, i.e. a careful composition of activity data and molecular descriptors in a training set, and the establishment of a statistically significant relationship between the biological activity and the molecular descriptors (usually characterized by a correlation coefficient, r^2 are part of almost each QSAR study.

The next step is the assessment how well the QSAR model can predict the activity of previously unseen compounds. Several strategies for this validation procedure have been published and discussed [85–87]. Cross-validation is the most common practice in internal model validation. In a single cross-validation step, either one (leave-one-out cross-validation, LOO) or several samples (leave-group-out cross-validation, LGO) are omitted from the data set. Their activity is predicted with a model which was generated from the remaining training samples. After each compound has been predicted once, a cross-validated correlation coefficient q^2 between predictions and observations is calculated. Especially, q^2 values are considered as an estimate of the predictive power of a model.

The division into a training and test set can be driven by rational criteria such as training set diversity or optimum coverage of the target range [85,88–90] or simply by random. A more objective way to assess the predictivity of a model is the application on a test set which was not used for model generation, often referred to as external validation. The ultimate test for a QSAR model, however, is the comparison of predictions to new data which were collected for instance during ongoing screening campaigns and were not available at the time of model generation.

In the following, we will denote the correlation between experiment and prediction for any data set which was not part of the training set by r^2_{test}. For such a data set, it has to be checked whether the model, in principle, is able to predict these compounds. Despite its paramount significance, the model applicability domain has been addressed in only few QSAR studies [75,78,81,85,91].

4. APPLICATIONS

4.1 Single-parameter relationships

In his pioneering work on structure–activity relationships in drug metabolism, Corwin Hansch could show that compound lipophilicity plays an essential role in the speed of metabolic degradation of drugs [92]. This view was further pursued by Lewis and coworkers, who investigated the influence of hydrophobicity on substrate binding for a wide range of CYP enzymes using relatively small sets of structurally diverse substrates [93–96]. In most cases, high correlations between favorable entropic changes upon desolvation of the CYP active site, expressed in terms of the calculated log P, and substrate affinities were observed. Different slopes and intercepts in the resulting regression equations were attributed to different degrees of hydrophobicity in the heme environment and to contributions of additional interactions such as hydrogen bonding or π–π stacking interactions, thus providing a hint towards substrate selectivity.

Although high correlations between compound lipophilicity could be observed for both substrates and inhibitors in some homologous series [97,98], this can not be generalized. For instance, a series of eight benzodiazepines revealed no significant correlation between inhibition of CYP2C19 and compound lipophilicity [99]. Especially the inclusion of structurally more diverse compounds often leads to a significant decrease of the correlation [97,100,101]. One explanation is that there is probably more than one linear lipophilic relationship for each CYP enzyme [95]. This was investigated in more detail by Riley and coworkers [102] for the interaction between CYP3A4 and a diverse set of 30 inhibitors with IC$_{50}$ values between 0.05 and 2700 µM. Using compound lipophilicity, estimated as log $D_{7.4}$, as a single descriptor, a correlation coefficient between observed and recalculated IC$_{50}$ values of $r^2 = 0.69$ was obtained. In a second step, the authors split the data set into two subsets, depending on the presence or absence of sterically unhindered N-containing heterocycles such as pyridines, imidazoles, or triazoles. Within both subsets, stronger relationships between IC$_{50}$ values and compound lipophilicities were observed. The authors attribute this to an additional binding energy contribution from the interaction between the heme iron and the lone pair on the nitrogen atoms. These single-parameter relationships were applied to ongoing research projects at AstraZeneca, yielding a correlation between observed and predicted IC$_{50}$ values of $r^2_{test} = 0.76$.

4.2 Models based on alignment-independent molecular descriptors

Regression equations using only a few calculated properties could be derived from small series of CYP substrates and inhibitors [19,93,103–105]. In general, the inhibitory potency or affinity was found to depend on the calculated log P, the molecular mass, frontier orbital energies, dipole moments or shape descriptors. In some cases, descriptors derived from modeled protein ligand–complexes contributed to the final linear equation. Typical examples are the number of hydrogen bonds or π–π stacking interactions formed between the ligand and active site residues [103–105].

In a study on CYP1A2 inhibition, Fuhr et al. correlated electrostatic and volume descriptors for a series of 38 quinolone-like compounds with their inhibitory potency [106]. Stepwise linear regression analysis revealed that the type of core structure, volume and charge characteristics, and lipophilicity at two substitution points established a regression equation with an r^2 of 0.56.

Several studies on the effect of flavenoids on CYP1A1 [107] and CYP1A2 mediated metabolism [76,107,108] have been published. In these MLR models, good correlations between observed and calculated IC_{50} values could be obtained in all cases, and similar key molecular properties of the flavenoids were identified to promote inhibition of CYP1A enzymes. Moon and coworkers also investigated the predictivity of the regression equations. Their final model was chosen according to the best q^2 value and the entire data set was split into a training set (14 compounds) for model development and a test set (5 compounds) for validation. With the best MLR model, the inhibitory potency of the test set compounds could be predicted with $r^2_{\text{test}} = 0.63$. Four out of five compounds were predicted with residuals of less than 0.5 log units.

Wanchana and coworkers investigated a series of 53 structurally diverse drugs which inhibit CYP3A4 [109]. The IC_{50} values covered almost six orders of magnitude (9 nM to 2 mM). After omitting compounds with low solubility or compounds which activate substrate metabolism, a final data set of 44 compounds remained. The compounds were characterized by Molconn-Z descriptors. After randomly picking nine compounds as a test set, PLS analysis with preceding descriptor selection was used to establish a correlation between the IC_{50} values and a subset of the descriptors. Descriptor selection was performed employing a genetic algorithm using q^2_{LOO} values as fitness function. The final model was constructed from 20 descriptors and showed a cross-validation performance of $q^2_{\text{LOO}} = 0.57$. The test set could be predicted with $r^2_{\text{test}} = 0.55$. The authors also analyzed the influence of single molecular properties on the PLS projection. The first latent component, which contributed 25.5% to the total variance of the biological data, was found to negatively correlate with the molecular mass. Because of the high correlation between hydrophobicity and molecular size [110], this result can be considered as additional support for the general observation that hydrophobicity plays an important role in CYP affinity.

Recently, Kriegl et al. used a set of 967 diverse drug-like compounds with IC_{50} values between 0.4 and 50 µM to derive multivariate models of CYP3A4 inhibition [100]. After removal of outliers, a D-optimal onion design strategy was used to split the data set into a training set of 551 and a test set of 379 compounds [88]. Several different descriptor sets were employed, ranging from physico-chemical properties to descriptors obtained from GRID MIFs and semi-empirical AM1 calculations. PLS models with q^2_{LGO} between 0.41 and 0.53 were obtained from the individual descriptor sets. Slightly better models could be generated if all descriptors were pooled ($q^2_{LGO} = 0.53$). A hierarchical model based on the scores of the individual models yielded a cross-validation performance of $q^2_{LGO} = 0.58$. The test set compounds could be predicted with r^2_{test} values between 0.48–0.57 (individual descriptor sets). All descriptors together or a hierarchical model yielded $r^2_{test} = 0.6$. Using the results of the regression model to classify the compounds into weak (IC$_{50}$ > 20 μM), moderate (2 μM < IC$_{50}$ < 20 μM), or strong inhibitors (IC$_{50}$ < 2 μM) gave an overall accuracy of about 60% correct classifications, and only few if any misclassifications over two classes.

Mao et al. investigated a data set of more than 2000 drug-like compounds with data for the inhibition of the metabolism of four different substrates of CYP3A4 [111]. The compounds were described by a battery of pharmacophore-based, surface area, topological and fragment descriptors. Descriptor selection with a genetic algorithm was followed by linear modeling, augmented by a model which describes the behavior of 'neighboring compounds'. The authors note that a single QSAR model of pIC50 which assumes a common binding mode for all inhibitors only leads to moderate predictivity ($r^2 = 0.57$, mean signed error = 0.51), in contrast to a previous model obtained for CYP2D6. It was suggested that better models can be obtained if one assumes that the data set can be subdivided into several classes, each adopting a distinct binding mode in the CYP3A4 active site. Based on the differential inhibition of pairs of the substrates, SVM models were trained to split the data set into subsets. Building separate QSAR models for each of the subsets considerably improved the predictions both in terms of r^2 and mean signed error. In the derivation of the models, all available data were used and no model validation with a test data set was performed. However, a small number of new data as an external validation set was predicted within the range of fitting errors.

Chohan et al. [75] used a diverse training set of 109 CYP1A2 inhibitors and 123 in-house descriptors to compare several linear and non-linear modeling techniques. Within the training set, all models faired equally well, with r^2 values between 0.7 and 0.85. All models suggest that lipophilicity, aromaticity, charge, and orbital energies are important features describing CYP1A2 inhibition, features also seen for CYP1A2 substrates. These models were applied to a diverse test set of 68 oral drugs. However, the obtained deviations between predictions and experimental results are larger than if the mean pIC$_{50}$ values of the training set would have been used to predict the test set activities. The authors attribute this to the fact that the test set is dissimilar to the training set, and find that the prediction error increases with the Euclidian distance in descriptor space.

Afzelius et al. [112] applied GRIND descriptors to generate quantitative and qualitative models for CYP2C9 inhibition. The resulting PLS model gave q^2_{LOO} and q^2_{LGO} values of 0.60 and 0.57, respectively. Application to an external test set of 12 compounds gave a standard deviation of prediction errors (SDEP) of 0.37 log units;

11 of the 12 compounds were predicted within 0.5 log units. This study was later extended to incorporate conformational flexibility of the ligands by using flexible GRID fields in the generation of the GRIND descriptors [113]. Here the same model accuracy could be achieved as in a previous paper where the conformational model was based on docking into CYP2C9 [114].

The MS-WHIM methodology was applied to inhibitors of CYP2A5 [52], 2C9 [115], 2D6 [116], and 3A4 [117], as well as to substrates of CYP2A5 [52], 2B6 [118], and 2D6 [119]. In all cases, several descriptor combinations were investigated to find the optimal one, and either LOO or LGO cross-validation was performed.

For the data set of 16 less diverse and rather rigid ligands of CYP2A5 (both inhibitors and substrates), a model with $q^2_{LGO} = 0.71$ could be obtained. For three diverse datasets of CYP2C9 inhibitors, inclusion of multiple conformers (4D-QSAR) generally improved the predictability of the models. However, only one of the three data sets yielded a predictive model as concluded from LGO cross-validation. In contrast to this, for two sets of CYP2D6 inhibitors, a 4D-QSAR model did not improve the predictability, which suggests that for these datasets a single conformation sufficiently described the conformational space. Both 3D and 4D models yielded a q^2_{LOO} of about 0.5. We note that using a LGO cross-validation approach, only one dataset still results in a predictive model ($q^2_{LGO} = 0.48$). Similar results were obtained for three data sets of CYP3A4 inhibitors. Only 4D-QSAR models yielded predictive models, with a best q^2_{LGO} value of 0.38.

For a set of 16 CYP2B6 substrates, a q^2_{LGO} value of 0.61 was reported. This model was then applied to a test set of 5 compounds. The K_m (apparent) of four of the five molecules could be predicted within one log unit. Literature K_m values for a set of 52 CYP2D6 substrates could be modeled with $q^2_{LOO} = 0.58$.

4.3 3D-QSAR models

In contrast to the descriptors discussed so far, MIF descriptors and the models based on them crucially depend on the alignment rule employed to superpose the compounds in the data set. Such 3D-QSAR models have been developed for ligands of CYP1A2 [120] and 2A5 [121], inhibitors of CYP1A2 [122], 2A5 [123,124], 2A6 [123,124], 2B1 [125], 2C9 [97,114,126–129], 2C19 [97,130] and 2D6 [131], as well as substrates of CYP2D [101] and 2E1 [132]. For a comprehensive survey, we refer to recent reviews [10,11,17,21,23,24]. Many studies were performed to identify important regions which promote or attenuate cytochrome-ligand interactions. Often the same molecular scaffold with different substitution patterns was used to constitute the training set. Although meaningful correlations between biological activity and MIFs as well as favorable internal validation statistics could be achieved in most cases, the models were tested in only few examples with an independent test set. This certainly also reflects the difficulty to obtain a general alignment rule for a set of diverse compounds. Recent publications attempt to overcome this problem by incorporating complementary computational approaches such as docking into homology models. In the following, we focus on selected studies which use diverse data sets and provide particular insights into the generalizability of the models.

The first quantitative QSAR study on CYP2C9 inhibition was published in 1996 by Jones et al. [126]. Manual alignment of 27 compounds (K_i = 0.1–70 µM), and a standard CoMFA analysis yielded a model with SDEP of 0.17 log units and a q^2_{LOO} of 0.70. The model predicted the experimental K_i values of both stereoisomers of warfarin within a factor of two. Later, this model was validated with 14 new compounds that had K_i values ranging from 0.1 to 48 µM [127]. While the initial training set was predominantly constituted of coumarin-like compounds, this validation set contained mostly sulfonamides. A further refinement through incorporation of a series of 17 benzbromarone analogues followed later [129]. The range of K_i values of this new, more diverse training set spanned now more than four orders of magnitude. Several CoMFA and CoMSiA models were generated both for the benzbromarone subset and the entire data set. The best model for the entire set was obtained from a CoMSiA model and gave a q^2 of 0.75. The inclusion of high-affinity ligands was shown to be important as this provided insight into key interaction regions which were not identified as significant in previous models [128].

Suzuki et al. investigated 24 hydantoin and barbiturate derivatives with respect to their potency to inhibit CYP2C9 [97]. These compounds showed K_i values between 5 and >1000 µM. A standard CoMFA model with a q^2_{LOO} value of 0.68 could be established. A CoMFA model derived from a structurally more diverse test set, which contained also 39 compounds from previous studies, showed a cross-validation performance of $q^2_{LOO} = 0.85$.

To work out the structural differences between CYP2C9 and CYP2C19 ligands, Locuson and coworkers extended their study on the CYP2C affinity of benzbromarones to CYP2C19 [130]. The K_i values of 15 compounds spanned a range from 0.3 to 13 µM. Although the tendency of CYP2C19 to prefer neutral benzbromarones could be derived from the inhibition data, no stable CoMFA models could be obtained, possibly due to the narrow range of K_i values. In parallel, the potency of 24 hydantoin and barbiturate analogues to inhibit CYP2C19 was investigated (K_i = 0.079 to >1000 µM) [97]. Among others, standard CoMFA fields were related to Ki, leading to cross-validation results of $q^2_{LOO} = 0.79$. From this series of structurally similar compounds, the importance of lipophilic and steric interactions in the CYP2C19 active site was deduced. The q^2 and CoMFA contours did not change significantly when the benzbromarones were added to the data set [130]. Interestingly, the model performance did not depend on the orientation of the benzbromarones in the alignment.

Afzelius et al. [114] reported a study that describes the generation of a GRID/GOLPE QSAR model for 21 diverse drugs which act as competitive CYP2C9 inhibitors with K_i values ranging from 0.5 to 245 µM. The compound alignment resulted from docking into a homology model of CYP2C9. With the MIFs generated by the DRY and OH probes, a model with $q^2_{LOO} = 0.73$ and $q^2_{LGO} = 0.67$ could be produced. Application of the model to an external test set of 8 compounds (K_i = 2.5–45 µM) gave a maximum error of 0.36 log units.

A first predictive 3D-QSAR model for substrate binding to CYP2D6 was developed by Haji-Momenian et al. [101]. They used a training set of 24 CYP2D6 substrates with K_m values from the literature (0.27–183 µM) to generate standard CoMFA models. Apart from a protonated nitrogen which was common to all compounds, the data set consisted of structurally diverse molecules. Manual alignment and the inclusion of multiple conformers led to $q^2_{LOO} = 0.55$. The activities of a 15 membered test set (K_m = 1.4–190 µM) could be predicted with r^2_{test} of 0.62. For 13 out of 15 substrates, the prediction error was less than 0.5 log units.

Recently, Vaz and coworkers reported a CoMSiA analysis of 36 aryloxypropanolamines which inhibit CYP2D6 with IC$_{50}$ values between 0.03 and >100 µM [131]. A training set of 30 compounds was docked into a homology model of CYP2D6. After superimposition of the energy-minimized conformers, a CoMSiA analysis was performed. The resulting PLS model exhibited a cross-validation performance of $q^2 > 0.6$ and was able to predict the IC$_{50}$ values of the five out of six test set compounds within less than one log unit.

A set of 46 CYP1A2 inhibitors (23 naphthalene derivatives, 23 lactones and similar compounds, IC$_{50}$ values between 2.3 µM and 40 mM) was subjected to CoMFA and GRID/GOLPE analysis by Korhonen et al. [122]. PLS models with $q^2_{LGO} = 0.69$ (CoMFA) and $q^2_{LGO} = 0.79$ (GRID/GOLPE) could be obtained. The CoMFA model was then applied to six small compounds similar to the training set molecules and six drug-like molecules which differ considerably from those used in the training set. The pIC$_{50}$ values for four out of six compounds exhibiting a pronounced similarity with members of the training set were predicted by less than 15% off. With one exception, the prediction error obtained for the drug-like molecules was less than a factor of two.

5. DISCUSSION AND OUTLOOK

Despite considerable progress over the last years, the accurate quantitative prediction of CYP inhibition or metabolism remains a considerable challenge. In our view, a number of factors contribute to the overall difficulty:

First of all, there seems a lack of large, diverse data sets on which to build models. Many of the QSAR studies on CYP interactions discussed in this article are restricted to rather small data sets of structurally similar compounds. Only recently data sets with ≫50 compounds were employed to construct predictive CYP models. In addition, a number of authors used data from several sources— either from the literature or from several in-house laboratories. This often induces additional noise, as it is not always clear if the assays are comparable [8]. Larger data sets are often available in pharmaceutical companies as measurements of e.g. CYP inhibition usually is part of early compound profiling. However, these data sets often only cover a very limited range of activities. Usually neither for strong inhibitors (IC$_{50}$ ≪ 1 µM) nor for very weak inhibitors (IC$_{50}$ ≫ 50 µM) an exact inhibition constant is determined. However, as shown by Locuson and Wahlstrom, especially the inclusion of high-affinity ligands in CYP-related QSAR analysis can improve the models [128].

A second reason for the difficulty to obtain good models may lie in the nature of the enzymes themselves: CYPs have evolved to exhibit broad substrate specificity and to metabolize a large range of compounds. Thus, different binding modes of inhibitors or substrates in the CYP active site may preclude a 3D-QSAR description via a single alignment rule. Alternatively, different compound classes may exhibit different interaction patterns with binding site residues. Then it might be difficult to build a model for all classes—several models for sub-classes would lead to better predictions [102,111]. While this is true for most enzymes, the special nature of CYPs would exacerbate this problem.

Independent of these problems, we consider a rigorous assessment of the model quality as a fundamental issue. Many of the published models were not validated by an (external) test set. Even then, internal validation with LOO cross-validated q^2 tends to overestimate the predictive power of a model [133]. On the other hand, LGO cross-validation may suffer from a chance bias if only one random split is employed. External validation can not be afforded in many cases due to limited training samples available for model generation.

Another issue that has attracted little interest in this area so far is that of model applicability: is the chemical space of the training set of a model close enough to the chemical space of the test set compounds so that a reasonably accurate prediction can be expected [134]? As is, for many models in the literature it is hard to judge their generalizability or transferability to other compound classes. Although these models provide insight into CYP-ligand interactions, they only work for the chosen training set, and their application to in-house compounds is not predictive.

An important feature of linear models over non-linear ones is that they usually are easier to interpret. However, the value of a model, expressed in terms of predictivity and interpretability, crucially depends on the type and number of descriptors which are incorporated. 3D-QSAR models based on MIFs allow to localize regions with favorable and less favorable interactions. Very often, predictive models are obtained by using multivariate approaches with many descriptors. The choice of descriptors determines the chemical variability which can be captured by a QSAR model [135]. During the model building process one has to find an appropriate balance between few, easily interpretable variables, and a broader multivariate characterization which may be more difficult to interpret.

Many of the more recent publications on *in-silico* assessments of CYP interactions employ non-linear approaches. They are well suited to be applied in cases were the model interpretability is less of an issue. In the same venue, classification models have been developed. Although coarse-grained, these models can be used as one of a number of filters to pick the most interesting compounds for subsequent experimental profiling. These classification models may also be a way to overcome the problem of noisy data, which are not reliable enough for the generation of a quantitative model [68–83,136].

Parallel to these trends, we expect information obtained from experimentally determined CYP structures (see [10] and references therein, [13,14]) to be combined with ligand-based methodologies. First examples include the use of docked ligand poses as alignment rules for subsequent 3D-QSAR models [114,120,131, 137]. Information about protein-ligand interactions (such as the number of hydrogen bonds formed [103–105] or interaction energy contributions [120,138] may be used as additional descriptors or may lead to compound-class specific models [102]. Another example is the program MetaSite [139–141] to predict the most probable site of metabolism in a ligand.

In addition to these combined approaches, ligand-based methodologies will continue to play an important role in the prediction of CYP inhibition and metabolism. From these computational tools, we expect a better understanding of the interaction between CYPs and their ligands and—hopefully—more reliable in-silico models which can be utilized in drug discovery.

REFERENCES

1. Weaver, R.J. Assessment of drug–drug interactions: concepts and approaches, Xenobiotica 2001, 31, 499–538.
2. Wienkers, L.C., Heath, T.G. Predicting in vivo drug interactions from in vitro drug discovery data, Nat. Rev. Drug Discov. 2005, 4, 825–33.
3. Lin, J.H., Lu, A.Y. Inhibition and induction of cytochrome P450 and the clinical implications, Clin. Pharmacokinet. 1998, 35, 361–90.
4. Hutzler, J.M., Messing, D.M., Wienkers, L.C. Predicting drug–drug interactions in drug discovery: where are we now and where are we going?, Curr. Opin. Drug Discov. Devel. 2005, 8, 51–8.
5. Kirchheiner, J., Seeringer, A. Clinical implications of pharmacogenetics of cytochrome P450 drug metabolizing enzymes, Biochim. Biophys. Acta 2007, 1770, 489–94.
6. Jenkins, K.M., Angeles, R., Quintos, M.T., Xu, R., Kassel, D.B., Rourick, R.A. Automated high throughput ADME assays for metabolic stability and cytochrome P450 inhibition profiling of combinatorial libraries, J. Pharm. Biomed. Anal. 2004, 34, 989–1004.
7. Ansede, J.H., Thakker, D.R. High-throughput screening for stability and inhibitory activity of compounds toward cytochrome P450-mediated metabolism, J. Pharm. Sci. 2004, 93, 239–55.
8. Zlokarnik, G., Grootenhuis, P.D., Watson, J.B. High throughput P450 inhibition screens in early drug discovery, Drug Discov. Today 2005, 10, 1443–50.
9. de Groot, M.J., Vermeulen, N.P. Modeling the active sites of cytochrome P450s and glutathione S-transferases, two of the most important biotransformation enzymes, Drug Metab. Rev. 1997, 29, 747–99.
10. de Graaf, C., Vermeulen, N.P., Feenstra, K.A. Cytochrome P450 in silico: an integrative modeling approach, J. Med. Chem. 2005, 48, 2725–55.
11. de Groot, M.J., Kirton, S.B., Sutcliffe, M.J. In silico methods for predicting ligand binding determinants of cytochromes P450, Curr. Top. Med. Chem. 2004, 4, 1803–24.
12. Maréchal, J.D., Sutcliffe, M.J. Insights into drug metabolism from modelling studies of cytochrome P450-drug interactions, Curr. Top. Med. Chem. 2006, 6, 1619–26.
13. Yano, J.K., Wester, M.R., Schoch, G.A., Griffin, K.J., Stout, C.D., Johnson, E.F. The structure of human microsomal cytochrome P450 3A4 determined by X-ray crystallography to 2.05 Å resolution, J. Biol. Chem. 2004, 279, 38091–4.
14. Yano, J.K., Hsu, M.H., Griffin, K.J., Stout, C.D., Johnson, E.F. Structures of human microsomal cytochrome P450 2A6 complexed with coumarin and methoxsalen, Nat. Struct. Mol. Biol. 2005, 12, 822–3.
15. Ekroos, M., Sjogren, T. Structural basis for ligand promiscuity in cytochrome P450 3A4, Proc. Natl. Acad. Sci. U.S.A. 2006, 103, 13682–7.
16. Rowland, P., Blaney, F.E., Smyth, M.G., Jones, J.J., Leydon, V.R., Oxbrow, A.K., Lewis, C.J., Tennant, M.G., Modi, S., Eggleston, D.S., Chenery, R.J., Bridges, A.M. Crystal structure of human cytochrome P450 2D6, J. Biol. Chem. 2006, 281, 7614–22.
17. Ekins, S., de Groot, M.J., Jones, J.P. Pharmacophore and three-dimensional quantitative structure activity relationship methods for modeling cytochrome P450 active sites, Drug Metab. Dispos. 2001, 29, 936–44.
18. de Groot, M.J., Ekins, S. Pharmacophore modeling of cytochromes P450, Adv. Drug Deliv. Rev. 2002, 54, 367–83.

19. Hansch, C., Mekapati, S.B., Kurup, A., Verma, R.P. QSAR of cytochrome P450, Drug Metab. Rev. 2004, 36, 105–56.
20. Jalaie, M., Arimoto, R., Gifford, E., Schefzick, S., Waller, C.L. Prediction of drug-like molecular properties, in Bajorath, J., editor. Chemoinformatics: Concepts, Methods, and Tools for Drug Discovery. Totowa: Humana Press; 2005, p. 449–520.
21. Crivori, P., Poggesi, I. Computational approaches for predicting CYP-related metabolism properties in the screening of new drugs, Eur. J. Med. Chem. 2006, 41, 795–808.
22. Refsgaard, H.H.F., Jensen, B.F., Christensen, I.T., Hagen, N., Brockhoff, P.B. In silico prediction of cytochrome P450 inhibitors, Drug Dev. Res. 2006, 67, 417–29.
23. Chohan, K.K., Paine, S.W., Waters, N.J. Quantitative structure activity relationships in drug metabolism, Curr. Top. Med. Chem. 2006, 6, 1569–78.
24. Arimoto, R. Computational models for predicting interactions with cytochrome P450 enzyme, Curr. Top. Med. Chem. 2006, 6, 1609–18.
25. Verras, A., Kuntz, I.D, Ortiz de Montellano, P.R. Cytochrome P450 enzymes: computational approaches to substrate prediction, in Spellmeyer, D.C., editor. Annual reports in Computational Chemistry. Amsterdam: Elsevier; 2006, p. 171–95. Ref. Type: Serial (Book, Monograph).
26. Madden, J.C., Cronin, M.T. Structure-based methods for the prediction of drug metabolism, Expert Opin. Drug Metab. Toxicol. 2006, 2, 545–57.
27. Danielson, P.B. The cytochrome P450 superfamily: biochemistry, evolution and drug metabolism in humans, Curr. Drug Metab. 2002, 3, 561–97.
28. Kumar, G.N., Surapaneni, S. Role of drug metabolism in drug discovery and development, Med. Res. Rev. 2001, 21, 397–411.
29. Rendic, S., Di Carlo, F.J. Human cytochrome P450 enzymes: a status report summarizing their reactions, substrates, inducers, and inhibitors, Drug Metab. Rev. 1997, 29, 413–580.
30. Guengerich, F.P. Common and uncommon cytochrome P450 reactions related to metabolism and chemical toxicity, Chem. Res. Toxicol. 2001, 14, 611–50.
31. Evans, W.E., Relling, M.V. Pharmacogenomics: translating functional genomics into rational therapeutics, Science 1999, 286, 487–91.
32. Goldstein, J.A. Clinical relevance of genetic polymorphisms in the human CYP2C subfamily, Br. J. Clin. Pharmacol. 2001, 52, 349–55.
33. de Groot, M.J., Alex, A.A., Jones, B.C. Development of a combined protein and pharmacophore model for cytochrome P450 2C9, J. Med. Chem. 2002, 45, 1983–93.
34. Wrighton, S.A., Schuetz, E.G., Thummel, K.E., Shen, D.D., Korzekwa, K.R., Watkins, P.B. The human CYP3A subfamily: practical considerations, Drug Metab. Rev. 2000, 22, 339–61.
35. Smith, D.A., Ackland, M.J., Jones, B.C. Properties of cytochrome P450 isoenzymes and their substrates. Part 2: properties of cytochrome P450 substrates, Drug Discov. Today 1997, 2, 479–86.
36. Smith, D.A., Ackland, M.J., Jones, B.C. Properties of cytochrome P450 isoenzymes and their substrates. Part 1: active site characteristics, Drug Discov. Today 1997, 2, 406–14.
37. Korzekwa, K.R., Krishnamachary, N., Shou, M., Ogai, A., Parise, R.A., Rettie, A.E., Gonzalez, F.J., Tracy, T.S. Evaluation of atypical cytochrome P450 kinetics with two-substrate models: evidence that multiple substrates can simultaneously bind to cytochrome P450 active sites, Biochemistry 1998, 37, 4137–47.
38. Hutzler, J.M., Tracy, T.S. Atypical kinetic profiles in drug metabolism reactions, Drug Metab. Dispos. 2002, 30, 355–62.
39. Hosea, N.A., Miller, G.P., Guengerich, F.P. Elucidation of distinct ligand binding sites for cytochrome P450 3A4, Biochemistry 2000, 39, 5929–39.
40. Schrag, M.L., Wienkers, L.C. Covalent alteration of the CYP3A4 active site: evidence for multiple substrate binding domains, Arch. Biochem. Biophys. 2001, 391, 49–55.
41. Wolff, T., Distlerath, L.M., Worthington, M.T., Groopman, J.D., Hammons, G.J., Kadlubar, F.F., Prough, R.A., Martin, M.V., Guengerich, F.P. Substrate specificity of human liver cytochrome P-450 debrisoquine 4-hydroxylase probed using immunochemical inhibition and chemical modeling, Cancer Res. 1985, 45, 2116–22.
42. Strobl, G.R., von Kruedener, S., Stockigt, J., Guengerich, F.P., Wolff, T. Development of a pharmacophore for inhibition of human liver cytochrome P-450 2D6: molecular modeling and inhibition studies, J. Med. Chem. 1993, 36, 1136–45.

43. Islam, S.A., Wolf, C.R., Lennard, M.S., Sternberg, M.J. A three-dimensional molecular template for substrates of human cytochrome P450 involved in debrisoquine 4-hydroxylation, Carcinogenesis 1991, 12, 2211–9.
44. Meyer, U.A., Gut, J., Kronbach, T., Skoda, C., Meier, U.T., Catin, T., Dayer, P. The molecular mechanisms of two common polymorphisms of drug oxidation—evidence for functional changes in cytochrome P-450 isozymes catalysing bufuralol and mephenytoin oxidation, Xenobiotica 1986, 16, 449–64.
45. Koymans, L., Vermeulen, N.P., van Acker, S.A., te Koppele, J.M., Heykants, J.J., Lavrijsen, K., Meuldermans, W., Donne-Op den Kelder, G.M. A predictive model for substrates of cytochrome P450-debrisoquine (2D6), Chem. Res. Toxicol. 1992, 5, 211–9.
46. Raunio, H., Rautio, A., Gullsten, H., Pelkonen, O. Polymorphisms of CYP2A6 and its practical consequences, Br. J. Clin. Pharmacol. 2001, 52, 357–63.
47. Lewis, D.F.V., Sams, C., Loizou, G.D. A quantitative structure–activity relationship analysis on a series of alkyl benzenes metabolized by human cytochrome P450 2E1, J. Biochem. Mol. Toxicol. 2003, 17, 47–52.
48. Pelkonen, O., Maenpaa, J., Taavitsainen, P., Rautio, A., Raunio, H. Inhibition and induction of human cytochrome P450 (CYP) enzymes, Xenobiotica 1998, 28, 1203–53.
49. Kubinyi, H. From narcosis to hyperspace: the history of QSAR, Quant. Struct.–Act. Relat. 2002, 21, 348–56.
50. Kubinyi, H. Quantitative structure–activity relationships in drug design, in v. Ragué Schleyer, P., editor. Encyclopedia of Computational Chemistry. Chichester: Wiley; 1998, p. 2309–20.
51. Todeschini, R., Consonni, V. Handbook of Molecular Descriptors, Weinheim: Wiley-VCH; 2000.
52. Bravi, G., Wikel, J.H. Application of MS-WHIM descriptors: 1. Introduction of new molecular surface properties and 2. Prediction of binding affinity data, Quant. Struct.–Act. Relat. 2000, 19, 29–38.
53. Bravi, G., Gancia, E., Mascagni, P., Pegna, M., Todeschini, R., Zaliani, A. MS-WHIM, new 3D theoretical descriptors derived from molecular surface properties: a comparative 3D QSAR study in a series of steroids, J. Comput.-Aided Mol. Des. 1997, 11, 79–92.
54. Greco, G., Novellino, E., Martin, Y.C. Approaches to three-dimensional quantitative structure–activity relationships, Rev. Comp. Chem. 1997, 11, 183–240.
55. Martin, Y.C. 3D QSAR: current state, scope, and limitations, Perspect. Drug Discov. Des. 1998, 12–14, 3–23.
56. Cramer III, R.D., Patterson, D.E., Bunce, J.D. Comparative molecular field analysis (CoMFA). 1. Effect of shape on binding of steroids to carrier proteins, J. Am. Chem. Soc. 1988, 110, 5959–67.
57. Kubinyi, H. Comparative molecular field analysis, in v. Ragué Schleyer, P., editor. Encyclopedia of Computational Chemistry. Chichester: Wiley; 1998, p. 448–60.
58. Klebe, G., Abraham, U., Mietzner, T. Molecular similarity indices in a comparative analysis (CoMSIA) of drug molecules to correlate and predict their biological activity, J. Med. Chem. 1994, 37, 4130–46.
59. Goodford, P.J. A computational procedure for determining energetically favorable binding sites on biologically important macromolecules, J. Med. Chem. 1985, 28, 849–57.
60. Cruciani, G., Watson, K.A. Comparative molecular field analysis using GRID force-field and GOLPE variable selection methods in a study of inhibitors of glycogen phosphorylase b, J. Med. Chem. 1994, 37, 2589–601.
61. Pastor, M., Cruciani, G., McLay, I., Pickett, S., Clementi, S. GRid-INdependent descriptors (GRIND): a novel class of alignment-independent three-dimensional molecular descriptors, J. Med. Chem. 2000, 43, 3233–43.
62. Cruciani, G., Pastor, M., Guba, W. VolSurf: a new tool for the pharmacokinetic optimization of lead compounds, Eur. J. Pharm. Sci. 2000, 11 Suppl. 2, S29–39.
63. Cruciani, G., Carrupt, P.A., Testa, B. Molecular fields in quantitative structure–permeation relationships: the VolSurf approach, J. Mol. Struct. (THEOCHEM) 2000, 503, 17–30.
64. Wold, S., Johansson, E., Cocchi, M. PLS–partial least-squares projection to latent structures, in Kubinyi, H., editor. 3D-QSAR in Drug Design. Leiden: ESCOM Science Publishers; 1993, p. 523–50.
65. Eriksson, L., Johansson, E., Lindgren, F., Sjøstrøm, M., Wold, S. Megavariate analysis of hierarchical QSAR data, J. Comput.-Aided Mol. Des. 2002, 16, 711–26.

66. Westerhuis, J.A., Kourti, T., MacGregor, J.F. Analysis of multiblock and hierarchical PCA and PLS models, J. Chemometrics 1998, 12, 301–21.
67. Wold, S., Kettaneh, N., Tjessem, K. Hierarchical multiblock PLS and PC models for easier model interpretation and as an alternative to variable selection, J. Chemometrics 1996, 10, 463–82.
68. Kriegl, J.M., Arnhold, T., Beck, B., Fox, T. A support vector machine approach to classify human cytochrome P450 3A4 inhibitors, J. Comput.-Aided Mol. Des. 2005, 19, 189–201.
69. Yap, C.W., Chen, Y.Z. Prediction of cytochrome P450 3A4, 2D6, and 2C9 inhibitors and substrates by using support vector machines, J. Chem. Inf. Model. 2005, 45, 982–92.
70. Molnár, L., Keserü, G.M. A neural network based virtual screening of cytochrome P450 3A4 inhibitors, Bioorg. Med. Chem. Lett. 2002, 12, 419–21.
71. Ekins, S., Berbaum, J., Harrison, R.K. Generation and validation of rapid computational filters for CYP2D6 and CYP3A4, Drug Metab. Dispos. 2003, 31, 1077–80.
72. Susnow, R.G., Dixon, S.L. Use of robust classification techniques for the prediction of human cytochrome P450 2D6 inhibition, J. Chem. Inf. Comput. Sci. 2003, 43, 1308–15.
73. Merkwirth, C., Mauser, H., Schulz-Gasch, T., Roche, O., Stahl, M., Lengauer, T. Ensemble methods for classification in cheminformatics, J. Chem. Inf. Comput. Sci. 2004, 44, 1971–8.
74. Kriegl, J.M., Arnhold, T., Beck, B., Fox, T. Prediction of human cytochrome P450 inhibition using support vector machines, QSAR Comb. Sci. 2005, 24, 491–502.
75. Chohan, K.K., Paine, S.W., Mistry, J., Barton, P., Davis, A.M. A rapid computational filter for cytochrome P450 1A2 inhibition potential of compound libraries, J. Med. Chem. 2005, 48, 5154–61.
76. Moon, T., Chi, M.H., Kim, D.H., Yoon, C.N., Choi, Y.S. Quantitative structure–activity relationships (QSAR) study of flavonoid derivatives for inhibition of cytochrome P450 1A2, Quant. Struct.-Act. Relat. 2000, 19, 257–63.
77. O'Brien, S.E., de Groot, M.J. Greater than the sum of its parts: combining models for useful ADMET prediction, J. Med. Chem. 2005, 48, 1287–91.
78. Arimoto, R., Prasad, M.A., Gifford, E.M. Development of CYP3A4 inhibition models: comparisons of machine-learning techniques and molecular descriptors, J. Biomol. Screen. 2005, 10, 197–205.
79. Kless, A., Eitrich, T. Cytochrome P450 classification of drugs with support vector machines implementing the nearest point algorithm, Lect. Notes Comput. Sci. 2004, 3303, 191–205.
80. Eitrich, T., Kless, A., Druska, C., Meyer, W., Grotendorst, J. Classification of highly unbalanced CYP450 data of drugs using cost sensitive machine learning techniques, J. Chem. Inf. Model. 2007, 47, 92–103.
81. Jensen, B.F., Vind, C., Padkjaer, S.B., Brockhoff, P.B., Refsgaard, H.H. In silico prediction of cytochrome P450 2D6 and 3A4 inhibition using gaussian kernel weighted k-nearest neighbor and extended connectivity fingerprints, including structural fragment analysis of inhibitors versus noninhibitors, J. Med. Chem. 2007, 50, 501–11.
82. Yap, C.W., Xue, Y., Li, Z.R., Chen, Y.Z. Application of support vector machines to in silico prediction of cytochrome P450 enzyme substrates and inhibitors, Curr. Top. Med. Chem. 2006, 6, 1593–607.
83. Korolev, D., Balakin, K.V., Nikolsky, Y., Kirillov, E., Ivanenkov, Y.A., Savchuk, N.P., Ivashchenko, A.A., Nikolskaya, T. Modeling of human cytochrome P450-mediated drug metabolism using unsupervised machine learning approach, J. Med. Chem. 2003, 46, 3631–43.
84. Fox, T., Kriegl, J.M. Machine learning techniques for in silico modeling of drug metabolism, Curr. Top. Med. Chem. 2006, 6, 1579–91.
85. Tropsha, A., Gramatica, P., Gombar, V.K. The importance of being earnest: validation is the absolute essential for successful application and interpretation of QSPR models, QSAR Comb. Sci. 2003, 22, 69–77.
86. Eriksson, L., Jaworska, J., Worth, A.P., Cronin, M.T., McDowell, R.M., Gramatica, P. Methods for reliability and uncertainty assessment and for applicability evaluations of classification- and regression-based QSARs, Environ. Health Perspect. 2003, 111, 1361–75.
87. Hawkins, D.M. The problem of overfitting, J. Chem. Inf. Comput. Sci. 2004, 44, 1–12.
88. Eriksson, L., Arnhold, T., Beck, B., Fox, T., Johansson, E., Kriegl, J.M. Onion design and its application to a pharmaceutical QSAR problem, J. Chemometrics 2004, 18, 188–202.
89. Golbraikh, A., Tropsha, A. Predictive QSAR modeling based on diversity sampling of experimental datasets for the training and test set selection, J. Comput.-Aided Mol. Des. 2002, 16, 357–69.

90. Wold, S., Josefson, M., Gottfries, J., Linusson, A. The utility of multivariate design in PLS modeling, J. Chemometrics 2004, 18, 156–65.
91. Sheridan, R.P., Feuston, B.P., Maiorov, V.N., Kearsley, S.K. Similarity to molecules in the training set is a good discriminator for prediction accuracy in QSAR, J. Chem. Inf. Comput. Sci. 2004, 44, 1912–28.
92. Hansch, C. Quantitative relationships between lipophilic character and drug metabolism, Drug Metab. Rev. 1972, 1, 1–14.
93. Lewis, D.F.V. On the recognition of mammalian microsomal cytochrome P450 substrates and their characteristics: towards the prediction of human P450 substrate specificity and metabolism, Biochem. Pharmacol. 2000, 60, 293–306.
94. Lewis, D.F.V., Modi, S., Dickins, M. Quantitative structure–activity relationships (QSARs) within substrates of human cytochromes P450 involved in drug metabolism, Drug Metab. Drug Int. 2001, 18, 221–42.
95. Lewis, D.F.V., Jacobs, M.N., Dickins, M. Compound lipophilicity for substrate binding to human P450s in drug metabolism, Drug Discov. Today 2004, 9, 530–7.
96. Lewis, D.F.V., Dickins, M. Baseline lipophilicity relationships in human cytochromes P450 associated with drug metabolism, Drug Metab. Rev. 2003, 35, 1–18.
97. Suzuki, H., Kneller, M.B., Rock, D.A., Jones, J.P., Trager, W.F., Rettie, A.E. Active-site characteristics of CYP2C19 and CYP2C9 probed with hydantoin and barbiturate inhibitors, Arch. Biochem. Biophys. 2004, 429, 1–15.
98. Venhorst, J., Onderwater, R.C., Meerman, J.H., Commandeur, J.N., Vermeulen, N.P. Influence of N-substitution of 7-methoxy-4-(aminomethyl)-coumarin on cytochrome P450 metabolism and selectivity, Drug Metab. Dispos. 2000, 28, 1524–32.
99. Lock, R.E., Jones, B.C., Smith, D.A., Hawksworth, G.M. Investigation of substrate structure activity relationships (SSAR) for cytochrome P450 2C19, Br. J. Clin. Pharmacol. 1998, 45, 511P.
100. Kriegl, J.M., Eriksson, L., Arnhold, T., Beck, B., Johansson, E., Fox, T. Multivariate modeling of cytochrome P450 3A4 inhibition, Eur. J. Pharm. Sci. 2005, 24, 451–63.
101. Haji-Momenian, S., Rieger, J.M., Macdonald, T.L., Brown, M.L. Comparative molecular field analysis and QSAR on substrates binding to cytochrome P450 2D6, Bioorg. Med. Chem. 2003, 11, 5545–54.
102. Riley, R.J., Parker, A.J., Trigg, S., Manners, C.N. Development of a generalized, quantitative physicochemical model of CYP3A4 inhibition for use in early drug discovery, Pharm. Res. 2001, 18, 652–5.
103. Lewis, D.F.V., Dickins, M. Substrate SARs in human P450s, Drug Discov. Today 2002, 7, 918–25.
104. Lewis, D.F.V., Modi, S., Dickins, M. Structure–activity relationship for human cytochrome P450 substrates and inhibitors, Drug. Metab. Rev. 2002, 34, 69–82.
105. Lewis, D.F.V. Quantitative structure–activity relationships (QSARs) within the cytochrome P450 system: QSARs describing substrate binding, inhibition and induction of P450s, Inflammopharmacology 2003, 11, 43–73.
106. Fuhr, U., Strobl, G., Manaut, F., Anders, E.-M., Sörgel, F., Lopez-de-Binas, E., Chu, D.T.W., Pernet, A.G., Mahr, G. Quinolone antibacterial agents: relationship between structure and in vitro inhibition of the human cytochrome P450 isoform CYP1A2, Mol. Pharmacol. 1993, 43, 191–9.
107. Iori, F., da Fonseca, R., Ramos, M.J., Menziani, M.C. Theoretical quantitative structure–activity relationships of flavone ligands interacting with cytochrome P450 1A1 and 1A2 isozymes, Bioorg. Med. Chem. 2005, 13, 4366–74.
108. Lee, H., Yeom, H., Kim, Y.G., Yoon, C.N., Jin, C., Choi, J.S., Kim, B.R., Kim, D.H. Structure-related inhibition of human hepatic caffeine N3-demethylation by naturally occurring flavonoids, Biochem. Pharmacol. 1998, 55, 1369–75.
109. Wanchana, S., Yamashita, F., Hashida, M. QSAR analysis of the inhibition of recombinant CYP 3A4 activity by structurally diverse compounds using a genetic algorithm-combined partial least squares method, Pharm. Res. 2003, 20, 1401–8.
110. Leo, A., Hansch, C., Jow, P.Y. Dependence of hydrophobicity of apolar molecules on their molecular volume, J. Med. Chem. 1976, 19, 611–5.
111. Mao, B., Gozalbes, R., Barbosa, F., Migeon, J., Merrick, S., Kamm, K., Wong, E., Costales, C., Shi, W., Wu, C., Froloff, N. QSAR modeling of in vitro inhibition of cytochrome P450 3A4, J. Chem. Inf. Model. 2006, 46, 2125–34.

112. Afzelius, L., Masimirembwa, C.M., Karlén, A., Andersson, T.B., Zamora, I. Discriminant and quantitative PLS analysis of competitive CYP2C9 inhibitors versus non-inhibitors using alignment independent GRIND descriptors, J. Comput.-Aided Mol. Des. 2002, 16, 443–58.
113. Afzelius, L., Zamora, I., Masimirembwa, C.M., Karlén, A., Andersson, T.B., Mecucci, S., Baroni, M., Cruciani, G. Conformer- and alignment-independent model for predicting structurally diverse competitive CYP2C9 inhibitors, J. Med. Chem. 2004, 47, 907–14.
114. Afzelius, L., Zamora, I., Ridderstrom, M., Andersson, T.B., Karlen, A., Masimirembwa, C.M. Competitive CYP2C9 inhibitors: enzyme inhibition studies, protein homology modeling, and three-dimensional quantitative structure–activity relationship analysis, Mol. Pharmacol. 2001, 59, 909–19.
115. Ekins, S., Bravi, G., Binkley, S., Gillespie, J.S., Ring, B.J., Wikel, J.H., Wrighton, S.A. Three- and four-dimensional-quantitative structure activity relationship (3D/4D-QSAR) analyses of CYP2C9 inhibitors, Drug Metab. Dispos. 2000, 28, 994–1002.
116. Ekins, S., Bravi, G., Binkley, S., Gillespie, J.S., Ring, B.J., Wikel, J.H., Wrighton, S.A. Three and four dimensional-quantitative structure activity relationship (3D/4D-QSAR) analyses of CYP2D6 inhibitors, Pharmacogenetics 1999, 9, 477–89.
117. Ekins, S., Bravi, G., Binkley, S., Gillespie, J.S., Ring, B.J., Wikel, J.H., Wrighton, S.A. Three- and four-dimensional quantitative structure activity relationship analyses of cytochrome P-450 3A4 inhibitors, J. Pharmacol. Exp. Ther. 1999, 290, 429–38.
118. Ekins, S., Bravi, G., Ring, B.J., Gillespie, T.A., Gillespie, J.S., Vandenbranden, M., Wrighton, S.A., Wikel, J.H. Three-dimensional quantitative structure activity relationship analyses of substrates for CYP2B6, J. Pharmacol. Exp. Ther. 1999, 288, 21–9.
119. Snyder, R., Sangar, R., Wang, J., Ekins, S. Three-dimensional quantitative structure activity realtionship for CYP2D6 substrates, Quant. Struct.–Act. Relat. 2002, 21, 357–68.
120. Lozano, J.J., Pastor, M., Cruciani, G., Gaedt, K., Centeno, N.B., Gago, F., Sanz, F. 3D-QSAR methods on the basis of ligand-receptor complexes. Application of COMBINE and GRID/GOLPE methodologies to a series of CYP1A2 ligands, J. Comput.-Aided Mol. Des. 2000, 14, 341–53.
121. Poso, A., Juvonen, R., Gynther, J. Comparative molecular field analysis of compounds with CYP2A5 binding affinity, Quant. Struct.–Act. Relat. 1995, 14, 507–11.
122. Korhonen, L.E., Rahnasto, M., Maehoenen, N.J., Wittekindt, C., Poso, A., Juvonen, R.O., Raunio, H. Predictive three-dimensional quantitative structure–activity relationship of cytochrome P450 1A2 inhibitors, J. Med. Chem. 2005, 48, 3808–15.
123. Poso, A., Gynther, J., Juvonen, R. A comparative molecular field analysis of cytochrome P450 2A5 and 2A6 inhibitors, J. Comput.-Aided Mol. Des. 2001, 15, 195–202.
124. Rahnasto, M., Raunio, H., Poso, A., Wittekindt, C., Juvonen, R.O. Quantitative structure–activity relationship analysis of inhibitors of the nicotine metabolizing CYP2A6 enzyme, J. Med. Chem. 2005, 48, 440–9.
125. Lesigiarska, I., Pajeva, I., Yanev, S. Quantitative structure–activity relationship (QSAR) and three-dimensional QSAR analysis of a series of xanthates as inhibitors and inactivators of cytochrome P450 2B1, Xenobiotica 2002, 32, 1063–77.
126. Jones, J.P., He, M., Trager, W.F., Rettie, A.E. Three-dimensional quantitative structure–activity relationship for inhibitors of cytochrome P450 2C9, Drug Metab. Dispos. 1996, 24, 1–6.
127. Rao, S., Aoyama, R., Schrag, M., Trager, W.F., Rettie, A., Jones, J.P. A refined 3-dimensional QSAR of cytochrome P450 2C9: computational predictions of drug interactions, J. Med. Chem. 2000, 43, 2789–96.
128. Locuson, C.W., Wahlstrom, J.L. Three-dimensional quantitative structure–activity relationship analysis of cytochromes P450: effect of incorporating higher-affinity ligands and potential new applications, Drug Metab. Dispos. 2005, 33, 873–8.
129. Locuson, C.W., Rock, D.A., Jones, J.P. Quantitative binding models for CYP2C9 based on benzbromarone analogues, Biochemistry 2004, 43, 6948–58.
130. Locuson, C.W., Suzuki, H., Rettie, A.E., Jones, J.P. Charge and substituent effects on affinity and metabolism of benzbromarone-based CYP2C19 inhibitors, J. Med. Chem. 2004, 47, 6768–76.
131. Vaz, R.J., Nayeem, A., Santone, K., Chandrasena, G., Gavai, A.V. A 3D-QSAR model for CYP2D6 inhibition in the aryloxypropanolamine series, Bioorg. Med. Chem. Lett. 2005, 15, 3816–20.
132. Waller, C.L., Evans, M.V., McKinney, J.D. Modeling the cytochrome P450-mediated metabolism of chlorinated volatile organic compounds, Drug Metab. Dispos. 1996, 24, 203–10.

133. Golbraikh, A., Tropsha, A. Beware of q2!, J. Mol. Graph. Model. 2002, 20, 269–76.
134. Netzeva, T.I., Worth, A., Aldenberg, T., Benigni, R., Cronin, M.T., Gramatica, P., Jaworska, J.S., Kahn, S., Klopman, G., Marchant, C.A., Myatt, G., Nikolova-Jeliazkova, N., Patlewicz, G.Y., Perkins, R., Roberts, D., Schultz, T., Stanton, D.W., van de Sandt, J.J., Tong, W., Veith, G., Yang, C. Current status of methods for defining the applicability domain of (quantitative) structure–activity relationships. The report and recommendations of ECVAM Workshop 52, Altern. Lab Anim. 2005, 33, 155–73.
135. Norinder, U. In silico modelling of ADMET-a minireview of work from 2000 to 2004, SAR QSAR Environ. Res. 2005, 6, 1–11.
136. Zuegge, J., Fechner, U., Roche, O., Parrott, N.J., Engkvist, O., Schneider, G. A fast virtual screening filter for cytochrome P450 3A4 inhibition liability of compound libraries, Quant. Struct.-Act. Relat. 2002, 21, 249–56.
137. Masimirembwa, C.M., Ridderstroem, M., Zamora, I., Andersson, T.B. Combining pharmacophore and protein modeling to predict CYP450 inhibitors and substrates, Method. Enzymol. 2002, 357, 133–44.
138. Ortiz, A.R., Pisabarro, M.T., Gago, F., Wade, R.C. Prediction of drug binding affinities by comparative binding energy analysis, J. Med. Chem. 1995, 38, 2681–91.
139. Boyer, S., Zamora, I. New methods in predictive metabolism, J. Comput.-Aided Mol. Des. 2002, 16, 403–13.
140. Zamora, I., Afzelius, L., Cruciani, G. Predicting drug metabolism: a site of metabolism prediction tool applied to the cytochrome P450 2C9, J. Med. Chem. 2003, 46, 2313–24.
141. Cruciani, G., Carosati, E., De Boeck, B., Ethirajulu, K., Mackie, C., Howe, T., Vianello, R. MetaSite: understanding metabolism in human cytochromes from the perspective of the chemist, J. Med. Chem. 2005, 48, 6970–9.
142. Shimada, T., Yamazaki, H., Mimura, M., Inui, Y., Guengerich, F.P. Interindividual variations in human liver cytochrome P-450 enzymes involved in the oxidation of drugs, carcinogens and toxic chemicals: studies with liver microsomes of 30 Japanese and 30 Caucasians, J. Pharmacol. Exp. Ther. 1994, 270, 414–23.
143. Lewis, D.F.V. Guide to Cytochromes P450 Structure and Function, Boca Raton: CRC Press; 2001.
144. Guengerich, F.P. Human cytochrome P450 enzymes, in Ortiz de Montellano, P.R., editor. Cytochrome P450: Structure, Mechanism, and Biochemistry. New York: Plenum Press; 1995, p. 473–535.
145. Lewis, D.F.V., Dickins, M. Quantitative structure–activity relationships (QSARs) within series of inhibitors for mammalian cytochromes P450 (CYPs), J. Enzym. Inhib. 2001, 16, 321–30.

Section 3
Chemical Education

Section Editor: Theresa Zielinski

Department of Chemistry, Medical Technology
and Physics
Monmouth University
400 Cedar Avenue
West Long Branch
NJ 07764-1898, USA

CHAPTER 6

Observations on Crystallographic Education

Phillip E. Fanwick[*]

Contents			
	1.	Introduction	85
	2.	Objectives for Teaching Crystallography	86
		2.1 Be familiar with the input and output for the X-ray crystallographic experiment	86
		2.2 Analyze and work with crystallographic results	87
		2.3 Solve and refine routine structures	88
		2.4 Collect data on routine crystal structures	88
		2.5 Collect and solve non-routine structures	88
		2.6 Create new methods and software for crystallography	88
		2.7 Develop new crystallographic instruments	89
	3.	Bragg's Law	89
	4.	Relating Crystallography to Chemistry	90
	5.	Creativity	96
	6.	Conclusions	98
	References		98

1. INTRODUCTION

Since the first crystallographic diffraction experiment was reported in 1912 [1], crystallography has been a major tool in the development of a variety of chemical sciences. Over the past several decades it has been the second most important analytical method after nuclear magnetic resonance spectroscopy. In chemistry, crystallography has been basic to the development of fields as diverse as organometallic chemistry and medicinal chemistry. It has had a pivotal role in the advancement of solid state chemistry. Diffraction studies are also now entrenched as a major research tool in structural biology.

Given the utility, diversity and importance of X-ray diffraction it is surprising how little is taught about the technique in the typical chemistry curriculum

[*] Department of Chemistry, Purdue University, West Lafayette, IN 47907, USA

Students are exposed to a hodge-podge of equations and facts that provide little insight into the workings and power of crystallography. Little education in the needs and output of a modern crystallography laboratory are provided. It is a testament to the intrinsic usefulness of crystallographic studies that the use of crystallography has grown. This has occurred despite how little most chemists understand about the method.

In this manuscript I hope to begin a discussion on what has to be done to improve crystallographic education. I will present some ideas on how to improve the teaching and also make some suggestions about what topics should be covered.

2. OBJECTIVES FOR TEACHING CRYSTALLOGRAPHY

A common complaint from students and researchers alike is that most attempts to provide instruction in crystallography appear to provide instruction in much of the theory of crystallography. Unfortunately this theory provides little insight into the actual use of the method. In most general chemistry text books there are discussions of Bragg's Law, closest packed structures and other topics that do not provide any orderly picture of the X-ray diffraction experiment or its results. The confusion continues in physical chemistry where the student is introduced to Miller Planes and Indices, and typically powder diffraction. Finally, there are the specialized texts which frequently begin with lessons on Fourier theory, unit cells, and space group symmetry. While all of these topics are important to understanding crystallography they fail to provide the student with an overview of the technique.

Therefore, if crystallographic education is to improve it is important to ask what the objectives in teaching the topic are. These should be organized in a hierarchy from simple to difficult. Like all educational objectives they should be [2]
Relevant
Unequivocal
Feasible
Logical
Observable
Measurable

Below are seven broad objectives for crystallography. The list is quite general and obviously there is considerable overlap between the objectives. The objectives are listed in order of increasing difficulty. Each objective can be then divided into more specific objectives which would meet all of the individual requirements.

2.1 Be familiar with the input and output for the X-ray crystallographic experiment

Before any chemist can use crystallographic methods he must be familiar with the requirements for the experiment and what results are to be obtained. First, knowledge of the sample requirements must be carefully understood. X-ray crystallography can only be done on samples that are single crystals (though more modern equipment can handle multiple crystals). The required size (ideally about 0.3 mm on an edge) and purity of the sample should be known. The concept of what a single crystal is must be introduced. Likewise, the concept that the experiment is not greatly affected by impurities on the surface of the crystal or coatings of glue or grease needs to be emphasized.

The output of the crystallographic study also needs to be understood. The experiment determines the dimensions of a unit cell which are used as basis vectors in a crystallographic coordinate system, the internal symmetry of this cell, the coordinates of the unique part of the unit cell in terms of the basis vectors and the atomic displacement parameters of each atom. While parameters such as bond distances and angles and molecular drawings can be determined from this data they are not the raw output of the experiment. It seems almost silly to mention that the results especially confirmations may only apply to the crystal studied and not to the isolated molecule or in solution.

2.2 Analyze and work with crystallographic results

Most chemists are not interested in the raw parameters from the crystallographic experiment but rather those that can be derived. Therefore, it is important to be able to obtain the derived results. Of special importance are the statistical descriptors. Every parameter has a standard uncertainty (su). This is usually reported as a number in parentheses and represents the uncertainty in the last digit. For su's between 1.0 and 1.95 the result is reported to two places. In general, anything differing by less than 3 su's is considered to be statistically indistinguishable. Therefore it is important when comparing parameters to not only focus on the difference in the values but also their statistical significance.

A second important skill is the ability to produce and manipulate molecular drawings derived from crystallographic data. Most chemists want to "see" their structures so the can readily observe the connectivity, isomer, and packing. Today there are many elegant programs that will run on personal computers and provide high quality graphics. Even better, many are free for academic users. A good introduction to what is available can be found on the CCP14 web site. Of special importance are the shapes of the atoms when thermal ellipsoids are plotted. These ellipsoids are referred to as the atomic displacement parameters (adp's) and result ideally from the vibrational motion of the atoms in the crystal. However, they can also reflect any systematic errors or mistakes in the structure. Unusually large or small adp's may reflect an incorrect element assignment. Unusually shaped ellipsoids can indicate disorder or other problems.

Today, the entire description and the results of the crystallography experiment are contained in the Crystallographic Information File (cif). This file contains a description of the experimental procedures used, the raw data and the derived parameters. All chemists should be familiar with the cif format and how to work with the file. The file consists of data headings followed by values. The data headings are rigorously defined in a dictionary which provides both a definition and range of acceptable values. Therefore, working with cif's can be a useful way of educating students about the crystallographic experiment.

The goal of all of the above is to enable students to use and critically evaluate structural data. Contrary to the beliefs of many chemists, crystallographic results have been published which have been incorrect. While many of these reports had technical problems with the crystallography, there have also been reports of totally incorrect compounds. All chemists not just crystallographers need the ability to evaluate the correctness of a structure. This is especially true today since most chemistry manuscripts containing crystallographic results are no longer reviewed by a crystallographer.

The two major objectives above should be included in the crystallographic education every chemistry student receives. They can be taught without the use of expensive equipment, specialized computer programs or dedicated text books. Meeting these first two objectives would place the knowledge of crystallographic methods at about the same level as NMR. Sadly, most undergraduate chemistry students and even many graduate students do not meet these criteria.

2.3 Solve and refine routine structures

Obviously working with raw crystallographic data provides further insight and requires greater skill than just using the results. To meet this objective requires evaluation of the reported unit cell, determination of the space group, solution and then refinement of the structure. Again much of the required software is available at no cost and will run on most personal computers. The data could be acquired from most diffraction labs so again no expensive X-ray equipment would be needed.

2.4 Collect data on routine crystal structures

To accomplish this objective a working diffractometer is required. However, there are currently several consortium's set up for remote access to the most modern instrumentation. To complete the objective the student would have to operate the diffractometer, determine and evaluate the unit cell, and collect a complete data set. It can be assumed that the data collection instrument works correctly and so instruction on how it works is not required.

2.5 Collect and solve non-routine structures

There are crystals that pose unusual difficulties in working with them. Among the problems encountered are disorder, twinning and non-commensurate structures. While all can be handled using modern instruments and software, they are beyond the norm for the typical chemist. This is the realm of the professional crystallographer.

2.6 Create new methods and software for crystallography

Again this is well beyond what is expected for a well trained chemist. However, there is a need for researchers with these advanced skills.

2.7 Develop new crystallographic instruments

This requires yet another increase in knowledge and a melding of crystallography, machinery and electronics.

3. BRAGG'S LAW

Nothing illustrates more clearly the problem of modern crystallographic education than Bragg's Law. It is the one piece of knowledge every chemist has regarding X-ray diffraction. This is not surprising in that the law and typically its derivation are found in almost every level of chemistry text from general chemistry to graduate texts. Yet, most students have no idea how this law is related to determining a crystal structure.

Bragg's Law is represented by the equation

$$2d \sin(\theta) = n\lambda.$$

In this equation d represents the distance between Bragg Planes, θ the angle of incidence and reflection of the X-ray beam, λ the wavelength of the X-rays, and n the order of diffraction. The law was derived by Sir W. H. Bragg and his son Sir W. L. Bragg and first reported in 1912 [3]. The Braggs were awarded the Noble Prize in physics in 1915 for their use of this equation to solve the structures of simple salts.

There is no question that Bragg's Law is very important in the history and development of science. This one equation gave proof to the electromagnetic nature of X-rays, the organization of crystals, and definitive support to the atomic theory. The equation allowed for the creation of the crystal spectrometer that was used by Henry Moseley to directly determine atomic numbers. Much of crystallography follows from this equation.

Besides its historical significance, Bragg's Law is very useful pedagogically. First it provides an equation with three (four if you count n) variables which can be the source of many practice and test questions. Even better, it also involves a trigonometric function. Beyond that it is one of the few laws that can be derived from first principals even in general chemistry. Thus there are good reasons for its ubiquitous presence.

However, there is much about Bragg's Law that places it outside the mainstream. It provides no insight what so ever into how a crystal structure is determined. In the end it tells where the data is to be found but gives no information about what the data is or how it is related to the crystal structure. It does not directly relate to any of the first three criteria listed above. If the workings of diffractometers are not to be taught then it is also not need for criterion 4.

Even more confusing is the typical derivation of Bragg's Law which is illustrated in Figure 6.1. It is inferred that the derivation represents the actual physics of the X-ray diffraction experiment with the X-rays reflecting off of various planes. This immediately brings up the question of what these planes are and why they reflect the X-ray beam. To demystify this problem it has become common to place

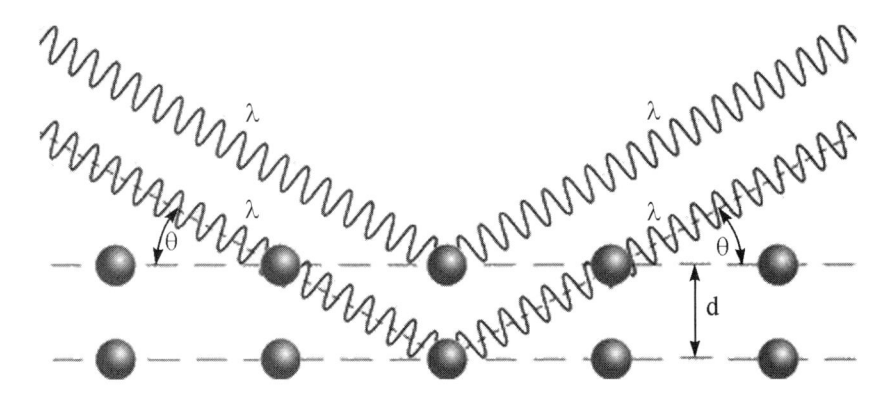

FIGURE 6.1 Typical derivation of Bragg's law.

the atoms on these planes. This is not correct. In the end the derivation of Bragg's Law frequently confuses students more than it provides useful information about crystallography.

There is one positive in deriving Bragg's Law. The derivation suggests that the diffracted intensity occurs because of constructive and destructive interference of the X-ray radiation. With some guidance this can lead to the correct idea that diffraction is another example of a Fourier transform technique which can then be expanded upon.

One final irony in this whole situation is that crystallographic data are referred to as "reflections" thus reinforcing an incorrect model. Even the cif dictionary begins all items involving the raw data as reflections. It is no wonder that non-specialists are confused by Bragg's Law and its place in understanding crystallography.

4. RELATING CRYSTALLOGRAPHY TO CHEMISTRY

A major problem with crystallographic education is that crystallography is frequently approached as some unique science. Therefore, chemistry students do not apply their knowledge of chemistry to analyzing or learning crystallography. This creates a disconnection between chemistry and crystallography which is not natural.

The first thing that needs to be recognized is that most crystal structures are done as a form of chemical analysis. While the X-ray diffraction experiment is certainly capable of providing answers to many questions beyond chemical analysis, these are not the justifications for most structural studies. A comparison of the analytical capabilities of NMR and X-ray crystallography are given in Table 6.1. Just providing this information goes a long way to satisfying the first objective above.

As mentioned above crystallography is actually an example of a Fourier transfer technique. Therefore it is reasonable to use what students know about other such techniques to explain the diffraction experiment. The most familiar such technique is Fourier Transform Nuclear Magnetic Resonance (ft-nmr). A typical ft-nmr

TABLE 6.1 A comparison of crystallography and NMR as analytical methods

Analytical result	Crystallography	NMR
Qualitative analysis	Good for light atoms Bad for metals	Possible but rarely done
Quantitative analysis	Fair—depends on the skill of the crystallographer	Fair
Connectivity and isomer determination	Excellent	Very good
Nature of bonding	Good—requires interpretation	Excellent
Absolute configuration	Can be determined	Poor
Follow reactions	Time resolved crystallography at synchrotron sources only	Excellent

spectrum is given in Figure 6.2. The top figure is the data as collected in the Time Domain and is frequently referred to as the free induction decay (fid). The fid is the input to the transform. The resulting spectrum is in the frequency domain and plots intensity versus frequency.

In some respects the crystallographic experiment is the same. The desired domain is the electron density at any coordinate x, y, z. An assumption is made that the density can be accounted for by the sum of the densities of the component atoms and that regions of high density represent the locations of atoms. There is a corresponding domain in which the data are collected. This is typically referred to as reciprocal space. Unlike the data collection domain in ft-nmr it has no simple physical basis and in fact has the rather complicated units of reciprocal distance. However, just as in the case of ft-nmr the data in the collection domain is not easily analyzed until the Fourier Transform is carried out.

There are several important differences between the crystallographic and ft-nmr experiment. First, obviously the ft-nmr data are one dimensional involving time and amplitude. The crystallographic data is three dimensional and therefore must be described by three spatial coordinates and the appropriate value. This makes it more difficult to visualize. Secondly, the ft-nmr data is a continuous wave. The crystallographic data consists of distinct diffraction spots. Instead of using x, y, z to label these data, the Miller indices h, k, l are used as coordinates and can be related to the Bragg planes.

A big difference between crystallography and ft-nmr is in what is measured. Ft-nmr measures the amplitude of the wave and this is suitable for direct use in performing the Fourier transform. For experimental reasons, X-ray diffraction is only capable of measuring the intensity of the data which is the square of the amplitude. This means that the sign of the data is missing. Strictly this is only correct if the symmetry of the unit cell contains an inversion center. For cells without an inversion center the situation is even more complex. In theory the ft-nmr data could be assigned signs by observing the maxima and minima and assuming that each crossing through zero alternates the sign. Because the crystallographic data

(A)

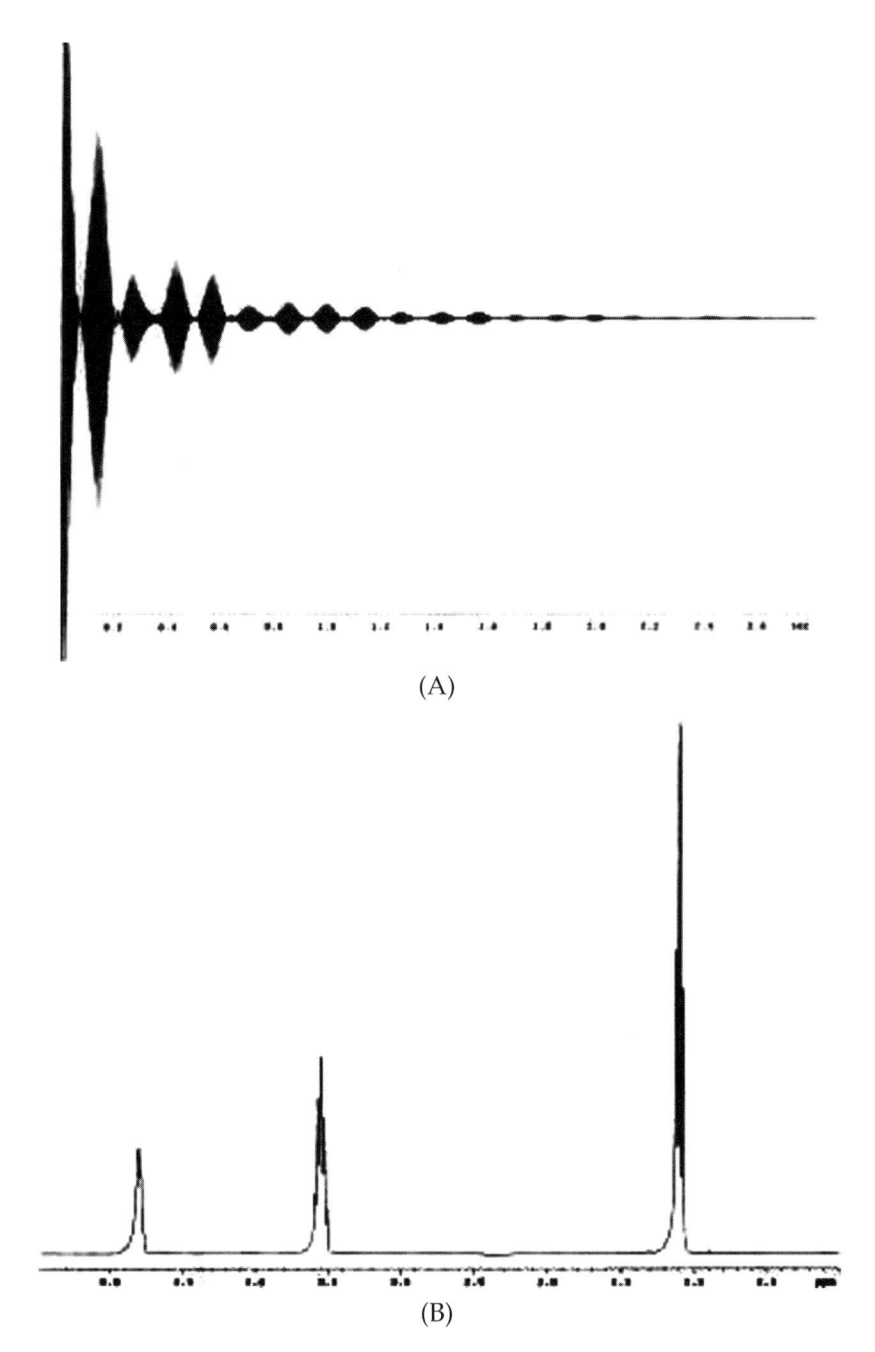

(B)

FIGURE 6.2 Ft-nmr spectra of neat ethanol. (A) The free inductive decay. (B) The resulting Fourier transform. (C) The Fourier transform using the first 7% of the FID.

(C)

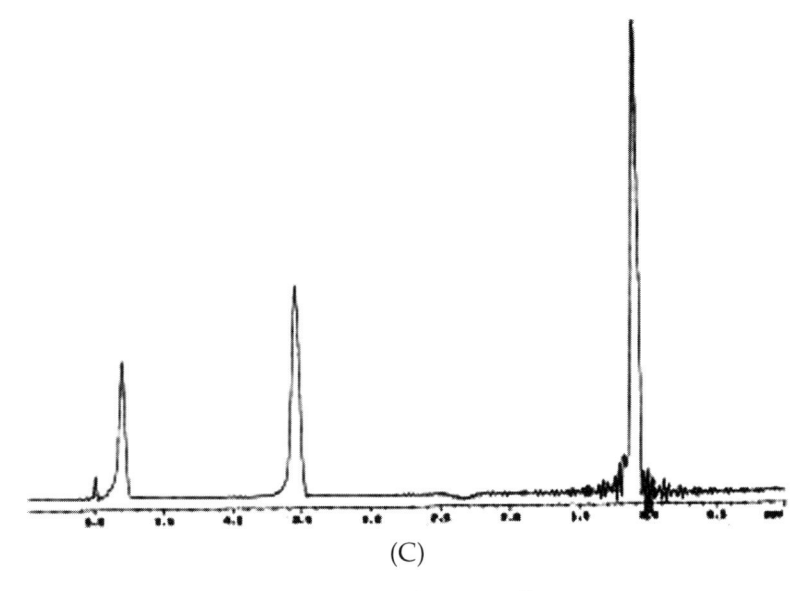

FIGURE 6.2 (*Continued.*)

is discrete, the saddle points are not likely to be found. The result is that the data collected cannot simply be Fourier transformed to the resulting electron density space. This is referred to as the phase problem and much of the processing of crystallographic data is a result of this problem.

One crystallographic concept that can readily be related to ft-nmr is resolution. In theory the fid extends forever while its amplitude decreases with time. Obviously, there is some time beyond which there is no reason to collect data because the signal is less than the noise level. For this data even multiple scans will not produce reliable values. However, whenever the data collection time is truncated then information will be lost. The lower spectrum in Figure 6.1 represents the Fourier transform of such a truncated collection. Obviously the crude details of the spectrum are still present but the fine structure of the splinting of these peaks is missing. The spectrum is said to lack resolution.

Similarly, crystallographic data could be collected over an infinite set of values of *hkl* but again the intensity decreases with increasing θ and no real additional information will be gained. However, truncating the data collection does result in some loss of information. Generally the resolution d is given in Angstroms and represents the distance at which two carbon atoms can be resolved in the electron density space. The smaller the value of d the greater the resolution is. This resolution can be converted into a maximum value of θ by entering the resolution as d in Bragg's Law. Most small molecule data collection is done to a resolution less than 1 Å. In protein crystallography the resolution frequently is limited to 2 Å or greater because of the large number of data to be collected and the large loss in intensity as the resolution is increased. A structure refined to such a low resolution may not allow the observation of individual atoms but clearly indicates the amino acids which make up the protein.

D4h

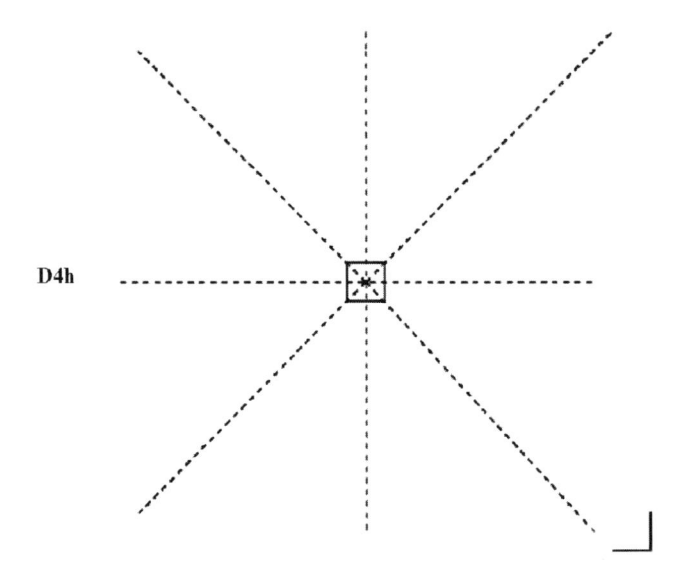

FIGURE 6.3 A minimal representation of D4h symmetry.

One last problem that must be noted is samples which provide a weak signal. If the signal is too weak then the resolution will always be high because beyond a certain resolution there will be no signal. This can be overcome by increasing the intensity of the X-ray beam and improving the signal-to-noise ratio of the data collection equipment.

A second area where students existing knowledge is not utilized when teaching crystallography is in introducing space group symmetry. Most undergraduates are introduced to point group symmetry and are already familiar with symmetry operations, groups, and using this information. It does not take much to extend this knowledge to space groups.

Consider a square planar molecule. This molecule belongs to the point group D4h. The group has sixteen symmetry operations. However, not all are unique and the drawing in Figure 6.3 illustrates the minimum to define this point group. The square in the middle is the 4-fold rotation axis. The dashed lines represent mirror planes that are perpendicular to the plane. The bracket at the lower right represents a mirror place perpendicular to the rotation axis. All the other operations can be derived from combining this minimal set. For example, it can be proved (it is a good homework exercise) that the line that is the intersection of two perpendicular mirrors is a 2-fold rotation axis.

Given this figure it is easy to place any single object containing D4h symmetry into the axis system. However, students are completely befuddled when asked to place water into such a symmetry drawing. They know that water possesses C2v symmetry which is considerably lower than D4h. However, the answer involves one additional concept. It is possible to place more than one molecule into the drawing. This is illustrated in Figure 6.4. On the pattern, a set of four molecules have been added. It is important to note that for this arrangement the 2-fold axis

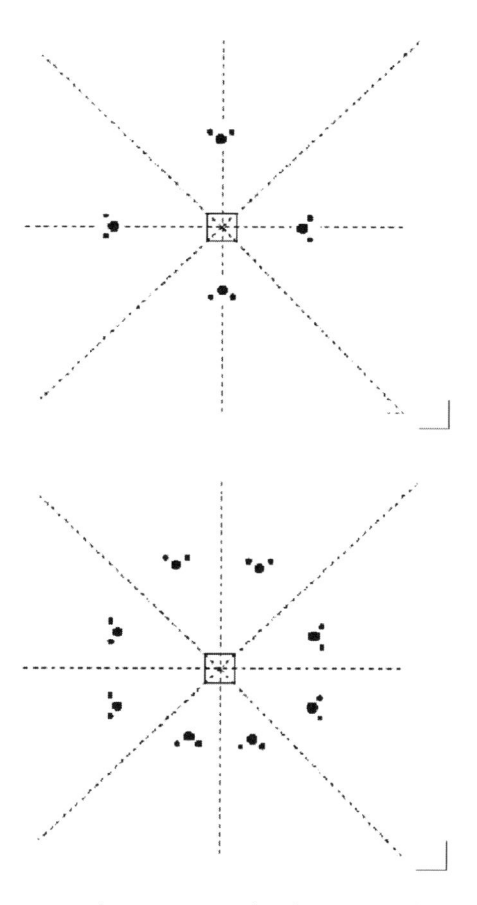

FIGURE 6.4 Placing water molecules into a D4h pattern.

in the water is coincident to an axis in the symmetry group and the same is true for the two mirror planes. This means that each water molecule has full C2v symmetry imposed upon it by the symmetry of the drawing.

Looking deeper into the drawing, assuming the horizontal mirror is in the $z = 0$ plane then the coordinates of one oxygen atom are $(x, 0, 0)$. The other three oxygen atoms are then produced by the 4-fold symmetry operation. This coordinate represents a special position within the unit cell with C2v symmetry. For anything to occupy this position the object must have C2v symmetry or have C2v as a sub-group of its point group. Since the water is located on this special position only the oxygen and one hydrogen atom are independent. The second hydrogen is related by the C2v symmetry of the special position.

The other drawing illustrates eight water molecules in this space group. In this case the space group only imposes the point group symmetry Cm on the water molecule. As shown, the oxygen and both hydrogen atoms are independent. However, the water could be rotated 90 degrees in which case the point group symmetry would still be Cm but now the two hydrogen atoms are related by the mirror plane.

In the most general case, sixteen identical items could be arranged in the illustrated pattern. In this case there would be two sets of eight items related by the horizontal mirror. There would be no restrictions on the point group symmetry of the item. Positions that do no contain any symmetry in the space group are called general positions. Note that there are sixteen symmetry operations in the point group D4h and that for objects on a general position each operation takes one molecule into a specific symmetry related one.

It is important to realize that this point group to space group approach is only an introduction to crystal symmetry. Other topics that need to be included are translation as a symmetry operation; combined operations such as glide planes; symmetry axes and planes that do not pass through the origin; use of the International Tables for Crystallography; and Hermann-Mauguin notation. Each of these ideas can be grafted onto the base formed from point group symmetry.

5. CREATIVITY

If crystallography is going to be taught in such a way that all chemistry students can understand and use it, then new methods for teaching the material must be found. All too often the material is taught as if the student already has a good understanding of the experiment when in reality they really do not understand what is going on. Unfortunately, in the past little creativity was applied to the presenting the material. However, there are exceptions.

Chapter 3 of Glusker and Trueblood [4] contains some of the most interesting pictures of diffraction. Through the use of optical transforms the effect of a translational repetition on the observed pattern is illustrated. Beyond this it also displays why the separation of the diffraction pattern is only a function of the distance between the objects. This is a very nice introduction to the diffraction experiment.

The Institute for Chemical Education at the University of Wisconsin [5] offers slides that allow one to illustrate the optical interference in the classroom. When a laser pointer is passed through the transform on the slide the diffraction pattern is observed. By changing the color of the laser the effect of wavelength can be demonstrated. The effect of patterns in the transform can readily be seen in the diffraction. This is a very useful way to illustrate how diffraction works beyond Bragg's Law.

Another nice example of creative crystallographic education is in an article by William Clegg [6]. He nicely explains the relationship between the amplitude of the diffracted wave and the electron density. Even better a Microsoft Excel macro downloadable from the journal website [7] is available which allows students to interactively play with the phase problem and Fourier transforms for a 1-dimensional crystal structure.

The Journal of Chemical Education offers a whole crystallographic education package [8]. At the heart of this is Marge Kastner's Crystallographic Courseware.

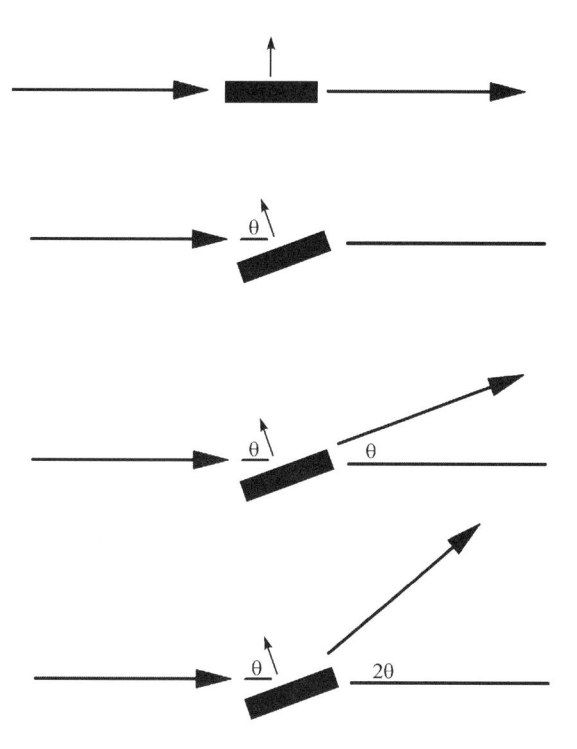

FIGURE 6.5 The moving crystal frame for setting the Bragg angles.

This is wonderful collection of lessons that illustrate everything from crystal growing techniques to many advance concepts. There are excellent sections on space group symmetry and adp's. It contains moving illustrations of many of the key concepts which are very helpful.

However, even in teaching simple concepts creativity should be applied. One question I have been asked is why Bragg's Law uses the angle θ but all diffractometers use 2θ. This can be answered in many ways but the simplest is the best to teach and is illustrated in Figure 6.5. It involves moving from a scheme in which the crystal is stationary and the beam and detector move to one where the beam is stationary and the crystal and detector move. This is much more realistic. In many cases it is not possible to move the beam while the small crystals used can easily be manipulated.

In the first step the crystal is aligned so the desired diffraction plane is parallel to the beam. The crystal is then rotated so it makes an angle θ with the incoming beam. Now the detector is moved. If it is just moved by the angle θ it then makes an angle of 0 with the plane. It must be moved another θ to obtain the desired geometry. Therefore the detector ends up offset an angle of 2θ from the incoming X-ray beam. This is what happens on nearly all data collection equipment.

6. CONCLUSIONS

First, crystallography is too important a tool for modern chemistry for it to be omitted from the average chemistry curriculum. It is widely used and needs to be more widely understood. However, it is also important to take into account the level of the students and what the objectives are for providing this instruction. Failure to set goals just leaves the students with a collection of unrelated and unusable facts.

If the teaching of crystallography is to increase then new approaches to the subject must be developed. Lessons must relate crystallography to what the students have already mastered. This must be done for both the analytical uses and for the science behind the method.

Lastly, more creativity must be brought to teaching the subject. The use of optical transforms, computer education and other modern educational methods can provide much clearer understanding of diffraction methods. Finding clearer ways to teach what are difficult concepts will provide more insight and allow students to better analyze and use crystallographic methods.

REFERENCES

1. Friedrich, W., Knipping, P., von Laue, M. Interferenz-Erscheinungen bei Rontgen-strahlen, Sitz. Bayer. Akad. Wiss. 1912, 303.
2. Guilbert, J.J. How to Devise Educational Objectives, Medical Education 1984, 18(3), 134.
3. Bragg, W.L. The diffraction of short electromagnetic waves by a crystal, Proc. Camb. Phil. Soc. 1913, 17, 43.
4. Glusker, J.P., Trueblood, K.N. Crystal Structure Analysis, A Primer, New York: Oxford University Press; 1985.
5. http://ice.chem.wisc.edu/catlogitems/ScienceKits.htm#OpticalTransform.
6. Clegg, W. An Excel spreadsheet for a one-dimensional Fourier map in X-ray crystallography, J. Chem. Educ. 2004, 81, 908.
7. http://www.jce.divched.org/Journal/Issues/2004/Jun/abs908.html.
8. http://www.jce.divched.org/JCESoft/Programs/Collections/CC/index.html.

CHAPTER 7

Achieving a Holistic Web in the Chemistry Curriculum

Henry S. Rzepa[*]

Contents

Introduction: The Impact of the Web on the Chemistry Curriculum	99
Background: The Trend Towards an Accumulation of Acrobat	101
1. The Properties of a PDF Collection	103
1.1 Existing solutions for structuring PDF collections	105
2. Formal Metadata Based Approaches	106
2.1 Standard meta-data: the Dublin core	106
2.2 Metadata: the resource description framework (RDF) approach	108
3. The Concept of Document Re-Use	111
4. Data as the Intel Inside	113
4.1 OpenSource software: Jmol	113
4.2 Molecular animations of pseudorotations	114
4.3 Data re-use from an enhanced web object	116
5. Towards the Holistic Approach: The Podcast	118
5.1 Selecting and viewing a Podcast	118
5.2 Authoring a Podcast	121
6. The Wiki	121
6.1 The Wiki as an information liberator	122
6.2 Example of course Wikis	129
6.3 Examples of chemical Wikipedia communities	129
Conclusion	131
References	132

INTRODUCTION: THE IMPACT OF THE WEB ON THE CHEMISTRY CURRICULUM

Since its emergence more than twelve years ago, the World-Wide Web has impacted upon undergraduate and graduate level chemical education in both obvious and more subtle ways. By liberating information from its physical containers, it has changed the manner in which a chemistry-based curriculum might be

[*] Department of Chemistry, Imperial College London, South Kensington Campus, London, SW7 2AZ, UK

designed, presented to students and utilized by them. The Web has also often provided the impetus to introduce new topics into the modern chemistry curriculum in a pro-active and visual manner using example and demonstration, such as molecular informatics and molecular modeling. It has probably been the catalyst (together with Powerpoint) for the widespread introduction of WiFi networking and computer data projection into lecture theaters and tutorial rooms. The present article does not represent any attempt at a comprehensive review of the myriad of Web based content management and learning technologies now available for deployment in such a curriculum. Rather the theme is explored of whether some of these technologies could be used as mechanisms for creating a more closely integrated knowledge environment. This applies not only in the area of computational chemistry, but across molecular science as a whole. Within this rather wider perspective, the article also reflects upon the scientific journal and the textbook, and what their holistic role might be.

The first review of the Web in chemistry, published in 1996 [1], covered its effective "big-bang" start (for chemistry at least) in 1993 and the adoption by this community during the period up to 1995. That article concluded by speculating whether the Web as an information medium might have the potential of one day enabling (fully or semi)-automated problem solving or pattern recognition, thus eventually allowing for the discovery of e.g. new pedagogic heuristics in chemistry. Automated elaboration of the Woodward–Hoffmann rules in organic chemistry was cited as one example of what might become possible. The present article provides a somewhat selective update in asking whether, more than a decade later, the current Web of 2007 has indeed evolved in such a semantic direction, and whether any of the new Web genres, known collectively as Web 2.0 [2] (this being a rather tongue-in-cheek term for various recent social and communal developments) are turning out to have such semantic potential. The question is a difficult one, since the Web (whether in the original vision or as version 2.0) now means so many things to so many people, depending on their relative viewpoint of being an information/knowledge creator, provider, curator or consumer. Most scientists of course adopt all four roles at some stage and to different degree. There is also another potential participant in the Web, whom I shall refer to by the (equally tongue-in-cheek) expression Human 2.0, a topic which will be discussed in the conclusion.

From a student's perspective as (initially at least) an information consumer, the Web is frequently synonymous with search engines such as Google, but accompanied (one suspects) by a feeling of frustration at frequently being unable to find high quality, trusted or relevant knowledge amongst the overwhelming mass of trivial information. Google may also be a convenient avoidance of the more onerous alternative of performing a skilled (or at least a trained) and rigorous search of available scientific data bases (including databases of reviewed scientific materials we could call books).

The perspective from an educator's point of view may be rather different. For them, the Web may be perceived as a convenient and ecologically friendly avoidance of the photocopier and the physical medium of paper. However, does electronic delivery of shrink-wrapped lecture-sized bundles of notes really liberate information? The reality is that most of these notes will probably end up being printed by the students anyway; it's what one does with PDF after all, and the growing bundle soon becomes reassuringly thick! The concept of information liberation has also not been actively promoted by the scientific publishing processes. Most publishers currently mandate that scientific articles be written in such a manner that facilities printing. Packaging the resulting reprint in the form of a PDF file was also designed largely for that purpose. The message that students get from this process is not pre-eminently one of information liberation!

BACKGROUND: THE TREND TOWARDS AN ACCUMULATION OF ACROBAT

This review starts from the premise that the wrapping of (chemical) content using portable document technologies (PDF) such as Acrobat, or as presentation slides (Powerpoint), and the deployment of such is now one ubiquitous component of the Web. The pragmatic issue is really whether information can still be liberated from such containers in a useful and chemically meaningful sense. Areas such as computational chemistry are both semantically and data-rich, and it is argued here, ideally suited for liberation.

As a preamble to this topic, the present reader is asked whether they have ever seen, or have contemplated adding information to the menus shown in Figure 7.1. These panes derive from the Adobe Acrobat 8 program interface to the document properties, a synonym for the document metadata. Part of the argument presented in this article is that metadata population should become an intrinsic part of the process of authoring and ultimately producing any document intended for Web dissemination. Current practice, it has to be noted, is that such formal population with metadata is still a rare occurrence. If the Web has indeed liberated information from its physical containers, it may have inadvertently re-imprisoned much of it into an electronic straightjacket. The question is whether can we find true liberation for it in the future?

Why does document metadata matter? Well, consider next the delivery mechanism for most PDF files (and the occasional Powerpoint presentation). The start point for curriculum based materials is often a so-called content management system (CMS) such as WebCT or Moodle [3], which have recently become popular in a teaching context. Content management is however something of a misnomer, since such systems do not unbundle and re-organize the actual content from within any Acrobat document. Instead their purpose is largely to provide institutional access controls (in modern parlance, a digital-rights-management or DRM mechanism), coupled with a parochial navigation tree for students to find their way to a document and its contextual relatives. The net outcome of this use of the Web for hosting a CMS is that the reader acquires an appropriately licensed and exact digital copy of the PDF file, which is then stored in their own document space. The subsequent life cycle of this document most probably involves printing, but currently at least, little other re-use. The re-use of information and data is a topic elaborated in more detail below.

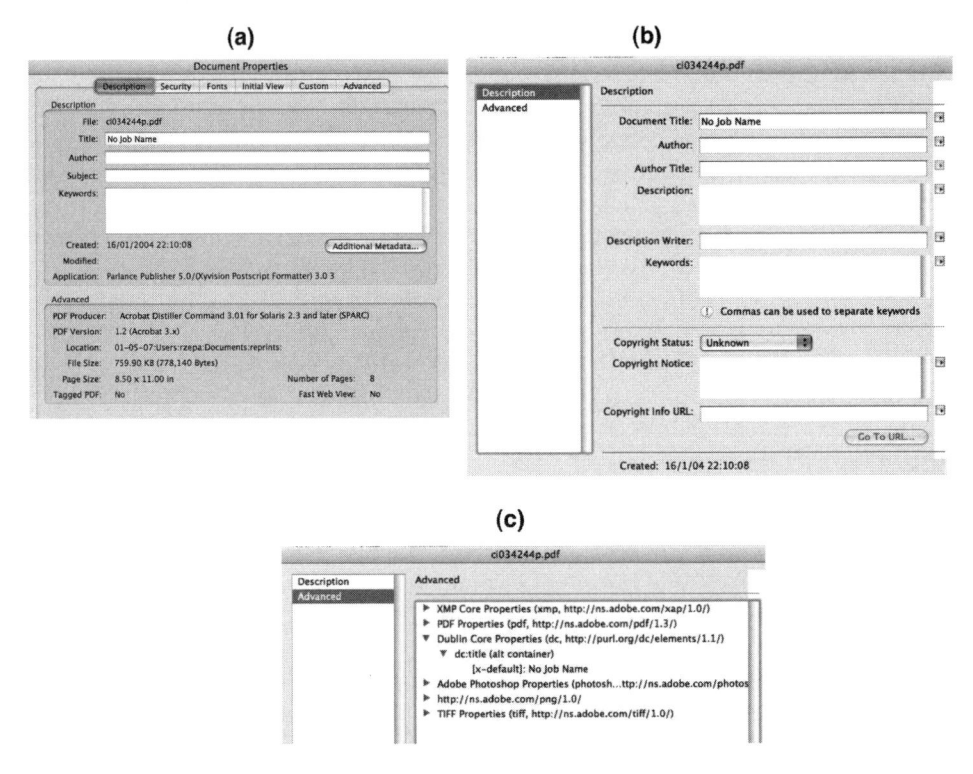

FIGURE 7.1 Menu pages in the Acrobat 8 Professional program, relating to document information and meta-data for (a) a typical journal reprint, with (b) the standard metadata schemas and (c) the advanced menu, where additional, more chemically relevant, metadata schemas can be added. The values, or more accurately, the absence of values for the various fields are typical for most Web-resident PDF files.

The other major source for student acquisition of PDF files is the primary literature in the form of journals. The modern electronic journal functions in much the same way as a content management system. Retrieval of individual articles has been vastly expedited by the recent introduction of the widely implemented digital-object-identifier (DOI) [4]. This provides 1-click access by means of a hyperlink embedded directly in other documents, or from e.g. the navigation tree of a HTML-based Web document. Although ostensibly introduced to provide rapid access to the published scientific literature, the DOI identifier is also the first of the technologies discussed here that has the potential to achieve the holistic and interlinked information structures implied in the title of this article.

One consequence of the introduction of the DOI for document delivery is that it facilitates the accumulation of significant numbers of PDF files onto a local storage device. Whilst there are no formal statistics available, an estimate of perhaps one thousand such documents acquired by a student during a typical 3–4 year study period seem not unreasonable. The present author for example has updated his own taught courses to include around one hundred DOI markers distributed within the podcasts, wiki and Web pages used by him to deliver content to students. These materials in turn constitute only a small percentage of the total curriculum content. An inspection of the hard drives on the computer this article is being authored on reveals the presence of 4675 PDF files. How many are unique, and what proportion actually relate to chemistry, is not easily discerned; indeed it is this very issue that is addressed in the present article.

1. THE PROPERTIES OF A PDF COLLECTION

A PDF collection has a number of characteristic features, several of which present real information issues.

- The original source for a document could be a published journal, a content management system, or increasingly a digital repository such as Fedora or DSpace [5] for the archival of preprints, Open-Access published reprints and other institutional materials. Such a source would almost certainly have provided context and metadata for that document. This might have taken the form of title, selected keywords, an abstract (often graphical), and list of authors, and the document identifier (DOI). Some chemistry journals also provide lists of molecules which may have been described within the body of the article, each with a unique molecule identifier such as the (proprietary) CAS registry number. A listing of related articles cited in the full text may also be part of the journal service. Because all this information is not formalized (in the sense that software rather than just a human could retrieve it) within the document itself, it risks being lost when the PDF file is separated from its source.
- For a typical reader, the PDF files themselves are likely to have been downloaded and distributed over time across a number of computers and their hard drives. From there, they may migrate to network storage areas, memory sticks and mobile devices such as iPods etc. Aggregating, removing duplicates and organizing these according to their content is non-trivial, and may well require substantial investment of human time and skills.
- The files themselves are unlikely to have any obvious, consistent, or even meaningful naming convention.
- Other than the filename itself, it is unlikely (in 2007) that *any* other substantive meta-data will have been added to the document itself during the authoring/editing/publishing process (c.f. Figure 7.1). Without such information, it becomes more difficult to answer even simple questions such as "is this document about chemistry" without actually opening and reading it. A "mouse-over" on an unopened document merely informs the reader what type of document they are dealing with; probing further with e.g. a "right-click" to reveal the document properties reflects merely the same undefined fields which were illustrated in Figure 7.1. Contrast this with what is possible. Shown in Figure 7.2 is a document where the publisher and authors have co-operated to populate the standard (Dublin Core schema) metadata fields. These fields in turn could be automatically exposed to operating-system based indexing engines (true of Mac

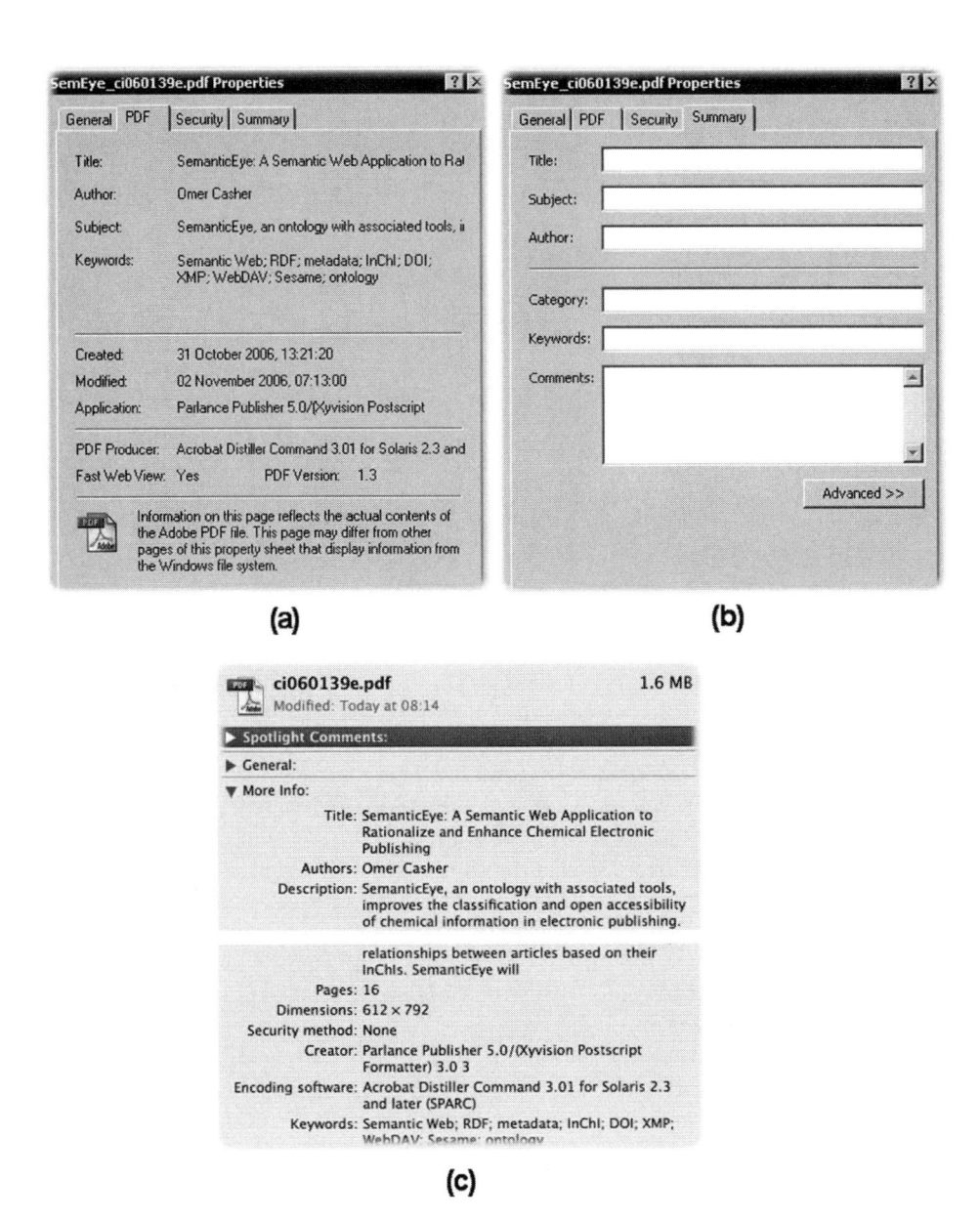

FIGURE 7.2 Document property panes from (a) Windows XP, revealing the PDF attributes. These correspond to the part of the Dublin Core metadata schema. (b) This contrasts with the attributes exposed to the XP operating system itself, which have not been inherited from the PDF attributes; third party tools, i.e. http://www.ifiltershop.com/xmpfilter.html are required to achieve this. The successor Vista system, in the January 2007 release, no longer reveals these (presumably still empty) fields. (c) Document properties exposed to the Mac OS X operating and the Spotlight search system include the full article abstract mapped to the Dublin Core Description field, as well as author, title and keyword fields.

OS X but in 2007 not yet of Microsoft Vista). By enabling appropriate weighting and categorization of the terms, the quality of resulting content searches (i.e. by field restriction to e.g. a named author) is improved. It is important to appreciate however that this process relates to the document as a whole; any finer grained internal structure, components or relationships which the document may contain, such as reference to discrete molecular species, cannot be easily captured at this level.

- It is not unusual for PDF files themselves to be copy protected, preventing data in the form of text from being "scraped" out of the document. It is quite possible that more elaborate, and restrictive, DRM may become common in the future.
- Standard document metadata is not designed to easily capture relationships between two or more documents, such as one document citing another, or two documents each covering the same concepts and terms but perhaps using synonyms for these. Whilst free text indexing and searching may reveal some such relationships, this is not a systematic or reliable process.
- Chemical relationships (defined by say having one or more molecules or even molecular substructures in common) may be particularly difficult to detect, even if individual documents are visually inspected, since molecules may not be represented in a consistent, and hence comparable, manner. Thus some molecules may merely be named, either trivially or systematically; others may be drawn as two-dimensional line diagrams with perhaps components such as the Markush R convention denoting a variable group which may be enumerated elsewhere. Three dimensional properties such as stereochemistry may be even less precisely or incompletely specified.
- Conceptual relationships in chemistry, which help provide a synoptic or holistic view of a topic, represent the ultimate challenge in perception. Traditionally, students would rely on a carefully and lucidly written text book to achieve this sort of perspective across a subject. The issue with texts books, or other static organizations of knowledge, is that they are not readily extensible, and new information cannot so easily be related to existing organized knowledge. Lecture handouts, it is true, can have an appended reading list, this indeed being part of the function of a lecturer giving a course, but it is still hard work for students to make the semantic connections between diverse sets of materials.

Perhaps a chemical analogy to this state of affairs can be drawn. The file storage systems would be the equivalent of a container filled with an inert gas. Being "inert", the individual atoms of this gas hardly interact with one another, and apart from a count of the total number of atoms, little further structure to the collection can easily be discerned. Molecules are certainly unlikely to form under these conditions, and nor will it be easy to *crystallize* this collection into something with a well defined structure and with internal relationships.

1.1 Existing solutions for structuring PDF collections

Partial solutions to some of these issues do exist.

- One might rely on the likes of Google (or the rather more selective Google Scholar [6]) to perform a free-text indexing of the content of any information object at its original site of publication, but, as noted above, with the limitation that heavy contamination with less relevant sources is likely. It is less likely that the Google robot would be allowed to access protected content in a local content/document management system, which probably has its own parochial index and organization. Such material remains isolated in its own information silo.

- If the PDF collection is consolidated onto a single user space, so-called desktop searching tools can be installed to index the free text-based content, and thus help to establish relationships between documents from diverse sources. However, tools such as e.g. Windows Desktop Search are not invariably deployed on staff and student computers, and even if present, the appropriate **iFilter** to extract content from e.g. PDF files may not have been installed. Even if successfully indexed, resulting searches will nevertheless rely on finding unique and common descriptors of suitable concepts in the body of the document. In the pedagogic example illustrated in more detail below, one will find terms such as *Berry pseudorotation* and *non dissociative ligand exchange* in the free-text. These in fact describe the same concept, but have no words in common, and it seems unlikely that any associations could be established by simple text searches.

- Higher order concepts, for example requiring an understanding of three dimensions, may be definable only with great difficulty. Describing the actual processes involved in a Berry pseudorotation using only simple words and definitions is quite a challenge! In the example elaborated below [7,8], we introduce two other ligand-exchange modes termed "lever" and "turnstile", and then proceed to show how some systems can have attributes of all three modes, which we call chimeric. If you are not familiar with all these terms, simply understanding the preceding sentence may difficult. Any indexing/search system might make a connection between these various concepts, but the outcome is likely to be quite variable in general, and may depend on how effective and up to date any dictionary of synonyms available to the indexing engine is. For example, how would associations and relationships for any newly created term such as *chimeric mode* be established?

- It is possible to restore some form of structure to a PDF document collection by using bibliographic tools such as *Reference manager* or *EndNote*. This in fact requires a great deal of effort on the part of the user. The lack of *self-identifying* metadata, as noted above (Figures 7.1, 7.2) makes this process relatively manual and arduous, and one few students would assiduously adopt over a degree course.

2. FORMAL METADATA BASED APPROACHES

2.1 Standard meta-data: the Dublin core

The lack of content demonstrated in Figure 7.1 served to emphasize how neglected meta-data is, and how even modern operating systems may not handle it at all! Simple fields such as document title, author names, creation dates, and subject are all too frequently omitted from all kinds of documents. This may be because the authors of such documents regard these fields as self-evident to anyone reading them. That would miss the point entirely. A human needing to scan 5000 documents, assuming 1 minute per document, would need to set aside 3.5 days to do so. As it happens, computers are rather good at doing automatically this highly repetitive task. Appropriately declared meta-data is simply a standard mechanism for exposing this information to a computer in a structured manner, enabling efficient subsequent indexing and retrieval. The so-called Dublin-core (DC) meta-data schema ensures cross-compatibility across a diversity of programs and documents, and is widely supported by many programs and some operating systems. Wherever possible, meta-data based on this schema should be declared in any document which forms part of curriculum content (c.f. Figure 7.2). In a PDF file, it is recommended the XMP-RDF approach be taken (discussed in the next two sections). In a conventional HTML page, such metadata might be transcluded by links of the type;

```
<link rel='schema.DC' href='http://purl.org/dc' />
<meta name='DC.Title' content='A Computational Study of the Non
dissociative Mechanisms that Interchange Apical and
Equatorial Atoms in Square Pyramidal Molecules' />
<meta name='DC.Identifier.citation'
content='http://dx.doi.org/10.1021/ic0519988' />
<meta name='DC.Subject' content='Pseudorotation, non-dissociative,
ligand exchange,' />
<meta name='DC.Creator' content='h.rzepa@imperial.ac.uk' />
<meta name='DC.Description' content='The lowest energy transition
state for the nondissociative
apical/equatorial atom exchange mechanism for three square pyramidal
AEX5 molecular
species was calculated (CCSD(T)/pVTZ; B3LYP/pVTZ, aug-cc-pV5Z) to
have a hemidirected geometry with Cs symmetry for BrF5, IF5, and
XeF5 ...' />
```

In this context, mention should be made of the open Office format introduced by Microsoft with Office 2007 [9] and known as ECMA standard 376. An Office 2007 document (suffix .docx) is essentially a container for around 12 XML-based documents (or parts) performing a variety of functions. One of these parts (core.xml) is used to define and store the document metadata or properties. An example populated with the same metadata as the previous HTML example is shown below.

```
<cp:coreProperties
xmlns:cp='http://schemas.openxmlformats.org/package/2006/metadata/core-
properties'
xmlns:dc='http://purl.org/dc/elements/1.1/'
xmlns:dcterms='http://purl.org/dc/terms/'
xmlns:dcmitype='http://purl.org/dc/dcmitype/'
xmlns:xsi='http://www.w3.org/2001/XMLSchema-instance'>
<dc:title>A Computational Study of the Non dissociative Mechanisms
that Interchange Apical and
Equatorial Atoms in Square Pyramidal Molecules</dc:title>
```

```
<dc:subject>Pseudorotation, non-dissociative, ligand
exchange</dc:subject>
<dc:creator>h.rzepa@imperial.ac.uk</dc:creator>
<dc:description>The lowest energy transition state for the
nondissociative apical/equatorial atom exchange mechanism for three square
pyramidal AEX5 molecular species was calculated (CCSD(T)/pVTZ; B3LYP/pVTZ,
aug-cc-pV5Z) to have a hemidirected geometry with Cs symmetry for BrF5,
IF5, and XeF5 ...
</dc:description>
<dcterms:created xsi:type='dcterms:W3CDTF'>2006-11-
20T14:04:00Z</dcterms:created>
</cp:coreProperties>
```

Highlighted above in bold is the attribute XSI of the last element, which defines a date and time datatype and is noted here in anticipation of the discussion of datatyping as a component of Wikis (section 6).

There is yet little experience about how docx documents might be populated with metadata and then queried using this information, or about how either the operating system, or third party tools might make use of this information to structure a docx collection according to its content.

2.1.1 Molecular meta-data: InChI

Unlike the bibliographic-derived DC metadata schema, general chemical metadata schemes are relatively underdeveloped. However, one particular type is worth noting at this stage; the so-called InChI string [10]. This *In*ternational *Ch*emical *I*dentifier enables a unique descriptor to be derived for a molecule, based purely on how the atoms are connected (but not specifying the type of bond that connects them). Although it is difficult to be precise, one can estimate that in a typical chemical curriculum, students may encounter around 1000 unique molecules exemplifying some aspect of their lectures, laboratories or problem solving tasks. Many of these molecules probably recur in more than one context, and many are associated with reactions and other properties. If each of these molecules were to be labeled with an InChI, then detecting instances of each molecule, and inferring reactions, would be greatly facilitated. InChI metadata can be embedded into HTML documents using for example RDF syntax.

```
<link type='application/rdf+xml' rel='meta'
href='http://www.ch.ic.ac.uk/rzepa/bpr/YX5.rdf'
title='InChI identifiers for YX5 species' />
```

2.2 Metadata: the resource description framework (RDF) approach

The type of metadata described above is a monodirectional descriptor, it merely declares that somewhere in the document it relates to can be found a particular type of data or information. It fails to provide an extensible framework for adding a controlled context to that information.

A more powerful expression of metadata has been developed to address this limitation and is known as RDF, or Resource Description Framework. "RDF was designed from the ground up to fulfill the role of an Internet architecture for meta-data. It is a foundation for processing metadata; it provides interoperability between applications that exchange machine-understandable information on the Web" [11]. RDF has three components, and for this reason is often also referred to as an information triple. These components are the *subject*, which makes an assertion or a *predicate* about an *object*. The predicate, which is normally defined in the form of a collection of related terms or attributes derived from a controlled dictionary or thesaurus (an RDF Schema) imparts a powerful context to the information, and allows a taxonomic classification of that subject. The object of one assertion can also be the subject of another (as for example when describing the product of one reaction transformation as forming the reactant for another). In theory at least, such declared properties could provide a powerful method of creating links between different types of information; a process which has been referred to as intertwingling. A visual way for expressing this intertwingling is to draw nodes for the subject and object and an arrow between them for the predicate. When another occurrence is encountered, the two pieces can be glued together at the common subject node. Generally such geometric/semantic objects are called labeled directed graphs. Until now, such mechanisms have hardly been exploited in chemistry, but they could form the basis for one future, more holistic approach to the subject.

For example, inspection of Figure 7.1c reveals a list of advanced document property types, each type identified using a so-called URI or namespace, the properties being encoded as RDF declarations. In this form, it can be queried by software via the XMP interface. This is what is referred to as self-identifying or queryable metadata. To refer back to the metaphor introduced of a container of argon gas, RDF can introduce strong intermolecular interactions, perhaps even leading to bonding (or in computer-speak, logical inferences or reasoning). The collection now becomes much easier to crystallize. An example of the use of RDF to populate a document with metadata can be found within the reprint [12] discussing its application.

The RDF declarations can be contained within an XML document; a small component is illustrated below;

```
<rdf:Description rdf:about='
 xmlns:UniqueIdentifier='http://www.doi.org/2004/DOISchema/'>
 <UniqueIdentifier:doi>
<rdf:Alt>
 <rdf:li>10.1021/ci060139e</rdf:li>
</rdf:Alt>
 </UniqueIdentifier:doi>
</rdf:Description>
<rdf:Description rdf:about='
 xmlns:molstruct='http://www.iupac.org/inchi/'>
 <molstruct:inchi>
<rdf:Seq>
 <rdf:li>InChI=1/C12H22O11/c13-1-4-6(16)8(18)9(19)11(21-4)23-12(3-
```

```
15)10(20)7(17)5(2-14)22-12/h4-11,13-20H,1-3H2/t4-,5+,6-,7+,8+,9-,10-,
11+,12-/m0/s1</rdf:li>
</rdf:Seq>
 </molstruct:inchi>
</rdf:Description>
```

This content is recognized by Adobe Acrobat via a technology for embedding metadata in documents referred to as XMP (eXtensible metadata platform). The XMP code above defines two information triples or RDF declarations. The subject of both is the PDF document itself [12]. The first predicate is the assertion that this document has an attribute known as a DOI identifier (defined by a URI http://www.doi.org/2004/DOISchema/) and the object of this assertion or attribute is the actual value of the DOI. The second predicate is the assertion that the document references a molecule having an InChI identifier (defined by the URI http://www.iupac.org/inchi/) with the value shown, a value which can be algorithmically recognized as a glucose molecule. This information manifests via the Adobe XMP interface (Figure 7.3).

Essentially, we have a formal (i.e. a logical, machine processable) declaration of two properties, one being a unique identifier for the document in question, and the other a unique identifier for one molecule mentioned within the body of its text, the latter being further processable to infer it has the trivial name (another predicate, if a rather fuzzy one!) of glucose. This knowledge can be inferred automatically, and would not need a human to accomplish.

This particular article [12] also illustrates in detail how the RDF declarations can be automatically extracted from a document, and used to populate a so-called triple store and displayed in graph form as a set of nodes and their connections. Armed with formal definitions of how to process each predicate (defined using e.g. another XML language known as OWL, or Web Ontology Language, and uniquely referenced using the URI for each predicate), such an environment could be used to perform context sensitive searches (using as search fields values of either the object, predicate or subject of each information triple), or even to use inference logic to automate the discovery of deeply nested relationships (or indeed logical contradictions) within the information.

Another tool which utilizes such metadata is Piggy Bank [13], which allows a communal repository of RDF to be created by users contributing to share the information they have collected by conventional Web browsing. The Piggy-bank extension achieves this by detecting any RDF associated with a Web page and banking it. If no RDF is present, it can utilize so-called semantic scrapers, pre-programmed with pattern recognition, to re-structure information on the fly and express it as RDF. A related concept is to be found in the Semantic Wiki, discussed in section 6 below.

Although the example described above relates to only a single document and the small number of RDF-based metadata declarations within it, in principle this could be scaled to well beyond the 5000 documents which might be owned by an individual, perhaps up to the tens of millions of documents which represent the current scientific literature. In another dimension, it might scale to the estimated thousands of potential chemical predicates or assertions, which taken together

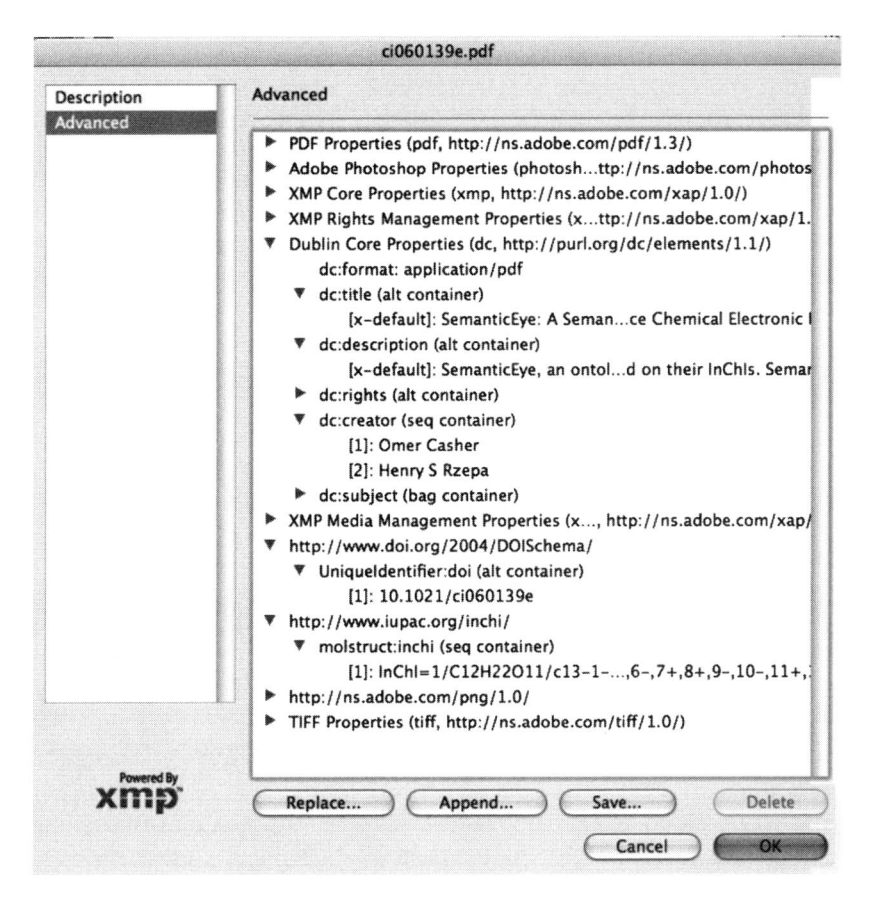

```
                            ci060139e.pdf
 Description        Advanced
 Advanced
                    ▶ PDF Properties (pdf, http://ns.adobe.com/pdf/1.3/)
                    ▶ Adobe Photoshop Properties (photosh...ttp://ns.adobe.com/photos
                    ▶ XMP Core Properties (xmp, http://ns.adobe.com/xap/1.0/)
                    ▶ XMP Rights Management Properties (x...ttp://ns.adobe.com/xap/1.
                    ▼ Dublin Core Properties (dc, http://purl.org/dc/elements/1.1/)
                          dc:format: application/pdf
                       ▼ dc:title (alt container)
                             [x-default]: SemanticEye: A Seman...ce Chemical Electronic
                       ▼ dc:description (alt container)
                             [x-default]: SemanticEye, an ontol...d on their InChIs. Seman
                       ▶ dc:rights (alt container)
                       ▼ dc:creator (seq container)
                             [1]: Omer Casher
                             [2]: Henry S Rzepa
                       ▶ dc:subject (bag container)
                    ▶ XMP Media Management Properties (x..., http://ns.adobe.com/xap/
                    ▼ http://www.doi.org/2004/DOISchema/
                       ▼ UniqueIdentifier:doi (alt container)
                             [1]: 10.1021/ci060139e
                    ▼ http://www.iupac.org/inchi/
                       ▼ molstruct:inchi (seq container)
                             [1]: InChI=1/C12H22O11/c13-1-...,6-,7+,8+,9-,10-,11+,
                    ▶ http://ns.adobe.com/png/1.0/
                    ▶ TIFF Properties (tiff, http://ns.adobe.com/tiff/1.0/)

     Powered By
     xmp           ( Replace... )  ( Append... )  ( Save... )     ( Delete )
                                                     ( Cancel )     ( OK )
```

FIGURE 7.3 Acrobat XMP interface to document metadata expressed as RDF information triples, including the two triples discussed above. Note the Append button for declaring (extending) further metadata information, and the Save button for extracting existing metadata information.

could constitute something like a complete ontological description of much of chemistry and its semantics. The reality of course is that we are currently a long way from such an achievement.

3. THE CONCEPT OF DOCUMENT RE-USE

With a strategy established for defining the context of a document and for identifying its properties, one can next consider what is meant by the concept of information or data re-use. It is important to emphasize that chemistry is a very data-rich subject, and that the center of much of this data can be considered to be the molecule. Indeed, the term datument has been coined [14] to describe a data-rich document, where the data is formally identified within the document structure. The concept of a data-rich document is also present in Web 2.0, in which context it is referred to as "Data as the Intel Inside", a reference to the heart of a computer.

The document may relate to precisely specified molecules and their properties, tabulated or graphed, which could be referred to as molecular data. The data in turn will be associated (implicitly or explicitly) with units; it will have the property of a datatype. The document may also contain components such as theory encapsulated in a mathematical expression, and more generally sets of relationships between molecules such as their reactions, their properties and the formal expressions describing these. All these various types of information could be described by a vocabulary of formal subject terms, and such vocabularies could belong to collections we can refer to as dictionaries, which in turn have some implied taxonomic organization. A human (and currently only a human) will be able to recognize much of this content (by virtue of their previous training) and might be able to select parts of the free text corresponding to such content in the original document, and transfer it to another software application, or document, in the operation known as copy/paste. It is quite probable that this will still need substantial editing or transformation so that its syntax suits the purpose intended for it.

A simple example should suffice to illustrate this (often instinctive) process. A document retrieved from a content management system asserts that a substance referred to as aspirin exhibits a type of interaction known as a hydrogen bond between two specific atoms, which may in turn impact upon some property (e.g. chromatographic behavior). Re-use of this information might involve creating the following tasks, formalized as a workflow.

- Firstly, the trivial chemical name *aspirin* has to be recognized as such, and then resolved into a full expression of the *molecular connectivity*, by querying the appropriate dictionary in which the term *trivial chemical name* is described, and then invoking the appropriate (Web) service to perform this function. A student (i.e. human) would probably recognize aspirin as a trivial chemical name without additional help, and then probably attempt to identify its molecular structure by querying Google. Although obtaining a pictorial representation of aspirin using Google is not that difficult (the Wikipedia entry for this substance will most likely come high on, if not top of the list of **16 million** or so apparent hits), finding a source which provides the connectivity in a **re-usable** form is more difficult. This is because most sources (including the Wikipedia article) give the structure only as a bit-mapped image, which a human then has to translate using appropriate software into a machine processable form. The challenge here of course is to create an infrastructure where software could itself be sufficiently guided to perform this task without human intervention. This is where the role of RDF-encoded metadata comes into play, and which would form the basis for the decisions that such software would need to make.
- Most probably at this stage, only a 2D structural representation of the connections between the component atoms of aspirin has now been made available. One has to perceive this of course, and then to recognize that of more value is actually a 3D representation of the spatial distribution of these atoms, and to infer the units of this distribution (which in fact are likely to be Å). Again, the

meaning of the term *3D molecular coordinates* has to be resolved, and again the appropriate query made to search for any 3D information consistent with the 2D connectivity. If more than one set of 3D coordinates is retrieved, a decision will need making as to which set is the most appropriate for the task at hand.
- These coordinates may then be used as the starting point for e.g. a theoretical computation of the structure, including modeling of the relevant hydrogen bond. Selecting the appropriate method for this again requires a great deal of experience, something only a human can really provide (although software could be primed with some default assumptions, which might include something like "first try a B3LYP Density functional calculation using the 6-31G(d) basis set").
- By now, one may have a quantitative measure of the strength and geometry of the hydrogen bond present in aspirin, but only of course if the correct conformer has been modeled. Perceiving that several conformations or tautomers might be possible, but that only some may sustain a hydrogen bond, is yet another challenge.
- The final step in this example is to map the computed structure onto an appropriate model which relates this to e.g. chromatographic behavior (if one such exists).

The above example serves two purposes. Firstly, to illustrate how much implicit understanding a human deploys during the every day tasks of re-using information. Secondly, it shows how much of this could nevertheless be formalized, and hence automatable, and by implication, scalable. The next section deals with how data-rich courseware can be created, and how some of the data can be re-used by students (although the complete, automated workflow described above is, in 2007, not yet fully achievable).

4. DATA AS THE INTEL INSIDE

The idea of using the Web to provide not merely information, but data, goes back to the earliest days. It was a point strongly made in the original review [1] and in a number of articles written during this period [15]. It is here illustrated via a project that owes its inspiration from a typical teaching scenario, but which evolved to eventually become a formal research article [8]. This in turn borrowed from its pedagogic origins to take advantage of the Web to augment the conventional figures in the printed version with *Enhanced Web Objects* in the online version, from which readily re-usable data could be extracted and subjected to the sort of process described in section 3 above. Before discussing this in detail, a short diversion to deal with Opensource software is appropriate.

4.1 OpenSource software: Jmol

No discussion of a data-rich environment can be complete without some mention of the software tools that will be to create and/or display the data. In the context of this article, we have made extensive use of a molecular visualization and scripting program called Jmol, which was specifically designed for Web use. Its origins go back to the 1980s, when (hugely expensive) national supercomputer centers were being set up, and the recognition that to persuade computational chemists to use these resources, open provision of good molecular visualization (and other) software would be required. Recognition that such tools and associated data needed to be Web-friendly was made in the form of the proposal of an Internet protocol known as chemical MIME [16]. The first package to support this protocol was based on the Rasmol visualization package and presented in a browser friendly form known as Chime (the name deriving from its support of the chemical MIME types). Chime however was not OpenSource in that no source code was openly released by the owner. The gradual cessation of Chime development was effectively therefore a business decision by the owner rather than a lack of interest by the community. This however provided the impetus for the community to resuscitate XMol, an early package developed within the supercomputer initiative. Re-written in the Web-friendly Java language. it was renamed as Jmol [17] and offered as OpenSource. Over the period of around 1998-present, a highly active set of interested individuals discussed and extended its capability. Now at version 11, Jmol offers a visually stunning array of features, including molecules rendered in a wide variety of representations, surfaces, wavefunctions, animations, and a very flexible scripting language, originally developed for Rasmol, but then fully implemented and extended in Jmol. One example serves to illustrate the awesome capability of this program, the BioMolecular Explorer 3D site [18].

4.2 Molecular animations of pseudorotations

The motivation for creating this example [7] arose because the present author was approached by a colleague who been asked to deliver a course for one year only whilst the original lecturer undertook a sabbatical. One aspect of this course involved discussing a phenomenon known as *pseudorotation*. The colleague had consulted two seemingly appropriate textbooks on this topic, which revealed several confusing aspects, along with the speculation of alternative modes termed *turnstile rotation* and *umbrella motion*. These explanations were presented in the form of static snapshots of the motion, projected into two dimensions. The surrogate lecturer felt that something more was needed to enable a class of students to appreciate the difference between e.g. Berry and turnstile modes (and they wanted to know whether the umbrella mode was even real). Conventionally, both lecturer and students would be advised to build 3D models from plastic kits, but given one is actually dealing with a dynamic process involving two or more transition state normal modes, such kits could provide only a small part of the required insight. The outcome of this request is shown (in part) in Figure 7.4.

The presentation was in fact deliberately couched to have the appearance of a journal article (which indeed it spun out into [8]), in part to emphasize to students what a holistic integration of journals and teaching materials might look like.

1. To gain an accurate representation of the process, we undertook *density functional* molecular orbital calculations for our systems. These held some surprises.

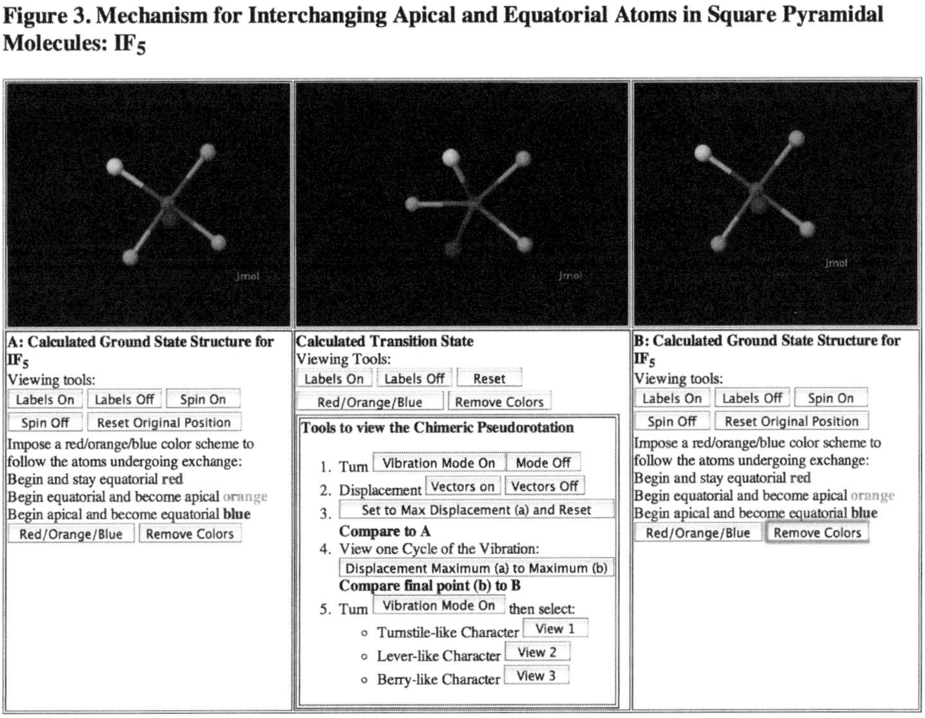

Figure 3. Mechanism for Interchanging Apical and Equatorial Atoms in Square Pyramidal Molecules: IF$_5$

A: Calculated Ground State Structure for IF$_5$	Calculated Transition State	B: Calculated Ground State Structure for IF$_5$
Viewing tools: [Labels On] [Labels Off] [Spin On] [Spin Off] [Reset Original Position] Impose a red/orange/blue color scheme to follow the atoms undergoing exchange: Begin and stay equatorial red Begin equatorial and become apical orange Begin apical and become equatorial blue [Red/Orange/Blue] [Remove Colors]	Viewing Tools: [Labels On] [Labels Off] [Reset] [Red/Orange/Blue] [Remove Colors] **Tools to view the Chimeric Pseudorotation** 1. Turn [Vibration Mode On] [Mode Off] 2. Displacement [Vectors on] [Vectors Off] 3. [Set to Max Displacement (a) and Reset] **Compare to A** 4. View one Cycle of the Vibration: [Displacement Maximum (a) to Maximum (b)] **Compare final point (b) to B** 5. Turn [Vibration Mode On] then select: ○ Turnstile-like Character [View 1] ○ Lever-like Character [View 2] ○ Berry-like Character [View 3]	Viewing tools: [Labels On] [Labels Off] [Spin On] [Spin Off] [Reset Original Position] Impose a red/orange/blue color scheme to follow the atoms undergoing exchange: Begin and stay equatorial red Begin equatorial and become apical orange Begin apical and become equatorial blue [Red/Orange/Blue] [Remove Colors]

FIGURE 7.4 One of the Enhanced-Web objects as a component of Ref. [8] and viewable at DOI: 10.1021/ic0519988.

2. For the molecule PF$_5$, we could find no evidence of *turnstile rotation* for PF$_5$, and concluded that *umbrella motion* was highly improbable.
3. Seeking to generalize the effect for students, we approached IF$_5$. During the course of viewing the Jmol-generated animations of this species, we concluded it approximated to an "upside-down" inverted version of pseudorotation. We struggled to find a name for this mode, and observed that depending on the angle the vibration was viewed from, it magically assumed differing characters. Finally, we settled on "chimeric pseudorotation". Chimeric stems from Chimera, the mythological Greek demon of the mixed and "monstrous" character of a goat, snake and lion. Recently, chimeric has become a byword for fabulous and fantastic—but utterly mixed-up—ideas, in this case ligand exchange via *Berry pseudorotation*, *turnstile rotation*, and a new mode called *lever rotation*. Ligand exchange in IF$_7$ proved equally challenging to describe using only words. The point of this discourse is that in extending how the concept of *pseudorotation* could be presented to an audience of students seeing it for the first time, we had rapidly discovered new and magical modes which had hardly been described in the chemical literature (and in which two well known text books actually made fundamental errors of description of these phenomena). By presenting the data from calculations of these effects in such an animated

manner, even these quite new and conventionally difficult concepts may become more accessible for students, and staff alike.
4. Further surprises were still in store for us. We found that the related BrF$_5$ could sustain not one but two distinct modes for interchanging the F atoms. We eventually found a convenient label for these two modes in a quite different area of chemistry; that of the stereochemistry of Pb(II) compounds. Such systems exhibit both *hemidirected* and *holodirected* coordinations. We had thus joined one aspect of the stable structures of lead systems with the modes adopted by the fluxional *transition states* of bromine molecules. A lot of concepts in chemistry have been covered by the last few sentences. Metadata mechanisms such as RDF would nicely capture these connections. Whereas we discovered the connection in part by accident (one of us attended a scientific conference and came across a student poster describing Pb(II) systems, which, some months later, "clicked" as they say), perhaps in the future the expression of these concepts as espoused above will allow a more systematic and less accidental mode of discovery!
5. Convinced that some of these insights were sufficiently novel to be worth publishing a formal research article, in our first submission we made the tactical mistake of referring to the pedagogic stimulation for the discoveries (a referee disparaged the work on this basis). Presenting the article in a different research-oriented context (might this be called "rebranding"?), was more successful; indeed the editor of the journal suggested that the figures be converted into Web-enhanced-Objects for the online version [8]. On that occasion, we were less successful in persuading the journal to enhance the PDF version of the article in the manner illustrated in Figure 7.2 (although the same publisher did shortly thereafter do this for another article [12]).

4.3 Data re-use from an enhanced web object

How might a student or a researcher re-use components of the above article? Figure 7.5 shows a small portion of the original Web-object shown in Figure 7.4, but with the addition of a menu derived from Jmol showing information about the data available for that model. This data is readily downloaded to the user's hard drive. A comment about the syntax of this data file is appropriate. This is shown as the original .XYZ data format developed for the Xmol program, representing the atomic numbers and Cartesian coordinates for the IF$_5$ transition state for ligand exchange. The XYZ format is something of a historical legacy; nowadays a more formally structured type based on the generic XML specifications, and implemented for chemistry as CML (Chemical Markup Language) would be considered more useful and future-proofed [19]. The astute reader may notice from the figure that IF$_5$ apparently contains 120 atoms and 100 bonds! The reality is of course that the file contains 20 coordinate snapshots corresponding to the molecular vibration. This does however nicely illustrate the need for a more structured representation of the information, where the data about the molecule, and 19 further coordinate sets, needs to clearly identified as such, information which older formats such as the XYZ carry implicitly and not explicitly.

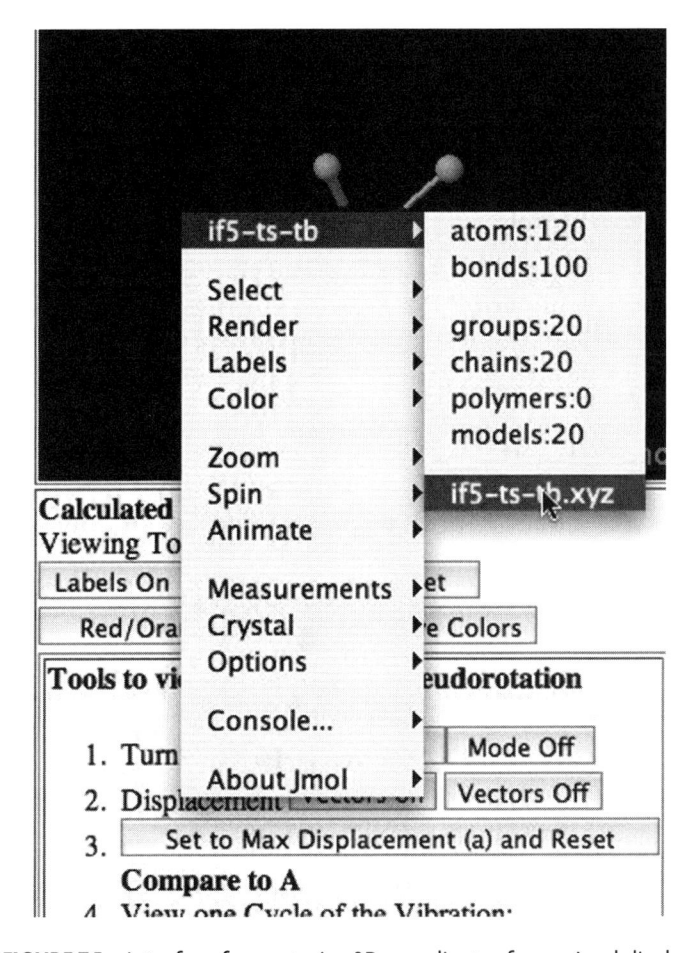

FIGURE 7.5 Interface for capturing 3D coordinates from a Jmol display.

In order to extract utility from the older .XYZ file, a converter (another highly useful opensource program known as OpenBabel, this being part of the more general Chemical Development toolkit) [20] has to be employed. This is part of the (currently human-driven) workflow outlined in section 3 above, and relies on OpenBabel having the built-in content model to correctly parse the .XYZ file, and if needed, to re-write it in another appropriate format. In our case, OpenBabel is actually implemented as part of the opensource Ghemical modeling program [21]. Let us, for arguments sake, say that a student or researcher wants to compute further properties associated with this particular transition state, such as a bond analysis not given in the original article. To achieve this, the file if5-ts-tb.xyz is transformed to (in this example) a Gaussian input file (.GJF) and subjected to further calculation. This process (Workflow) is actually described in a Wiki article developed for the purpose, and illustration of this is deferred to section 6 below.

Finally in this section one may ask why the student or researcher derives benefit from this re-use of data. The advantage for this example is that location of the transition state at the relatively expensive level of theory employed (B3LYP/aug-cc-pV5Z(5d,7f), itself a semantically rich acronym!) was non trivial, and required much expertise acquired by progress through a painful learning curve for the original researchers. There would really be little point in others repeating all of this again.

5. TOWARDS THE HOLISTIC APPROACH: THE PODCAST

The metaphor of data and metadata-rich, student learning resources merging with a journal article is one means of achieving a holistic overview of a taught curriculum. One aspect thus far entirely neglected is the audio-visual components of teaching. Might it be possible to somehow combine the two approaches? In the late 1990s, the *newsfeed*, more formally known as the RSS feed, started becoming popular. In its original incarnation, RSS carried **only** the meta-data describing the properties of a recently created or changed resource on the Web. This alerting mechanism has recently been adapted to cover broadcast and audio-visual materials, and in this guise has become known as a *podcast*. The most mature implementation of RSS for podcasts is by Apple Computer. Although proprietary (and not OpenSource) it is described here in some detail, so that some of its characteristic features can be identified and contrasted with the approaches described above.

5.1 Selecting and viewing a Podcast [22]

There are two software tools associated with podcasts. iTunes is an example of the first, a content aggregation service and management interface for acquiring podcast meta-data. This in turn provides information to the reader for browsing and accessing the audio and/or visual material of the podcast. A recent snapshot of the iTunes store relating to podcasts is shown in Figure 7.6. This illustrates the meta-data fields captured for each podcast, essentially again a sub-set of the Dublin Core schema. The search can be restricted by the user to match only the desired values of the fields, in this case restricting the category to Science and Medicine, and the title and description to have chemistry in their values. This has something of the features illustrated in Figure 7.2 (iTunes, perhaps unfortunately, does not handle PDF files).

With the information provided by the metadata, the user selects a podcast of interest, an action which induces a further set of meta-data to be downloaded. This provides them with more finely grained metadata about each **episode** of the podcast (Figure 7.7) such as its size and topic. This metadata is important, since it enables efficient management of an expensive resource, the download bandwidth (an individual episodes may well be 0.5 Gbytes or larger in size). Selection can be again filtered down by providing, now merely a single, search term for matching. In this example, the user is interested only in podcasts that pertain to the topic **mechanism**, which as a result of the earlier filtering, can be presumed to relate to chemical mechanism.

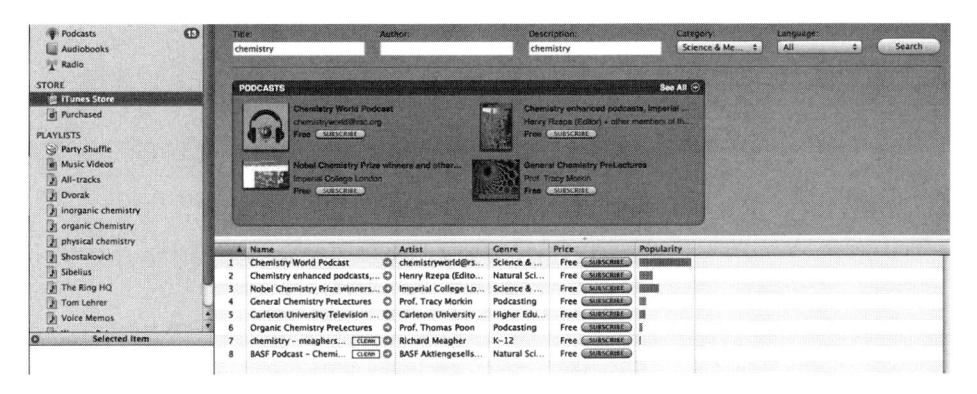

Achieving a Holistic Web in the Chemistry Curriculum   119

FIGURE 7.6 The podcast page in the Apple iTunes store, filtered using the search terms *chemistry* in the title and description.

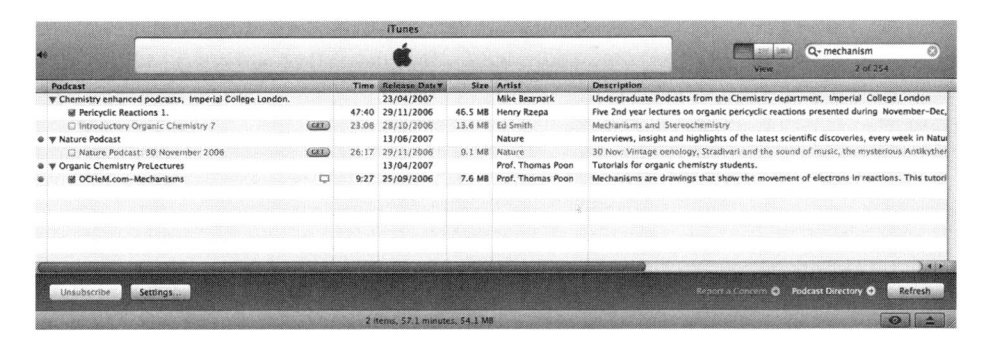

FIGURE 7.7 Metadata describing the individual episodes contained with one podcast feed. Here, they have been filtered by using the search term *mechanism*.

After the individual episode is downloaded (normally a process that can occur in the background, or if the files are big, overnight) the episode can be explored. Two types of content are allowed for an episode; enhanced audio, and video. In the current iTunes implementation, the latter has no further metadata associated with it other than the timecode. Enhanced audio however has two components, an audio track and an associated artwork. Each separate artwork component is known as a chapter to the episode, and this can have further metadata such as a title and an associated link to another web resource. Unfortunately, this additional metadata is only viewable, and is not currently searchable. The artwork links can be for example to an article DOI, or to a Jmol page with 3D content or animation. An example is shown in Figure 7.8.

Can a podcast be associated with the type of re-usable data described in section 4? Since an RSS feed is really just an XML document, any suitable namespaced components can be added to it. If these components are not recognized by the processing software (iTunes in this case), they are simply ignored. To demonstrate this, we added a <cml:>...</cml:molecule> namespace, which can function as a container for chemistry [23]. Into this container we defined an appropriate mole-

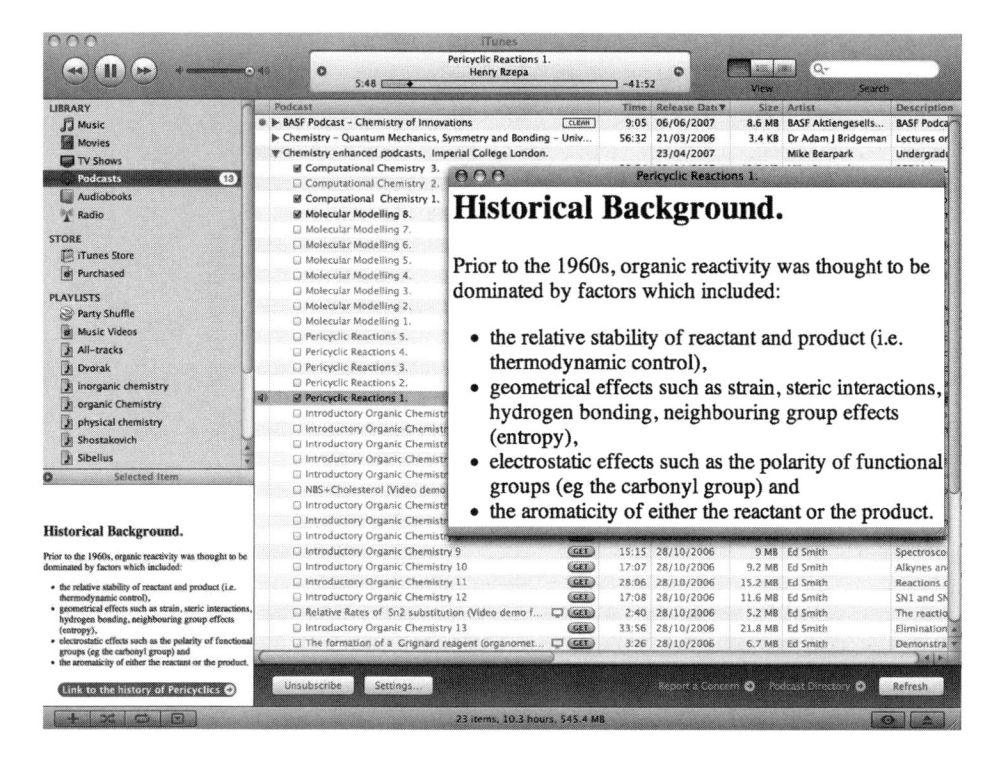

FIGURE 7.8 Components of a podcast, showing individual episodes, one of which is an enhanced audio podcast containing sub-sections known as chapters, one of which is expanded in the central pane, and which can be linked to further information resident on the Web.

cule relating to an episode of an audio podcast, with an InChI identifier added. Although iTunes itself will simply ignore this, an RSS-enabled program such as Jmol can extract this content and display it directly [23]. Other namespaces such as <mathml: /> (for mathematical content) and <svg: /> (for graphical content) could be inserted as well.

The podcast viewed in this manner becomes an interesting vehicle for accurately expressing a variety of types of information and data, with episodes of a podcast to give it context, and a timeline. Such content does not have to be viewed using iTunes. For example, the Firefox Web browser supports RSS feeds, and it can display the video, the enhanced audio, and additional components such as MathML and SVG directly within the browser window. A system known as Bioclipse [24] is essentially a more highly configurable browser which can be used to extract further molecular content. Finally, it is worth noting that the podcast need not have a human as its intended consumer; appropriate software primed with some heuristics for making its own decisions, could automatically aggregate, filter and extract components in a manner quite similar to that described above. A glimmer of a how a holistic approach to the Web might be achieved is starting to be seen!

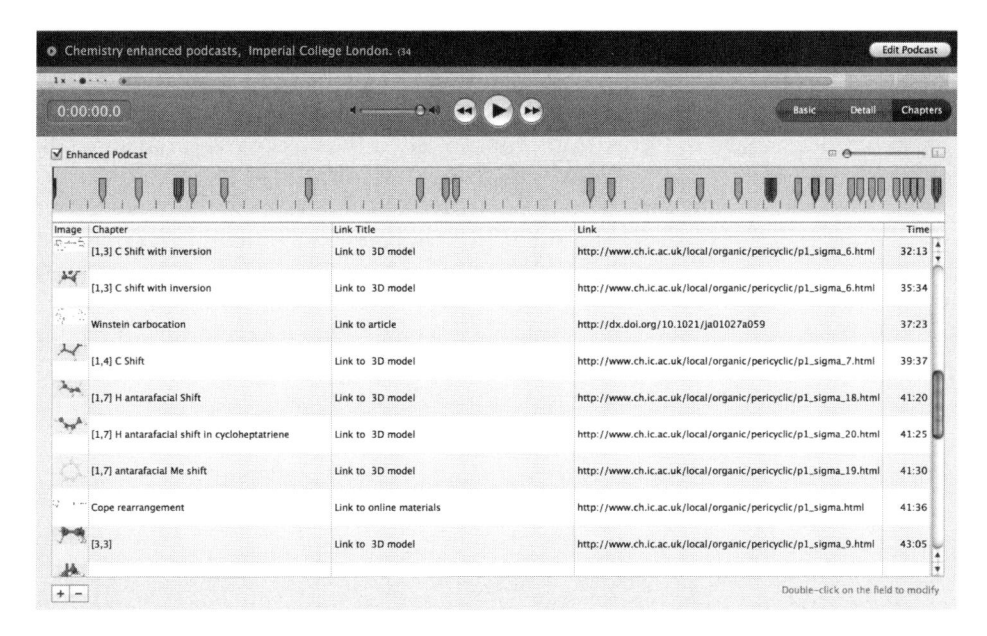

FIGURE 7.9 Interface for creating an enhanced podcast. Each *episode* contains a single audio file, with defined timepoints at which visual material (normally JPEG artwork, but it can also be an Acrobat file) is introduced as a so-called *chapter*. Each chapter can have a title which will be displayed, and a hyperlink to further materials such as original journal articles (via the DOI), 3D models (using Jmol) and other Web-resident information and data.

5.2 Authoring a Podcast

The structure of an enhanced podcast is illustrated in Figure 7.9, as revealed using an authoring program such as Podcast Maker. The program additionally provides interfaces (Basic and Detail) for populating the various metadata fields described above, some of which are controlled in the sense of allowing only certain values defined by a dictionary of terms. The category of podcasts for example comes from a controlled vocabulary defined by Apple Computer, which in turns allows them to customize their interface more carefully (c.f. Figure 7.6). This restriction means for example that there is no category such as **Chemistry** available.

6. THE WIKI

The Wiki is a relatively slow burner in the Web-world. The concept was first demonstrated in 1995 by Ward Cunningham, and as the story goes, was named after a shuttle bus in Honolulu airport, deriving from the Hawaiian language word for fast. It was envisaged as a solution to several issues relating to multi-author creation of Web pages, and was created as a user-interface to a database, the output of which was to be standardized XHTML for expressing content, stylesheets for expressing style, and metadata for describing the content. As such, it mandated no standards for this interface, the outcome being the development over the next twelve years of a variety of interfaces and styles. Since the purpose was fast and simple authoring of content, it is perhaps inevitable that many approaches will have evolved to achieving such simplicity. The purpose here is not to analyze this evolved diversity, but to explore what the concept can achieve in terms of the holistic ideal. To do this, the focus is on one particular flavor of Wiki known as **MediaWiki**. This actually has two advantages; it happens to be the Wiki chosen to deploy the Wikipedia, and secondly, this has been the chosen platform to further develop the Wiki semantically. The Wiki (another representative of Web 2.0) is both a simplification of the original Web concept and a return to the roots of the Web as originally envisaged by Berners-Lee in being a symmetrical medium, where people can act as both consumers and contributors.

The two main differences to a conventional Web page are:

1. The basic authoring syntax is designed to be simpler than HTML. The latter has around 60 *elements* and many attributes, which have to be interpreted and handled correctly. Wiki reduces this to a smaller number of simpler rules, expressed with minimal syntactic requirements. These are simple enough to require no user interface (other than text), although the latter can be built if desired (for chemical use for example, one can add an interface for the selection of Greek and other special characters). There is a smart processor (based on php) in the background that does the difficult work of interpreting these rules and storing the result in a MySQL relational database. If during the authoring of HTML an unclosed or improperly constructed element is defined, this will probably result in a error, severe or mild, at some unpredictable stage. Using a Wiki, the authored content either works properly or it is not a container; the underlying rule processor ensures it is not possible to store a broken element or container in the database.
2. This approach allows a Wiki to move away from the *write once, read many* conventional approach to a more symmetric *write many, read many* metaphor. *Write many* can either encompass the entire world, be constrained to say a local subset of students and teachers in order to address institutional concerns about legality etc., or extend to a federation of organizations. To achieve such control, **authentication** has always been at the core of most Wikis. MediaWiki for example supports such authentication protocols as LDAP and Shibboleth to achieve this.

6.1 The Wiki as an information liberator

A Wiki is very much a collaborative and communal medium. The database driven approach tends to reduce the error rate in the underlying syntax to a minimum, and hence strives to make the Wiki as user-friendly as it can. The communal ethos also means that Wiki has at its heart information liberation, rather than re-imprisonment, since it recognizes that most users are just as likely to be content creators as content consumers.

Wiki authoring syntax has two purposes. More trivially, the syntax is designed to achieve **appearance**. Thus in MediaWiki, enclosing a word ' "**thus**" ' will embolden it on the final Wiki page, whilst *"double quotes"* will italicize the word. The

Wiki rules transform this to CSS stylesheet entries on the final delivered XHTML (a syntax most people find too difficult to easily implement).

Of more interest are the structuring and semantic rules. Thus ==enclosing a title== allows a section heading to be defined, and this in turn is used to construct a table of contents for the document. More elaborate structure in the form of a table is imparted using:

    {|
    |cell| |cell
    |}

a syntax which somewhat simplifies the HTML original, but also does introduce some theoretical difficulties (which a markup purist might object to, along with blurring the distinction between content and appearance). In this case for example, the line break between the first {| and the second line is essential. This dependence on line breaks (and as it happens also on some white space) is a result of pragmatic decisions for simplicity, rather than achieving markup-rigor.

The most far-reaching structuring construct is that associated with *Linking*. In a multi-authored environment, links between different pages or sections of pages (I am trying to avoid using the term document here) are essential. As an author, you may recognize that a particular term you are using deserves further elaboration, and wish to highlight the term for others possibly more familiar with the subject to expand upon. Suppose you are writing something about a natural product synthesis, and one step involves what you recognize as an electrocyclic reaction. Rather than explain this term immediately, you flag it as being a particularly meaningful term worthy of further explanation elsewhere, and include in your prose the following string: [[electrocyclic reaction | A class of pericyclic reaction]]. The syntax of this example indicates a link to another page (on the same Wiki) with the **object** of this link being a page entitled *electrocyclic reaction*, and with metadata for this link comprising the value of the field following the | character. As the author, you do not need to know the location of this page, but you are expressing the need for it to be part of the (**implicit**) dictionary of terms being built by the community. The preceding two sentences each contain an emboldened word; respectively object and implicit. The first of these is highlighted to flag that **object** is being using in the same sense as one of the three components of an RDF declaration. The second is to introduce the difference between implicit (which a human can recognize) and explicit (which also software can process) semantic declarations. In section 6.1.3, the concept of a Semantic Wiki is introduced in which semantic meanings are made **explicit** (in this example, the term "electrocyclic reaction" is to be rendered with deliberate semantic meaning).

Consider next how this term might appear on the Wiki page itself to another reader. If in fact an object named **electrocyclic_reaction** does not already exist (this is reflected by rendering the term in red on the referring page, as opposed to blue if it does exist), then selecting the link will forward the reader to a blank editing page. They are invited to then become its author! This Janus-like role for the *Wikipedian* (an informal term for someone who both reads and contributes to the Wiki communal base of knowledge) is quite novel. It has the potential to symmetrize the relationship between and teacher and student, since both in principle could contribute terms and explanations about items they consider themselves knowledgeable.

An elaboration of this last construct is [[organic:electrocyclic reaction | A class of pericyclic reaction]]. The word preceding the: character defines a *category* and a provides a means of constructing a taxonomy of the subject by subdivision into categories, rather than just having a flat dictionary of terms. If the category is found useful by the community of users creating the Wiki pages, they will continue to use it and expand upon it. This communal approach is another Web 2.0 concept, the product being a **folksonomy** rather than a formal taxonomy. In this example, a community has agreed to construct a dictionary of terms relating to organic chemistry. A further (loose) analogy can be drawn to the RDF discussion above. Each link originates from an *subject* page and defines an *object* page. When the latter is created, it in turn becomes a new *subject* from which further *objects* can be referred. Viewed this way, the Wiki incorporates many of the concepts already discussed in section 2 above. Indeed, some Dublin Core (and an alternative scheme known as Creative Commons) metadata for a subject page can be expressed as RDF automatically, including the title, keywords, creation date and author identity (derived from any authentication process). The Wiki can be configured to expose this metadata as RDF by adding link statements to the generated HTML using the syntax shown in section 2.1.1 above, and these in turn can be detected by tools such as Piggy-Bank [13].

6.1.1 Wiki templates

A particularly innovative feature of the Wiki is the use of templates. These are pages which can be inserted into other pages via a process called transclusion. Templates contain defined parameters, which can either be defined with the call to the template, or evaluated using some logical procedure. A simple example is a DOI template: [http://en.wikipedia.org/wiki/Digital_object_identifierDOI]: [http://dx.doi.org/{{{1}}}{{{1}}}] where the term {{{1}}} is a parameter whose value will be supplied by the template call. {{{1}}} is used twice, once to append to the prefix http://dx.doi.org/ which serves to actually resolve the DOI and again to display its value on the screen to the user. Invoked as {{DOI | 10.1021/ci060139e}}, its transclusion produces the appearance DOI:10.1021/ci060139e, where the first component preceding the colon is a URL to the Wikipedia entry describing the DOI, and the second component following the colon is a URL which actually resolves the journal article.

The significance of the template is that the Wiki community can contribute these just as freely as they can articles. A number of chemistry-based templates have already been created, including examples such as {{Chem-Data}} and {{Drug-Box}} which are containers for a comprehensive selection of molecular properties and attributes. The advantages of transcluding such a template into a Wiki page is that the user's exposure to syntactic complexity is minimized, and more significantly that data can be presented (liberated) in a standard pattern which greatly facilitates the use of e.g. semantic data-scrapers such as Piggy-Bank or OSCAR [25].

6.1.2 Wiki extensions

One final data-liberating feature of the Wiki is the ability to define Extensions to its functionality. Five chemically useful examples are listed here, two of which will be examined in more detail below.

1. CharInsert, which enables transclusion of non-standard and Greek characters from any editing box and is particularly useful for chemistry.
2. Cite, which allows the Wiki text to be populated with citations, and which then collects and (re)numbers these at the bottom of the page in a standard pattern which again can facilitate the use of scrapers.
3. Parserfunctions, which provides a rich variety of mathematical and logical expressions used to enrich templates and content by dynamic evaluation. These are particularly useful when used in conjunction with Semantic templates (see below).
4. A Jmol enabler, in effect a template which when evaluated maps onto an invocation of the Jmol applet.
5. Semantic MediaWiki, which extends the Wiki with Semantic functions, and is used together with the Parserfunctions above.

A useful hint for identifying the extended functionality of any given (Media)Wiki is a URL of the form: http://site-name/wiki/index.php/Special:Version, which lists all the installed extensions, their versions, and associated hooks.

6.1.3 Wiki Jmol

The Jmol enabler can be invoked (in MediaWiki) thus:

```
<jmol>
  <jmolApplet>
<size>150</size>
<color>red</color>
<inlineContents><![CDATA[<cml:molecule xmlns:cml='http://www.xml-cml.org/schema/cml2/core'
 title='cyclohexane chair.mol'>
<cml:metadataList title='generated automatically from Openbabel'>
<cml:metadata name='dc:creator' content='OpenBabel version 1-100.1'/>
<cml:metadata name='dc:description' content='Cyclohexane'/>
<cml:metadata name='dc:identifier' content='InChI=1/C6H12/c1-2-4-6-5-3-1/h1-6H2'/>
<cml:metadata name='dc:rights' content='open'/>
<cml:metadata name='dc:type' content='chemistry'/>
<cml:metadata name='dc:contributor' content='rzepa'/>
<cml:metadata name='dc:creator' content='Openbabel V1-100.1'/>
<cml:metadata name='dc:date' content='Tue Jun 06 19:11:23 BST 2006'/>
<cml:metadata name='cmlm:structure' content='yes'/>
</cml:metadataList>
<cml:atomArray atomID='a1 a2 a3 a4 a5 a6 a7 a8 a9 a10 a11 a12 a13 a14 a15 a16 a17 a18'
elementType='C C C C C C H H H H H H H H H H H H'
formalCharge='0 0 0 0 0 0 0 0 0 0 0 0 0 0 0 0 0 0'
x3='1.688100 2.148400 2.936100 ....'
y3='2.157200 3.621700 3.912600 ....'
```

```
z3='-3.019300 -2.979700 -1.694200 ....'/>
<cml:bondArray atomRef1='a1 a1 a1 a1 a2 a2 a2 a3 a3 a3 a4 a4 a4 a5 a5
a5 a6 a6'
atomRef2='a2 a6 a7 a8 a3 a9 a10 a4 a11 a12 a5 a13 a14 a6 a15 a16 a17
a18' order='1 1 1 1 1 1 1 1 1 1 1 1 1 1 1 1 1 1'/>
</cml:molecule>]]>
</inlineContents>
  </jmolApplet>
</jmol>
```

The example is shown at some length to illustrate several points, but particularly the concept of re-use.

1. It shows the rather ad-hoc nature of how the Wiki user interface has evolved for extension. Thus to transclude a Jmol applet, XML markup is used (<jmol>...</jmol>) and an author has to cope with this via a simple text-editing interface. Another such example is Wiki re-use of some XHTML components, such as ^{superscript} to achieve styling of e.g. a molecular formula. The latter is in fact much better represented within CML.
2. The Jmol container in turn transcludes another XML component, defined by the namespace as CML. The Jmol parser reads this into memory, and displays any molecule it detects according to the coordinates it finds. The usual Jmol interface can in turn be used to retrieve the complete CML file (c.f. Figure 7.5).
3. The CML file in turn contains further metadata (according to the Dublin Core schema) and data, the predictable structure of which means that a data-scraper could be written to extract the information for re-use by e.g. Piggy-Bank.

6.1.4 The semantic Wiki

This is an early stage project to extend the conventional Wiki to being able to author what in computer science is referred to as an ABox statement. Expressed more simply, an ABox statement is an assertion component comprising simple facts about information objects, and would be used to enable more complete capture of relationships for an RDF declaration. To achieve this, the Wiki is extended with two new declarations which can be embedded in a page to achieve semantic annotation:

1. A subject can have a declared relation to an object. Thus a Wiki page with the subject of Mauveine can contain the following declaration: [[Oxidized to::Safranine|Mauveine can be oxidized to Safranine]]. In plain(er) English, this carries the information that a Wiki subject page identified as **Maueveine** is related to an object page identified as **Safranine** by the relationship **Oxidized to**, which too has to be a defined page on the Wiki. The text actually appearing on the page is what follows the | character. These three information objects are thus formally related by an RDF information triplet. This construct differs from a conventional Wiki in that the assertion (synonyms = relation or predicate) must also have a page which defines its meaning.
2. A subject can have a declared attribute rather than a relation, and that attribute can have a value. Thus the entry [[InChI:=1/C27H24N4/c1-17-9-11-20(12-10-

17)29-21-13-19(3)27-26(15-21)31(22-7-5-4-6-8-22)25-16-23(28)18(2)14-24(25)30-27/h4-16H,1-3H3,(H2,28,29)/p+1]]] contained in the Mauveine subject page states that one attribute of this molecule is that an InChI identifier has been declared, and that it has the value shown. A page belonging to the Wiki category **Attribute** (and declared as Attribute:InChI) elaborates the properties of the InChI attribute. The most important is that a datatype for the attribute (see section 2.3.1) must be declared, in this instance of type **String**. Here we see for the first time a formal mechanism for declaring what type of data any automated system would have to deal with. This datatype is declared using the syntax [[has type::Type:String]]. Attributes can themselves have attributes. For example, the InChI attribute can have a "See also" attribute declared [[**See also**:=http://www.iupac.org/inchi/|See IUPAC for full definitions]], the value of which itself has a datatype of URI, and which could be used for retrieval of further (machine processable) information relating to how to process an InChI string. Currently, the range of allowable datatypes is quite limited, and none relate specifically to chemistry. For example, one could envisage a datatype of *molecular formula*, which would be defined by rules expressing what might constitute an allowable formula, and which could be used to check for validity.

The formal properties of an environment created using relations and attributes as described above are both subtle and powerful, and largely beyond the scope of this article. Some of these features are merely listed here with minimal explanation. Thus multiple attributes can be related through chains of relations, an approach designated as multi-relational datamining. An in-lined query can also be embedded within a page, using syntax such as <ask limit="2">Attribute:+</ask> which results in a displayed listing of the first two known attributes associated with the subject. This is in fact a very powerful mechanism for dynamically updating the contents of any given page based on the relations and attributes of the entire Wiki. The <ask>...</ask> syntax can also be used within a Template to provide parameter values. Such a **Semantic Template** could have much potential for enabling semantic capture of information at the authoring stage, and thence as an information liberator.

How might this semantic content be exposed to the user (or to a software robot)? The most obvious difference between a conventional and a semantic Wiki is the automated collection of known facts about the subject page (Figure 7.10), and which can be exported from this environment into others appropriate for e.g. logical reasoning. Considered in this way, such a semantic article (a "Web 3.0" article) carries within its cellular structure the ability to exchange information about itself (its visiting card, to use an anthropomorphic analogy) not only with humans but with other such articles. It could be induced to do the latter by (automated) software processes. Such semantic articles would then not be limited by the attention span of a mere human; it is thought humans rarely read to completion any article they may come across in conventional form in a journal or book, scouring it merely for the salient fact they suppose to be relevant.

The search interface for a Semantic Wiki is also different (Figure 7.11). One can specify a search through any of the three information components, i.e. the subject, object and/or any relation between them. In the example below, a search

Facts about Mauveine — Click +⚲ to find similar pages.	RDF feed ⚛

	Relations to other articles
Member of	Imperial College Chemistry +⚲
Affiliation	Imperial College Chemistry +⚲
See also	10.1039/CT8793500717 +⚲, safranine +⚲, 10.1021/ja042233m +⚲, Schultz +⚲, and 10.1039/P19940000005 +⚲
Reaction with	Lead Dioxide +⚲, and Zinc +⚲
Reaction product	Parasafranine +⚲
Sub structure	HN-Phenyl +⚲, and HN-Tolyl +⚲
	Attribute values
Name	Henry Rzepa +⚲
Email	h.rzepa@imperial.ac.uk +⚲
Foaf:mbox	mailto:h.rzepa@imperial.ac.uk✉ +⚲
Homepage	http://rzepa.ch.ic.ac.uk/⌐ +⚲
Foaf:phone	2075945774 +⚲
See also	http://www.ch.ic.ac.uk/rzepa/rzepa.xrdf⌐ +⚲
Cites	http://dx.doi.org/10.1039/CT8793500717⌐ +⚲, http://dx.doi.org/10.1021/ja042233m⌐ +⚲, Schultz +⚲, and http://dx.doi.org/10.1039/P19940000005⌐ +⚲
MF	$C_{27}H_{24}N_4$ +⚲
Number Atoms	55 +⚲
MW Measured	406.4 +⚲
MW Calculated	404.2 +⚲
MW diff	2.2 +⚲
Oxidized group	HN-aryl +⚲
Reduced group	NO_2 +⚲
MF Diff	C_7H_6 +⚲, and C_6H_4 +⚲
InChI	InChI=1/C27H24N4/c1-17-9-11-21(12-10-17)31-26-15-22(28)18(2)13-24(26)30-25-14-19(3)23(16-27(25)31)29-20-7-5-4-6-8-20/h4-16H,28H2,1-3H3/b29-23- +⚲, and InChI=1/C27H24N4/c1-17-9-11-20(12-10-17)29-21-13-19(3)27-26(15-21)31(22-7-5-4-6-8-22)25-16-23(28)18(2)14-24(25)30-27/h4-16H,28H2,1-3H3/b29-21+ +⚲
SMILES	CC1=C(N)C=C(N(C5=CC=C(C)C=C5)C(C4=N3)=C/C(C(C)=C4)=N\C2=CC=CC=C2)C3=C1 +⚲, and CC1=C(N)C=C(N(C5=CC=CC=C5)C(C4=N3)=C/C(C=C4C)=N\C2=CC=C(C)C=C2)C3=C1 +⚲

Categories: Person I Phenazinium I Dyestuff I Molecular tectonics	

FIGURE 7.10 The explicit facts known about a Semantic Wiki page, as shown on screen, and available as RDF. These include information about the author as well as about the molecule mauveine.

Simple semantic search

Subject article:	Relation name:	Object article:	
Mauveine	Reaction product	Parasafranine	(Search Relations)
	Attribute name:	Attribute value:	
	InChI	InChI=1/C27H24N4/c1-1	(Search Attributes)

Search results (attributes)

Mauveine   InChI   InChI=1/C27H24N4/c1-17-9-11-20(12-10-17)29-21-13-19(3)27-26(15-21)31(22-7-5-4-6-8-22)25-16-23(28)18(2)14-24(25)30-27/h4-16H,28H2,1-3H3/b29-21+ +⚲

FIGURE 7.11 A semantic Search interface, showing the components of an information triple.

has been conducted on the value of an attribute, being in fact the InChI identifier for mauveine. The search results (there is only one hit in this example) show all three components of the information triple. Consider how much more powerful

2.1 Conformational analysis I: Chair and Boat-like conformations of Cyclohexane [edit]

1. Construct chair and boat-like conformations of cyclohexane. Compare the energies of both forms.
2. Check carefully if your boat really is a boat, or whether it has any apparent distorsion.
3. Try changing one or more of the CH_2 groups into an oxygen and see if that affects things.
4. For the record, the point group symmetries of the various species which may be involved are D_{3d} for the chair conformation, C_{2v} for a boat form (if it exists?), and D_2 for any twisted boat form.

2.1.1 References [edit]

1. The first suggestion of two forms for cyclohexane goes as far back as H. Sachse, *Chem. Ber*, 1890, **23**, 1363 and *Z. Physik. Chem.*, 1892, 10, 203. This is nicely explained here. E. Mohr, *J. Prakt. Chem.*, 1918, **98**, 315 and *Chem. Ber.*, 1922, **55**, 230, translated Sachse's argument into a pictorial one.
2. The article that put conformational analysis on the map: D. H. R. Barton and R. C. Cookson, *The principles of conformational analysis*, Q. Rev. Chem. Soc., **1956**, *10*, 44. DOI:10.1039/QR9561000044
3. Wikipedia article

FIGURE 7.12 Section of a molecular modeling course expressed as a Wiki. Features to note are the edit links, the linked to other pages presence of a 3D molecular model containing re-usable data, and the use of the DOI template to provide 1-click connection to a literature article.

such a search could be compared to conventional retrievals. It is of course also possible to pipe the result of one search into further searches to refine the results. There is some resemblance of this operation to the filtering demonstrated with Podcasts in section 5.1.

6.2 Example of course Wikis

There are numerous excellent chemical examples of this type of presentation, indeed too many to try to even summarize here. Instead, the examples shown below are from the author's own materials and are selected to demonstrate the components in action. Figure 7.12 shows an article which was written by the author and several colleagues and how each contributed different components from their own areas of expertise. The course was actually delivered "locked" to the students, although they were allowed to post self-help comments on each item.

The next example (Figure 7.13) shows a follow up interface, allowing students to extract the coordinates of any molecule shown in the course notes, and submit a calculation to a High-performance-cluster pool for further elaboration. The outcome of this step in the workflow could be deposition into a digital repository for more permanent archival.

Figure 7.14 demonstrates the communal aspects of a Wiki. Students were presented with three choices to annotate a chemistry project page. They could populate suggested themes for molecules (Figure 7.14a), or propose molecules of their own (Figure 7.14b) or extend the list of available templates to enhance the other pages. All three proved popular!

6.3 Examples of chemical Wikipedia communities

Another illustration of the community approach that the Wiki espouses can be seen in the diversity of communities that have sprung up with the objective of

JOBS IN PROJECT 3RD YEAR

Job ID	Application	Description	Submission Time	Wall Time	Status	Input files	Output files	Delete	Repository
753	Gaussian	res cavity mono	2007-04-19 07:12:58	0:19:33	Finished	Gaussian Input file / Download	Gaussian Log File / Download	Delete	View
735	Gaussian	cavity only	2007-04-17 21:20:09	27:42:31	Finished	Gaussian Input file / Download	Gaussian Log File / Download	Delete	View
694	Gaussian	res inside only	2007-04-14 17:43:57	0:22:42	Finished	Gaussian Input file / Download	Gaussian Log File / Download	Delete	View
691	Gaussian	res cavity only	2007-04-14 15:54:32	Not reported	Finished	Gaussian Input file / Download	Gaussian Log File / Download	Delete	Publish
495	Gaussian	SR	2007-03-04 13:10:21	13:9:0	Finished	Gaussian Input file / Download	Gaussian Log File / Download	Delete	View
494	Gaussian	RS	2007-03-04 13:10:04	31:35:58	Finished	Gaussian Input file / Download	Gaussian Log File / Download	Delete	View
491	Gaussian	chair-5me-iso	2007-03-04 12:17:10	0:22:11	Finished	Gaussian Input file / Download	Gaussian Log File / Download	Delete	View
490	Gaussian	boat-5me-iso	2007-03-04 11:47:51	0:21:54	Finished	Gaussian Input file / Download	Gaussian Log File / Download	Delete	View
488	Gaussian	chair-5me	2007-03-04 11:33:44	0:22:9	Finished	Gaussian Input file / Download	Gaussian Log File / Download	Delete	View
431	Gaussian	nor-3	2007-02-25 15:56:27	0:37:15	Finished	Gaussian Input file / Download	Gaussian Log File / Download	Delete	View

FIGURE 7.13 An interface between a course Wiki and a computing pool allowing annotation of further molecular properties.

Main Project Page (a)

Please do not edit this page itself. Click on one of the titles to start editing.

Project Number	General Keywords
01	Lignocaine (used in dentistry as a "local")
02	Piperine (active ingredient of both black and white pepper)
03	Rapamycin (prevents transplant rejection)
04	Gossypol (male birth control)
05	Gentamicin A (aminoglycoside antibiotic)
06	Herceptin (topical anticancer drug)
07	Zingerone (the characteristic smell of ginger)
08	Sucralose (non-metabolizable sweetening agent)
09	Bufotoxin (active component of the toad *Bufo vulgaris*)
10	Roaccutane (treatment for severe acne)
11	Sibutramine (appetite suppresor)
12	Anandamide (the "feel-good" factor in chocolate)
13	Ammonia-borane: H_3N-BH_3 (Hydrogen storage molecule?)

Supplemental Project Page (b)

This area is for people who wish to create their own projects if none of the above appeal to them. Click on the **Edit** button to the right to open up an editable page, then add an entry below as follows

- [[t:name_of_project|Descriptive name of intended project]]
- This will produce the effect: Descriptive name of intended project

- Sitagliptin
- Retinal, molecule of sight
- Tamoxifen, breast cancer treatment
- Morphine, painkiller
- Pelargonidin, colouring in nature
- Propanil, weedkiller
- Ranitidine, antiulcerative
- chlorphentermine, anorectic

Wiki Templates (c)

Template:DOI and Template:Doi-inline are providea as (protected) templates for your use. Many other templates exist, often to be found on e.g. Wikipedia pages. You may decide one of these is of particular use, or of interest. If so, you can install it on the wiki here for you and others to use. Add below a line that looks like Template:Template-name, save, and click on the red text to create the new template. If you prefer the task of adding useful templates to that of adding information about molecules, then you will be given full credit for performing this valuable service for others! --Rzepa 14:41, 20 October 2006 (BST)

Template:Chem-Data

Template:Drug-Box - For pharmaceutical drugs just copy variable names and code generates tables

Template:Chembox supplement - to be linked to from the supplementary section of the table in the template above, for usage see here

Template:NFPA_704 - for notes on how to use, see here

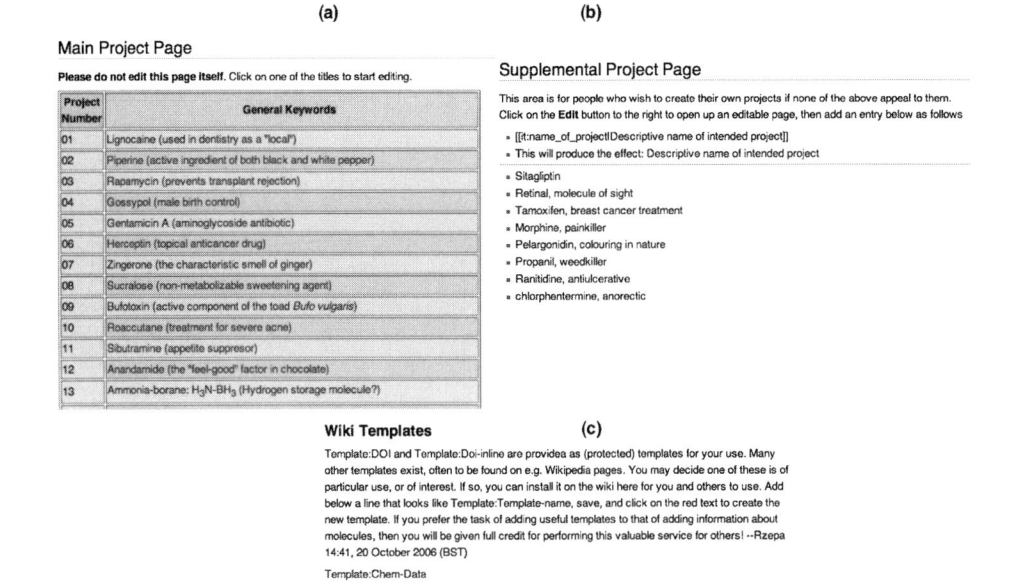

R & S Phrases

FIGURE 7.14 A Communal chemistry Wiki project page.

enhancing or annotating the Wikipedia itself. Some of the ones active in 2007 and pertaining to chemistry are listed below.

1. http://en.wikipedia.org/wiki/Wikipedia:WikiProject_Chemistry handles general articles on chemistry. The discussion part of this page has an active group
2. http://en.wikipedia.org/wiki/Portal:Chemistry is the main portal for users
3. http://en.wikipedia.org/wiki/Wikipedia:WikiProject_Chemicals handles chemical compounds
4. http://en.wikipedia.org/wiki/Wikipedia:WikiProject_Chemicals/Style_guidelines enumerates the standard formats deployed in the other pages
5. http://en.wikipedia.org/wiki/Dictionary_of_chemical_formulas comprises a listing of molecules with some standard properties.

CONCLUSION

I noted at the start that the Web, by liberating information from its traditional physical containers, had created the potential for entirely new metaphors for its movement, manipulation and re-purposing. In reality however, what we have seen is an increasing transfer of the world's scientific and chemical knowledge into only marginally less restrictive electronic containers such as PDF files or Powerpoint presentations. Less thought appears to have been given to how these new objects might be subsequently used and integrated into a more holistic whole. Publishers and authors alike have solutions in their own hands; they should start taking metadata far more seriously, and start adding it to PDF files in a manner which is properly exposed and queryable by software. Some types of electronic container, such as the podcast, already have much of this incorporated into their design, and tools for creating podcasts already allow metadata creation. Perhaps the most natural medium for storing information in a liberated manner is the Wiki, of which the best exemplars are the Wikimedia family [26] spun out from Wikipedia and which now includes WikiBooks, Wikiversity, Wiktionary and WikiSource. These all strongly espouse the Open movement, and although Wikipedia is currently the subject of much discussion in terms of its reliability and quality [27], the concept in its maturity will surely represent a strong challenge to the unliberated medium of the printed book. To emphasize just one limitation of the latter, the author feels impelled to note that this article was written in January 2007, and in the manner of the fast-moving digital world, some aspects may have already evolved substantially by the time you read the words imprisoned on this printed page.

An extension to the Wiki concept is called the Semantic Wiki. This fully assimilates modern approaches such as the information triple and the idea of a community of like-minded participants who are not merely information consumers but who also contribute to the community. Whereas Web 2.0 brought us communities and social, collaborative software, combining this with a holistic semantic environment is already being described as the evolutionary Web 3.0.

It seems appropriate to conclude this review by introducing onto the stage one actor who presence has been implied repeatedly above. The community can include not just humans, but also machines in the form of intelligent software that can also absorb the liberated information and reason using it, on a vast scale if necessary since unlike humans it is unlikely to get bored (although it might well get very confused!). One of the more provocative, stimulating and certainly controversial predictions of the future is Kurzweil's [28] concept of Human v2.0, in which the existing biological organism will be supplemented by a symbiosis with ultra-fast computers. This composite entity, it is argued, will enable human evolution to suddenly proceed at a far faster pace than it has in the past, a moment in time referred to as the *singularity*. This new being will be a voracious information consumer, but of the semantic type described in this article. Leaving aside the host of other issues this raises, how capable would Human v2.0 be at creative chemical discovery? Perhaps the answer will be revealed as soon as 2030!

REFERENCES

1. Rzepa, H.S. Chemistry and the World-Wide-Web, in Bachrach, S., editor. The Internet: A Guide for Chemists. American Chemical Society; 1996.
2. Boulos, M.N.K., Maramba, I., Wheeler, S. Wikis, blogs and podcasts: a new generation of Web-based tools for virtual collaborative clinical practice and education, BMC Medical Education 2006, 6, 41. DOI:10.1186/1472-6920-6-41.
3. Gregorius, R.M. Various learning environments and their impact on student performance. Part I: Traditional versus PowerPoint and WebCT augmented classes, Chemical Educator 2005, 10, 72–7. DOI:10.1333/s00897050880a.
4. Paskin, N. Digital object identifiers for scientific data, Data Science Journal 2005, 4, 12–20. DOI:10.2481/dsj.4.12. A chemical analogue to the DOI has recently been proposed;
   Rzepa, H.S. Semantic wiki as a model for an intelligent chemistry journal. Abstracts of Papers, 233rd ACS National Meeting, Chicago, IL, United States, 2007 CINF-053.
5. Kajosalo, E., Branschofsky, M. DSpace as an institutional repository, Abstracts of Papers, 228th ACS National Meeting, Philadelphia, PA, United States, 2004, CINF-033.
6. Marx, W., Schier, H. Hermann, CAS versus Google, Nachrichten aus der Chemie 2005, 53, 1228–32. See also http://scholar.google.com/.
7. Cass, M.E., Hii, K.K., Rzepa, H.S. Mechanisms that interchange axial and equatorial atoms in fluxional processes: illustration of the berry pseudorotation, the turnstile, and the lever mechanisms via animation of transition state normal vibrational modes, J. Chem. Ed. 2006, 83, 336. Webware: http://www.jce.divched.org/JCEDLib/WebWare/collection/reviewed/JCE2006p0336_2WW/.
8. Rzepa, H.S., Cass, M.E. A computational study of the nondissociative mechanisms that interchange apical and equatorial atoms in square pyramidal molecules, Inorg. Chem. 2006, 3958–63. DOI:10.1021/ic0519988.
9. For details, and specifications, see http://en.wikipedia.org/wiki/Microsoft_Office_Open_XML and http://www.ecma-international.org/publications/standards/Ecma-376.htm.
10. Heller, S.R., Stein, S.E., Tchekhovskoi, D.V. InChI: open access/open source and the IUPAC International Chemical Identifier, Abstracts of Papers, 230th ACS National Meeting, Washington, DC, United States, Aug. 28–Sept. 1, 2005, CINF-060;
    Coles, S.J., Day, N.E., Murray-Rust, P., Rzepa, H.S., Zhang, Y. Enhancement of the chemical semantic web through the use of InChI identifiers, Org. Biomol. Chem. 2005, 3, 1832–4. DOI:10.1039/b502828k.
11. The best source for information about RDF is http://www.w3.org/RDF/. For information on how the Qualified Dubin Core is expressed as RDF/XML, see http://dublincore.org/documents/dcq-rdf-xml/. For a chemical application of RDF, see Taylor, K.R., Gledhill, R.J., Essex, J.W., Frey, J.G., Harris, S.W., De Roure, D.C., Bringing chemical data onto the semantic web, J. Chem. Inf. Model.,

2006, 46, 939–952. DOI:10.1021/ci050378m. For an article specifically relating to learning technology, see Nilsson, M., The semantic web: How RDF will change learning technology standards, 2001, http://www.cetis.ac.uk/content/20010927172953.
12. Casher, O., Rzepa, H.S. SemanticEye: a semantic web application to rationalise and enhance chemical electronic publishing, J. Chem. Inf. Mod. 2006, 46, 2396–411. DOI:10.1021/ci060139e.
13. Huynh, D., Mazzocchi, S., Karger, D. Piggy Bank: experience the semantic web inside your web browser, International Semantic Web Conference (ISWC), 2005.
14. Murray-Rust, P., Rzepa, H.S. The next big thing: from hypermedia to datuments, J. Digital Inf., 2004, 5, article 248, 2004-03-18.
15. Rzepa, H.S., Whitaker, B.J., Winter, M.J. Chemical applications of the World-Wide-Web, J. Chem. Soc., Chem. Commun., 1994, 1907. DOI:10.1039/C39940001907;
Casher, O., Chandramohan, G., Hargreaves, M., Leach, C., Murray-Rust, P., Sayle, R., Rzepa, H.S., Whitaker, B.J. Hyperactive molecules and the World-Wide-Web information system, J. Chem. Soc., Perkin Trans 1995, 2, 7. DOI:10.1039/P29950000007.
16. Rzepa, H.S., Murray-Rust, P., Whitaker, B.J. The application of chemical multipurpose Internet mail extensions (Chemical MIME) Internet standards to electronic mail and World-Wide Web information exchange, J. Chem. Inf. Comp. Sci. 1998, 38, 976–82. DOI:10.1021/ci9803233.
17. See the Jmol site at http://jmol.sf.net/. For a typical application of Jmol, see Herraez, A. Biomolecules in the Computer. Jmol to the rescue, Biochem. Mol. Biology Education 2006, 34, 255–61. DOI:10.1002/bmb.2006.494034042644.
18. Martz, E. 3D molecular visualization with protein explorer, Intro. Bioinformatics 2003, 565–86; Reichsman, F., Martz, E. Biomolecular Explorer site, at http://www.umass.edu/molvis/bme3d/.
19. Murray-Rust, P., Rzepa, H.S. Chemical Markup, XML and the Worldwide Web. Part 4. CML Schema, J. Chem. Inf. Comp. Sci. 2003, 43, 757–72;
Holliday, G.L., Murray-Rust, P., Rzepa, H.S. Chemical Markup, XML and the Worldwide Web. Part 6. CMLReact; An XML Vocabulary for Chemical Reactions, J. Chem. Inf. Mod. 2006, 46, 145–57. DOI:10.1021/ci0502698.
20. Guha, R., Howard, M.T., Hutchison, G.R., Murray-Rust, P., Rzepa, H.S., Steinbeck, C., Wegner, J., Willighagen, E.L. The Blue Obelisk—interoperability in chemical informatics, J. Chem. Inf. Mod. 2006, 991–8. DOI:10.1021/ci050400b.
21. Acton, A., Banck, M., Bréfort, J., Cruz, M., Curtis, D., Hassinen, T., Heikkilä, V., Hutchison, G., Huuskonen, J., Jensen, J., Liboska, R. Rowley, C. http://www.uku.fi/~thassine/projects/ghemical/.
22. Morgan, R. Science in the City: the podcasts, vodcasts, and blog activities, Abstracts, 19th Rocky Mountain Regional Meeting of the American Chemical Society, Tucson, AZ, United States, 2006, RM-223;
Garritano, J.R., Eisert, D.B. On the go with CHM 125, ECON 210, PHYS 218, and BIOL 205: Coursecasting at a large research university, Abstracts of Papers, 231st ACS National Meeting, Atlanta, GA, United States, 2006, CINF-005.
23. Murray-Rust, P., Rzepa, H.S., Williamson, M.J., Willighagen, E.L. Chemical Markup, XML and the Worldwide Web. Part 5. Applications of Chemical Metadata in RSS Aggregators, J. Chem. Inf. Comp. Sci. 2004, 44, 462–9. DOI:10.1021/ci034244p.
24. Bioclipse; see http://www.bioclipse.net/ and http://sourceforge.net/project/showfiles.php?group_id=150681.
25. Adams, S., Kuhn, S., Murray-Rust, P., Steinbeck, C., Townsend, J.A., Waudby, C.A. Reviving analytical data of the past with open submission databases and text mining tools, Abstracts of Papers, 232nd ACS National Meeting, San Francisco, CA, United States, 2006, CINF-101.
26. See the following URLs: http://www.wikimedia.org/, http://en.wikibooks.org/wiki/, http://en.wikiversity.org/wiki/, http://en.wikisource.org/, http://en.wiktionary.org/.
27. Giles, J. Internet encyclopedias go head to head, Nature 2006, 438, 900–1. DOI:10.1038/438900a; Walker, M.A. Wikipedia: Social revolution or information disaster?, Abstracts of Papers, 231st ACS National Meeting, Atlanta, GA, United States, 2006, CINF-007;
Yager, K. Wiki ware could harness the Internet for science, Nature 2006, 440, 278. DOI:10.1038/440278a.
28. Kurzweil, R. Our human-machine civilization, Science 1999, 285(5426), 339–40;
Kurzweil, R. Human life: The next generation, New Scientist 2005, 187(2518), 32-7. See also http://www.kurzweilai.net/.

Section 4
Materials and Polymers

Section Editor: Jeffry Madura

Duquesne University
Department of Chemistry and Biochemistry
Center for Computational Sciences
600 Forbes Ave.,
Pittsburgh, PA 15282
USA

CHAPTER 8

The Role of Long-Time Correlation in Dissipative Adsorbate Dynamics on Metal Surfaces

Jeremy M. Moix[*] and **Rigoberto Hernandez**[*]

Contents		
	1. Introduction	137
	2. Classical Surface Diffusion	140
	3. Molecular Dynamics Simulations and Projective Models	143
	3.1 Atomistic models	143
	3.2 Projective models	144
	4. Summary and Conclusions	146
	Acknowledgements	147
	References	147

1. INTRODUCTION

The motion of adsorbates on metal surfaces remains an area of active research as a result of its technological implications for a diverse set of processes such a catalysis, epitaxial growth, and self-assembly, among many others. Aside from the relevance to practical applications, fundamental scientific interest remains strong because of the rich physics governing these processes and the challenges facing both the experimental and theoretical communities to develop accurate techniques that are capable of investigating and describing the various phenomenon. Nevertheless, these communities have made great advances over the past two decades. Thus a full discussion of all of the recent developments is beyond the scope of this report. Numerous others have focused on various aspects and the curious reader is directed to several excellent reviews of recent contributions in experimental, theoretical and computational methodologies, and their interpretation from experimental [1–6] or theoretical [6–12] perspectives, although obviously neither can be

[*] Center for Computational Molecular Science and Technology, School of Chemistry and Biochemistry, Georgia Institute of Technology, Atlanta, GA 30332-0400, USA
E-mail: hernandez@chemistry.gatech.edu

completely devoid of the other. For a review discussing the challenges and recent successes in modeling surface dynamics across very large length and time scales while retaining atomic level details as needed, see Ref. [13]. Alternatively, algorithmic developments attempting to accelerate molecular dynamics simulations of surface diffusion are reviewed in Ref. [14].

The primary aim of this report is to demonstrate that classical simulations of adsorbate dynamics on surfaces can be described by reduced-dimensional models and that such models can provide insight into their qualitative—and sometimes quantitative—behavior. In principle, one would prefer not to resort to such coarse-grained models in favor of the inclusion of all degrees of freedom explicitly. However, such models remain computationally infeasible for many interesting chemical systems given the current state of computing capabilities and will remain so for some time in the near future given the foreseeable performance increases. Aside from their affordability, reduced-dimensional models also provide a lens on which modes of a system are important to the detailed dynamics as well as those that can be successfully ignored and are evidently not important. For example, in the cases illustrated in Figure 8.1, the all-atom simulations require the integration of the dynamics not just for the atoms, molecules, or polymers adsorbed on the surface, but of the much larger underlying surface. As the number and complexity of the adsorbates grows, it becomes increasingly useful to be able to ignore the detailed dynamics of the underlying surface, and this necessarily requires a projection. The latter is not a rigorous coarse-graining because the length scale of the remaining adsorbates has not been changed. However, these length scales may also need to be coarse-grained in order to assess macroscopic properties of the surface [13].

There is, of course, one very important additional level of complexity in these systems that arises when the length and energy scales are such that classical mechanics no longer offers the correct equations of motion. (The Electronic dynamics are necessarily such a case.) Consequently, the correct inclusion of quantum mechanical degrees of freedom into the description of surface events and molecular simulations in general remains a topic at the forefront of current theoretical efforts. Unfortunately, such simulations are prohibitively expensive for processes occurring on all but the shortest of length and time scales. For reviews on the subject within the context of surface science one is referred to Refs. [9,15–18]. While the present report will not focus on such approaches, it is perhaps worthwhile to point out that one potential advantage of the identification of reduced-dimensional models lies in the possibility that the subsystem can be quantized independently of the projected classical models [13].

Presently simulations represent a compromise between the level of detail one would like to include and that which is capable of reasonably being included. This is particularly relevant when long-time correlations exist between the adsorbate and the surface. In this context, long-time scale implies long on the molecular scale but it should be noted this is usually very short on the experimental scale. The use of simulation in this instance has been instrumental in interpreting the complicated experimental results that typically represent an ensemble average over many adsorbates completing multiple hops between observations. One could cer-

FIGURE 8.1 Much of the method development discussed here is focused on the equations of motion for a single particle on a surface as illustrated in the top graphic. However, the projective methods become all the more useful as one begins to consider many adsorbates (as in the middle graphic) or polymers (as in the bottom graphic). If the atomistic dynamics of the underlying surface can be subsumed through a single effective stochastic representation, then the dissipative MD can become tractable even for very large extended coatings.

tainly include ipso facto the long-time correlations through the use of an all-atom model that is sufficiently large as to include all of the requisite local modes, surface phonons, and bulk phonons while maintaining this classical-like prescription. However, the key aim here is the use of reduced-dimensional models that allow for the inclusion of the long-time correlations without recourse to such detailed models. Meanwhile such reduced-dimensional models, while projecting the bulk modes into dissipative terms can also project the electronic modes into the so-called electronic friction [9,19–21]. Thus, in limits where the former has been seen to be effective, then the models described here with the appropriate friction terms can also account for quantum electronic effects. An essential requirement of any reduced-dimensional approach is that some transferability of the coarse-grained description will exist over the relevant parameter range of interest. Fortunately, this requirement is often satisfied. Obviously, one should not project out degrees of freedom of the system that will be important to the process of interest. However, once one has an acceptable model for surface diffusion with appropriate long-time correlations then it can presumably be used to study various questions concerned with surface transport.

It is perhaps also helpful to briefly note a few experimental techniques in surface science and their relationship to the corresponding microscopic dynamics as elucidated through an appropriate reduced-dimensional or atomistic model. This duality can be exploited to assist and validate experimental or computational interpretations of complicated systems, and several reviews have discussed this topic in detail [1–5,10,11,22]. For slow diffusion and low coverage, scanning tunneling microscopy (STM) and field ion microscopy (FIM) are able to directly follow diffusive events and thus provide a description of the microscopic time-dependent probability distribution [6,23–25]. However, in more general cases, these methods lack the temporal resolution needed to observe faster hopping rates and it also becomes difficult to unambiguously identify the unique diffusion events as coverage increases. In such cases, quasi-elastic helium atom scattering (QHAS) becomes the preferred method although it provides only an indirect measure of adsorbate diffusion [22,26–28]. The measured structure factor from these experiments is related to the microscopic hopping rates as determined for example, from a microscopic master equation.

2. CLASSICAL SURFACE DIFFUSION

Surface diffusion consists of a series of jumps or events between minima on the energy landscape. Each of these jumps is inherently a rare event (on the molecular scale) due to the large barriers impeding the hopping process. These barriers are typically on the order of a few tenths of an eV and generally an order of magnitude larger than the thermal energy [29,30]. At relevant temperatures, such activated processes occur on time scales that lend themselves only to computationally-taxing classical molecular dynamics simulations or more feasible coarse-grained Langevin simulations. Taking into account that this is an activated process, the diffusion rate can usually be well represented by an Arrhenius-like form

$$D(T) = D_0(T) \exp(-\beta E_A) \tag{1}$$

where E_A is the classical activation energy, $\beta = 1/k_B T$, and $D_0(T)$ is the pre-exponential factor whose temperature dependence is often weak [31]. This formalism provides one with the ability to extract the activation energy and prefactor from a series of simulations performed over different temperatures. However the simple interpretation is not quite adequate. At normal temperatures, an adsorbate would never have enough energy to surmount a barrier since typically $E_A \gg k_B T$. This elementary analysis makes it clear that the surface is a vital component of the diffusion process serving as an effective bath that is constantly exchanging energy with the adsorbates. The particle surmounts a barrier when it has accumulated a sufficient amount of energy from the surface and continues to diffuse across the barriers until the energy has been transferred back. This type of argument readily gives rise to a Langevin description of surface diffusion and has been invoked frequently in the past [32–37]. The possibility of correlated hops to sites farther than the nearest wells during an excursion depends on the rate of energy relaxation. Because the probability of such events is usually low, correlated hops have often been ignored or included implicitly. However when the temperature is large or the adsorbate is weakly bound, then correlated hops have been seen to play an important role [37–39].

While the diffusion constant is dominated by the equilibrium barrier height E_A, all of the dynamical contributions are contained in the prefactor. Unfortunately, brute force atomistic simulations have only been computationally feasible for relatively high temperatures. (Of course, this constraint is constantly relaxing.) Molecular dynamics simulations are now possible for long runs of hundreds of nanoseconds or alternatively many short runs of several hundred picoseconds. However, in the case that the ratio of the barrier height to the thermal energy is large, acquiring sufficient hopping statistics through either approach becomes impractical at an exponential rate. As a result, one often takes advantage of the Arrhenius behavior and studies simulations at higher temperatures and then relies on an extrapolation to lower temperature regimes. In light of this difficulty alternative methods such as Monte Carlo simulations are often preferred [8,29,40–42]. In this situation one relies on the assumption that hops are rare and uncorrelated events, and hence the entire trajectory can be taken as a series of independent (Markovian) processes. The diffusion coefficient may thus be cast as the product of a rate constant based on (canonical) transition state theory and the length of the hop (lattice constant), l. That is,

$$D = \frac{1}{2\alpha} k_{\text{TST}} l^\alpha, \qquad (2)$$

where α represents the dimensionality of the system. Within this approximation, the elegant techniques developed from transition state theory may be employed thereby allowing one to extract very accurate rates with substantial computational savings [43–47]. However, the Markovian approximation of independent hops has recently been tested in the self-diffusion of Cu on Cu(1 1 1) and it was found that correlations exist between hops even at low temperatures when the assumption is expected to be valid [48]. This raises serious questions about blind applications of the Monte Carlo average of canonical TST rates, but clearly more work is needed given its success thus far.

Further caution is warranted given that not all of the elementary mechanisms comprising surface diffusion are fully understood. Traditional hops to nearest neighbors may be accompanied by many other processes including site exchange whereby the adsorbate and a surface atom exchange positions resulting in a net displacement [49,50], long jumps to sites further than nearest-neighbors, and possibly even sub-surface diffusion. These various complex mechanisms make reactive flux approaches difficult to employ when the transition path is not known or perhaps contains multiple geometrically disparate exit channels.

The situation is compounded by the fact that in order to achieve an accurate description of the bulk, many layers of surface atoms must be included in simulations [32]. In principle, the only way to know when one has included a large enough underlying surface is to continue to increase the number of layers until a clear convergence of the target observable has been reached. To circumvent this costly exercise, several investigations have allowed only those atoms within the vicinity of the adsorbate to move while others are pinned to their equilibrium positions [33,34,51–53]. While this results in a dramatic computational savings, it overly dampens the long time scale processes of the surface. Furthermore, one must average over many trajectories in order to acquire reasonable statistics of the hopping process and properties of interest such as the hopping distribution. All of these considerations lead to state-of-the-art simulations of surface diffusion in which thousands (or more) surface atoms are subdivided into varying hierarchies of dynamical treatments, while nevertheless requiring substantial computing power in order to integrate several thousand trajectories each on the order of a nanosecond. Unfortunately most molecular dynamics packages are not designed for surface dynamics and therefore lack the ability or needed generality to correctly and accurately construct these hierarchies, and this leads many researchers to write their own simulation codes.

Beyond the dilute adsorbate limit that is the focus of this report, the role of step edges, coverage, sticking and scattering in adatom diffusion has been explored by a cornucopia of computational approaches ranging from coarse-grained to ab initio methods [7–10,12,54]. For example, it might seem that surface coverage effects can be completely described by a simple extension of the theory for single adatom hopping. However nothing could be farther from the truth as the diffusion now often includes concerted motions and may proceed by completely different mechanisms resulting in large qualitative changes in the diffusion rates [2–4,26]. In such cases, the degree to which long-time correlations persist plays an important role in determining diffusion rates. Similar complications arise in the case of gas-surface scattering. Of particular note, within the context of reduced-dimensional descriptions, the "washboard model"—wherein the atomistic surface is replaced by an effective washboard potential—has been shown to qualitatively capture the energetic and orientational dependencies of the process [55,56]. In cases with low energy gas-surface collisions, the sticking probability increases thereby leaving behind a transient ballistic adsorbate species that can not be described by the standard equilibrium diffusion description [2]. A more extensive discussion of the role of these features and many other relevant factors is beyond the scope of this review and will not be discussed further. Indeed, the main focus of this review is on those regimes where the surface diffusion can be adequately characterized by a reduced-dimensional model in which the primary modes that are explicitly retained are sufficient to describe the adsorbate(s) motion parallel to the surface.

3. MOLECULAR DYNAMICS SIMULATIONS AND PROJECTIVE MODELS

3.1 Atomistic models

Despite the difficulties facing computational efforts, their use has been essential in unraveling the complicated dynamics associated with surface diffusion and in determining the fundamental dissipation pathways. The seemingly simple transition from one adsorption site to its nearest neighbor is deceptively complex and is governed by a number of competing elementary mechanisms. As has been emphasized in the previous sections, the many-body coupling between the surface and adsorbate dominates surface diffusion and it governs when an adsorbate will hop and when it will stop. The primary coupling pathways can be represented by mechanical dephasings resulting from the classical interactions to the phonon bath of the surface and the quantum interactions with electron-hole pair excitations in the electronic continuum of the surface [2,9,19–21,40,57]. Generally the ratio of the vibrational frequencies of the adsorbate to that of the substrate dictates whether the phonon or electronic pathway will dominate with the latter being favored when the ratio is far from unity, although in general both mechanisms will be active to some extent [4,9]. Within a classical framework, the bulk can be represented within a simulation as an external harmonic bath—whose spectrum is equivalent to the phonon spectrum of the underlying bulk surface up to some desired level of accuracy—to which the surface atoms are coupled [33,34]. However an accurate theory is needed to account for the coupling of the surface to the bulk metal's electronic states. The inherent nature of the process brings into question the validity of the Born–Oppenheimer approximation given that the adsorbate does not interact with a single adiabatic potential energy surface, but rather a continuum of delocalized states associated with the metal, and relatively small changes in energy can result in excitations of electron-hole pairs [21]. As a result, Carr–Parinello dynamics involving electronic processes occurring on surfaces may not be applicable in general. Provided the coupling is weak, one possible approach that has been proposed is the "molecular dynamics with electronic friction" method designed to take electron-hole pair contributions into account resulting in a Langevin-type equation of motion for each explicit degree of freedom [19]. That is, an additional Langevin bath is introduced to describe the electronic response to adsorbate motion, albeit with its own characteristic electronic spectrum and perhaps also an effective electronic temperature that is not equal to the bulk temperature.

Several other theoretical and computational studies have been carried out that focus on particular aspects of surface diffusion and demonstrate that many factors can be important under certain circumstances. The role of lattice vibrations in the hopping process has been explored in detail by Rahman and co-workers [31, 58–60]. They note that although the rate is largely determined by the activation energy, which can in turn be accurately obtained from static high level ab initio calculations, the vibrational (entropic) contributions to the prefactor on flat surfaces and along step edges may be large and should not be ignored. In addition, a description of the diffusion of light atoms across metal surfaces must incorporate quantum mechanical effects. Hydrogen migration on a variety of substrates has been explored within the context of quantum transition state theory in which it was found that the motion of the hydrogen normal to the surface results in a significant number of barrier recrossings leading to substantial errors in the transition state theory approximation [61–65]. Several computational studies aimed at elucidating the effect of surface geometry on the hopping mechanisms and the ability of simulations to reproduce experimental diffusion coefficient values have been explored in detail [66–70]. Systematic studies of the role of coverage and long jumps have shown that both can have a dramatic impact on bulk diffusion rates [42,63, 71–73]. These and other results have provided the foundation for general scaling arguments for the magnitude of the self diffusion constant based on macroscopic properties of the solid [29,30].

A fundamental requirement on all of the computational studies on metal surface dynamics is the need to perform simulations with realistic potentials and in a feasible amount of time. To this end, the temperature-accelerated dynamics method [14,74,75] has arisen as a possible approach for reaching the latter limit. With the exception of quantum simulations, most classical simulations are based on semiempirical potentials derived either from the embedded atom method or effective medium theory [76–78]. However a recent potential energy surface for hydrogen on Cu(1 1 0) based on density functional theory calculations produced qualitatively different results from those of the embedded atom method including predictions of different preferred binding sites [79].

This brief report is in no way a complete list of all the simulations of surface diffusion that have been performed to date, but rather serves to illustrate the range of simulations that can typically be performed at present and the physical properties they can address, as well as an indication of the challenges facing current and future efforts. Moreover, it should be clear that there are substantial advantages to developing and implementing new algorithms in which one can reduce the number of atoms that must be treated explicitly within these models while retaining accuracy and the ability to calculate the requisite observables. This is precisely where projective methods can play a role, though they do not always retain the latter two requirements.

3.2 Projective models

As has been described, classical molecular dynamics simulations provide a detailed description of surface diffusion but are remiss of all important quantum effects and sit on the cusp of computationally feasible simulations. As a result, many techniques for constructing coarse-grained simulations have been developed ranging from purely phenomenological descriptions to more rigorous approaches. The crux of these models lies in constructing projection operators to remove the fast degrees of freedom that are irrelevant to the process of interest [80–82]. In this manner, many modes of the surface are reduced to one effective mode influencing the surface diffusion. Usually this is described within a generalized Langevin framework for each degree of freedom of the adsorbate whose equation of motion is of the form

$$m\dot{v} = -m\int_0^t \gamma(t-t')v(t')\,dt' + \xi(t) - \nabla U(q), \tag{3}$$

where $\gamma(t-t')$ is a time-dependent friction term including the entire history of the trajectory, $\xi(t)$ is a zero-mean stochastic process and $U(q)$ is the deterministic potential of mean force experienced by the adsorbate. This equation of motion is equivalent to a Hamiltonian system in which the primary degrees of freedom are coupled bilinearly to a bath of harmonic oscillators [81–83]. The origin of the stochastic force in the Langevin equation is related to the undetermined initial conditions of the bath in the Hamiltonian system. The former is connected to the friction by a fluctuation-dissipation relation that ensures the system approaches thermal equilibrium in the steady state [84]. Equation (3) is in no way more tractable than the Hamiltonian approach until one assumes a form for the friction kernel that is generally short-lived and physically corresponds to a quick response of the bath to the system motion. Calculations have shown that this approximation is usually justified particularly in the context of surface diffusion [2, 32,37,85]. Following this prescription, Tsekov and Ruckenstein have developed an exact Langevin description of the surface within the harmonic approximation for the coupling of the surface atoms [35,36]. Additionally in a series of papers, Tully and co-workers have developed a model in which a small number of surface atoms closest to the adsorbate are treated explicitly [33,34,51–53]. These are then coupled to a Langevin bath constructed to represent the bulk phonon spectrum of the surface.

These studies have paved the way for more phenomenological approaches that although less rigorous have nonetheless provided useful insights into the diffusion process. It is generally accepted that the potential of mean force $U(q)$ on which the adsorbate moves is corrugated and can be adequately approximated by a periodic function whose period is related to the lattice spacing of the surface [2,27,32]. The barrier heights impeding diffusion are derived from the activation energy determined either experimentally or from a higher level of theory. Furthermore, the friction kernel typically decays on a very short time scale on the order of a few hundred femtoseconds [32] and the magnitude lies in the low to intermediate friction regime [27,37]. The fast decay is much shorter than any time scale related to surface diffusion and so the generalized Langevin equation can be approximated by the simplified memoryless Langevin equation

$$m\dot{v} = -m\gamma v + \xi(t) - \nabla U(q). \tag{4}$$

This assumption is similar to the adiabatic approximation in which the bath is assumed to instantaneously respond to the motions of the subsystem.

At this level of description, the simulations are so inexpensive that it is possible to explore the full range of the parameters and obtain converged results. Often here one is not concerned with making quantitative comparisons with experimental values but rather in extracting general qualitative trends or in searching for new phenomenon [86–89]. Using such models, several groups have employed phenomenological Langevin equations to probe the role of friction and how the particular structure of the potential of mean force can influence surface diffusion [90–95]. Alternatively, as for example in Refs. [26] and [27] through comparison with experimental data, the authors were able to deduce the form of the potential and friction so as to reproduce experimental results providing valuable insights on the nature of the system. Within this formalism, the subject of multiple hops has received considerable attention. It has been shown that the probability of observing a long jump can be strongly influenced by both the form of the potential as well as the friction [37–39,86,90,92,94–99]. The ability of these models to probe a large range of parameters provides bounds for which to look for new phenomenon experimentally or at a higher level of simulation, as well as an intuitive and simplified picture of the process which can often be obscured by the sheer amount of data produced by all-atom simulations.

As has been emphasized throughout the article there are many length and time scales that influence adsorbate diffusion from couplings with electron-hole pairs to surface phonon modes all of which have been shown to be non-negligible in certain instances [9,21,31,57,59,100]. However, in most coarse-graining procedures all of these effects are averaged into a single effective mode governed by a simple (generalized) Langevin equation with a static potential of mean force. A central question remains as to whether this is an adequate description. If not, how should one go about constructing a reduced dimensional model that is capable of retaining some of these features. Recent advances involving the coupling of a particle to multiple or time-dependent baths leading to modifications of the friction kernel and noise term may provide some insight to the problem [101,102]. Another possibility is to incorporate the multi-scale details in the potential of mean force by allowing for time-dependent variations in the structure as have been explored by several authors [103–108].

4. SUMMARY AND CONCLUSIONS

Admittedly this report only provides a biased glimpse of the current methods available to model surface diffusion and the problems which have been addressed with them. In particular, high-level quantum simulations have largely been ignored here although they must be taken into account for an accurate description of any electronic processes—such as in conduction or light absorption/emission—and their role has thus far been invaluable in understanding surface phenomenon. As has been demonstrated above, in general there are multiple and varied dynamical processes simultaneously occurring at metal surfaces. This poses great challenges for future simulations to consistently incorporate all of these simultaneous events in a unified approach. Moreover, while the separation of time and space scales on which some of these varied processes evolve sometimes allows for the use of adiabatic approximations, it also makes it a potentially daunting task to include them all within a single simulation.

Current research efforts have largely been driven by experimental methods capable of creating and probing structures at ever smaller scales. Soon it should be possible to perform detailed classical simulations of systems that are of the same macroscopic size as those that can be created in the laboratory. Unfortunately, while the length scales obtainable by theory and experiment may become comparable, there will still be a large disparity in the accessible time scales. That is, without major algorithmic and theoretical advances, standard molecular dynamics will not be able to reach physically-relevant macroscopic time scales in the near future. As was outlined in this report, coarse-graining on some level will continue to be necessary. The use of projective (stochastic) models—e.g. the Langevin model—in a rigorously defined approach allows for a consistent treatment of the remaining modes whose detailed dynamics are of interest. Although the diminution of the computational effort to model surface dynamics using projective and coarse-graining approaches does not solve all of the outstanding problems in surface dynamics, it is nevertheless a useful tool and hence is part of the increasing effort in the multiscaling community.

ACKNOWLEDGEMENTS

This work was partly supported by the US National Science Foundation and by the Alexander von Humboldt-Foundation. RH is grateful to Prof. Peter Saalfrank for his hospitality and fruitful discussions during the writing of this report. The computational facilities at the CCMST have been supported under NSF grant CHE 04-43564.

REFERENCES

1. Gomer, R. Diffusion of adsorbates on metal surfaces, Rep. Prog. Phys. 1990, 53, 917.
2. Barth, J.V. Transport of adsorbates at metal surfaces: From thermal migration to hot precursors, Surf. Sci. Rep. 2000, 40, 75.
3. Naumovets, A.G., Zhang, Z. Fidgety particles on surfaces: how do they jump, walk, group, and settle in virgin areas?, Surf. Sci. 2002, 500, 414.
4. Naumovets, A.G. Collective surface diffusion: An experimentalist's view, Physica A 2005, 357, 189.
5. Graham, A.P. The low energy dynamics of adsorbates on metal surfaces investigated with helium atom scattering, Surf. Sci. Rep. 2003, 49, 115.
6. Kellogg, G.L. Field-ion microscope studies of single-atom surface-diffusion and cluster nucleation on metal-surfaces, Surf. Sci. Rep. 1994, 21, 1.
7. Tully, J.C. Theories of the dynamics of inelastic and reactive processes at surfaces, Annu. Rev. Phys. Chem. 1980, 31, 319.
8. Doll, J., Voter, A.F. Recent development in the theory of surface diffusion, Annu. Rev. Phys. Chem. 1987, 38, 413.
9. Tully, J.C. Chemical dynamics at metal surfaces, Annu. Rev. Phys. Chem. 2000, 51, 153.
10. Ala-Nissila, T., Ferrando, R., Ying, S.C. Collective and single particle diffusion on surfaces, Adv. Phys. 2002, 51, 949.
11. Miret-Artés, S., Pollak, E. The dynamics of activated surface diffusion, J. Phys.: Condens. Matter 2005, 17, S4133.
12. Feibelman, P.J. Theory of adsorbate interactions, Annu. Rev. Phys. Chem. 1989, 40, 261.
13. Starrost, F., Carter, E.A. Modeling the full monty: baring the nature of surfaces across time and space, Surf. Sci. 2002, 500, 323.
14. Voter, A.F., Montalenti, F., Germann, T.C. Extending the time scale in atomistic simulations of materials, Annu. Rev. Mater. Res. 2002, 32, 321.
15. Haight, R. Electron dynamics at surfaces, Surf. Sci. Rep. 1995, 21, 275.

16. Radeke, M.R., Carter, E.A. *Ab initio* dynamics of surface chemistry, Annu. Rev. Phys. Chem. 1997, 53, 248.
17. Brivio, G.P., Trioni, M.I. The adiabatic molecule–metal surface interaction: Theoretical approaches, Rev. Mod. Phys. 1999, 71, 231.
18. Saalfrank, P. Quantum dynamical approach to ultrafast molecular desorption from surfaces, Chem. Rev. 2006, 106, 4116.
19. Head-Gordon, M., Tully, J.C. Molecular dynamics with electronic frictions, J. Chem. Phys. 1995, 103, 10137.
20. Kindt, J.T., Tully, J.C., Head-Gordon, M., Gomez, M.A. Electron–hole pair contributions to scattering, sticking, and surface diffusion: CO on Cu(100), J. Chem. Phys. 1998, 109, 3629.
21. Wodtke, A.M., Tully, J.C., Auerbach, D.J. Electronically non-adiabatic interactions of molecules at metal surfaces: Can we trust the Born–Oppenheimer approximation for surface chemistry?, Int. Rev. Phys. Chem. 2004, 23, 513.
22. Hofmann, F., Toennies, J.P. High-resolution helium atom time-of-flight spectroscopy of low-frequency vibrations of adsorbates, Chem. Rev. 1996, 96, 1307.
23. Swartzentruber, B.S. Direct measurement of surface diffusion using atom-tracking scanning tunneling microscopy, Phys. Rev. Lett. 1996, 76, 459.
24. Zambelli, T., Trost, J., Wintterlin, J., Ertl, G. Diffusion and atomic hopping of N atoms on Ru(0001) studied by scanning tunneling microscopy, Phys. Rev. Lett. 1996, 76, 795.
25. Pedersen, M.Ø., Österlund, L., Mortensen, J.J., Mavrikakis, M., Hansen, L.B., Stensgaard, I., Lægsgaard, E., Nørskov, J.K., Besenbacher, F. Diffusion of N adatoms on the Fe(100) surface, Phys. Rev. Lett. 2000, 84, 4898.
26. Ellis, J., Graham, A.P., Hofmann, F., Toennies, J.P. Coverage dependence of the microscopic diffusion of Na atoms on the Cu(001) surface: A combined helium atom scattering experiment and molecular dynamics study, Phys. Rev. B 2001, 63, 195408.
27. Graham, A.P., Hofmann, F., Toennies, J.P., Chen, L.Y., Ying, S.C. Experimental and theoretical investigation of the microscopic vibrational and diffusional dynamics of sodium atoms on a Cu(001) surface, Phys. Rev. B 1997, 56, 10567.
28. Jardine, A., Ellis, J., Allison, W. Quasi-elastic helium-atom scattering from surfaces: experiment and interpretation, J. Phys.: Condens. Matter 2002, 14, 6173.
29. Agrawal, P.M., Rice, B.M., Thompson, D.L. Predicting trends in rate parameters for self-diffusion on FCC metal surfaces, Surf. Sci. 2002, 515, 21.
30. Feibelman, P.J. Scaling of hopping self-diffusion barriers on fcc (100) surfaces with bulk bond energies, Surf. Sci. 1999, 423, 169.
31. Kürprik, U., Kara, A., Rahman, T.S. Role of lattice vibrations in adatom diffusion, Phys. Rev. Lett. 1997, 78, 1086.
32. Gershinsky, G., Georgievskii, Y., Pollak, E., Betz, G. Microscopic and macroscopic estimates of friction: application to surface diffusion of copper, Surf. Sci. 1996, 365, 159.
33. Tully, J.C., Gilmer, G.H., Shugard, M. Molecular dynamics of surface diffusion. I. The motion of adatoms and clusters, J. Chem. Phys. 1979, 71, 1630.
34. Tully, J.C. Dynamics of gas-surface interactions: 3D generalized Langevin model applied to fcc and bcc surfaces, J. Chem. Phys. 1980, 73, 1975.
35. Tsekov, R., Ruckenstein, E. Stochastic dynamics of a subsystem interacting with a solid body with application to diffusive processes in solids, J. Chem. Phys. 1994, 100, 1450.
36. Tsekov, R., Ruckenstein, E. Two-dimensional Brownian motion of atoms and dimers on solid surfaces, Surf. Sci. 1995, 344, 175.
37. Georgievski, Y., Kozhushner, M.A., Pollak, E. Activated surface diffusion: Are correlated hops the rule or the exception?, J. Chem. Phys. 1995, 102, 6908.
38. Pollak, E., Bader, J., Berne, B.J., Talkner, P. Theory of correlated hops in surface diffusion, Phys. Rev. Lett. 1993, 70, 3299.
39. Georgievski, Y., Pollak, E. Long hops of an adatom on a surface, Surf. Sci. Lett. 1996, 355, L366.
40. Wahnström, G. Role of phonons and electron-hole pairs in hydrogen diffusion on a corrugated metal surface, Chem. Phys. Lett. 1989, 163, 401.
41. Sholl, D.S., Skodje, R.T. Kinetic phase transitions and bistability in a model surface reaction. II. Spatially inhomogeneous theories, Surf. Sci. 1995, 334, 305.

42. Brown, D.E., Sholl, D.S., Skodje, R.T., George, S.M. Surface diffusion of H and CO on Cu/Ru(001): evidence for long-range trapping by copper islands, Chem. Phys. 1996, 205, 23.
43. Chandler, D. Statistical mechanics of isomerization dynamics in liquids and the transition state approximation, J. Chem. Phys. 1978, 68, 2959.
44. Voter, A.F., Doll, J.D. Dynamical correction to transition state theory for multistate systems, J. Chem. Phys. 1985, 82, 80.
45. Zhang, Z., Metiu, H. Adsorbate migration on a solid surface: The connection between hopping dynamics and the atom-surface interaction energy, J. Chem. Phys. 1990, 93, 2087.
46. Hänggi, P., Talkner, P., Borkovec, M. Reaction-rate theory: fifty years after Kramers, Rev. Mod. Phys. 1990, 62, 251, and references therein.
47. Pollak, E., Talkner, P. Reaction rate theory: What it was, where is it today, and where is it going?, Chaos 2005, 15, 026116.
48. Ferrón, J., Gómez, L., de Miguel, J.J., Miranda, R. Nonstochastic behavior of atomic surface diffusion on Cu(111) down to low temperatures, Phys. Rev. Lett. 2004, 93, 166107.
49. Kellogg, G.L., Feibelman, P.J. Surface self-diffusion on Pt(001) by atomic exchange mechanism, Phys. Rev. Lett. 1990, 64, 3143.
50. Feibelman, P.J. Diffusion path for an Al adatom on Al(001), Phys. Rev. Lett. 1990, 65, 729.
51. Nitzan, A., Shugard, M., Tully, J.C. Stochastic classical trajectory approach to relaxation phenomena. II. Vibrational relaxation of impurity molecules in Debye solids, J. Chem. Phys. 1978, 69, 2525.
52. Tully, J.C., Cardillo, M.J. Dynamics of molecular motion at single-crystal surfaces, Science 1984, 223, 445.
53. Adelman, S.A., Doll, J.D. Generalized Langevin equation approach for atom/solid-surface scattering: Collinear atom/harmonic chain model, J. Chem. Phys. 1974, 61, 4242.
54. Rettner, C.T., Auerbach, D.J., Tully, J.C., Kleyn, A.W. Chemical dynamics at the gas–surface interface, J. Phys. Chem. 1996, 100, 13021.
55. Tully, J.C. Washboard model of gas-surface scattering, J. Chem. Phys. 1990, 92, 680.
56. Yan, T., Hase, W.L., Tully, J.C. A washboard with moment of inertia model of gas-surface scattering, J. Chem. Phys. 2004, 120, 1031.
57. Tully, J.C., Gomez, M., Head-Gordon, M. Electronic and phonon mechanisms of vibrational relaxation: CO on Cu(100), J. Vac. Sci. Technol. 1993, 11, 1914.
58. Yildirim, H., Kara, A., Durukanoğlu, S., Rahman, T.S. Calculated pre-exponential factors and energetics for adatom hopping on terraces and steps of Cu(100) and Cu(110), Surf. Sci. 2006, 600, 484.
59. Durukanoğlu, S., Kara, A., Rahman, T.S. The role of lattice vibrations in adatom diffusion at metal stepped surfaces, Surf. Sci. 2005, 587, 128.
60. Al-Rawi, A.H., Kara, A., Rahman, T.S. Comparative study of anharmonicity: Ni(111), Cu(111), and Ag(111), Phys. Rev. B 2002, 66, 165439.
61. Haug, K., Wahnström, G., Metiu, H. Hydrogen motion on a Cu surface: A model study of the rate of single and double site-to-site jumps and the role of the motion perpendicular to the surface, J. Chem. Phys. 1989, 90, 540.
62. Haug, K., Wahnström, G., Metiu, H. Hydrogen motion on a rigid Cu surface: The calculation of the site to site hopping rate by using flux-flux correlation functions, J. Chem. Phys. 1990, 92, 2083.
63. Haug, K., Metiu, H. Quantum simulation of hydrogen migration on Ni(100): The role of fluctuations, recrossing, and multiple jumps, J. Chem. Phys. 1991, 94, 3251.
64. Rick, S.W., Lynch, D.L., Doll, J.D. The quantum dynamics of hydrogen and deuterium on the Pd(III) surface: A path integral transition state theory study, J. Chem. Phys. 1993, 99, 8183.
65. Carter, E.A. Adsorption and diffusion energetics of hydrogen atoms on Fe(110) from first principles, Surf. Sci. 2003, 547, 85.
66. Evangelakis, G.A., Papanicolaou, N.I. Adatom self-diffusion processes on (001) copper surface by molecular dynamics, Surf. Sci. 1996, 347, 376.
67. Kallinteris, G.C., Evangelakis, G.A., Papanicolaou, N.I. Molecular dynamics study of the vibrational and transport properties of copper adatoms on the (111) copper surface; comparison with the (001) face, Surf. Sci. 1996, 369, 185.
68. Papanicolaou, N.I., Papathanakos, V.C., Papageorgiou, D.G. Self-diffusion on Al(100) and Al(111) surfaces by molecular-dynamics simulation, Physica B 2001, 296, 259.

69. Prévot, G., Cohen, C., Schmaus, D., Pontikis, V. Non-isotropic surface diffusion of lead on Cu(110): a molecular dynamics study, Surf. Sci. 2000, 459, 57.
70. Bulou, H., Lucas, O., Kibaly, M., Goyhenex, C. Long-time scale molecular dynamics study of Co diffusion on the Au(111) surface, Comp. Mat. Sci. 2003, 27, 181.
71. Ramirez-Pastor, A.J., Nazzarro, M., Riccardo, J.L., Pereya, V. Surface diffusion of dimers: I repulsive interactions, Surf. Sci. 1997, 391, 267.
72. Papathanakos, V., Evangelakis, G.A. Structural and diffusive properties of small 2D Au clusters on the Cu(111) surface, Surf. Sci. 2002, 499, 229.
73. Jacobsen, J., Jacobsen, K.W., Sethna, J.P. Rate theory for correlated processes: Double jumps in adatom diffusion, Phys. Rev. Lett. 1997, 79, 2843.
74. Sørensen, M.R., Voter, A.F. Temperature-accelerated dynamics for simulation of infrequent events, J. Chem. Phys. 2000, 112, 9599.
75. Miron, R.A., Fichthorn, K.A. Accelerated molecular dynamics with the bond-boost method, J. Chem. Phys. 2003, 119, 6210.
76. Daw, M.S., Baskes, M.I. Semiempirical, quantum mechanical calculation of hydrogen embrittlement in metals, Phys. Rev. Lett. 1983, 50, 1285.
77. Daw, M.S., Baskes, M.I. Embedded-atom method: Derivation and application to impurities surfaces and other defects in metals, Phys. Rev. B 1984, 29, 6443.
78. Jacobsen, K.W., Nørskov, J.K., Puska, M.J. Interatomic interactions in the effective-medium theory, Phys. Rev. B 1987, 35, 7423.
79. Bae, C., Freeman, D.L., Doll, J.D., Kresse, G., Hafner, J. Energetics of hydrogen chemisorbed on Cu(110): A first principles calculations study, J. Chem. Phys. 2000, 113, 6926.
80. Mori, H. Transport, collective motion, and Brownian motion, Prog. Theor. Phys. 1965, 33, 423.
81. Zwanzig, R. Ensemble method in the theory of irreversibility, Phys. Rev. 1961, 124, 983.
82. Zwanzig, R. Statistical mechanics of irreversibility, in Brittin, W.E., Downs, B.W., Downs, J., editors. Lectures in Theoretical Physics (Boulder), vol. 3. New York: Wiley-Interscience; 1961, p. 106–41.
83. Zwanzig, R. Nonequilibrium Statistical Mechanics, London: Oxford University Press; 2001.
84. Kubo, R. The fluctuation-dissipation theorem, Rep. Prog. Phys. 1966, 29, 255.
85. Moix, J., Hernandez, R., unpublished results.
86. Bao, J.D., Abe, Y., Zhuo, Y.Z. Inhomogeneous friction leading to current in periodic system, Physica A 1999, 265, 111.
87. Zhang, X.P., Bao, J.D. Stochastic resonance in multidimensional periodic potential, Surf. Sci. 2003, 540, 145.
88. Talkner, P., Hershkovitz, E., Pollak, E., Hänggi, P. Controlling activated surface diffusion by external fields, Surf. Sci. 1999, 437, 198.
89. Moix, J.M., Shepherd, T.D., Hernandez, R. A phenomenological model for surface diffusion: Diffusive dynamics across incoherent stochastic aperiodic potentials, J. Phys. Chem. B 2004, 108, 19476.
90. Chen, L.Y., Baldan, M.R., Ying, S.C. Surface diffusion in the low-friction limit: Occurence of long jumps, Phys. Rev. B 1996, 54, 8856.
91. Cucchetti, A., Ying, S.C. Memory effects in the frictional damping of diffusive and vibrational motion of adatoms, Phys. Rev. B 1996, 54, 3300.
92. Moro, G.J., Polimeno, A. Multi-barrier crossing regulated by the friction, Chem. Phys. Lett. 1992, 189, 133.
93. Vega, J.L., Guantes, R., Miret-Artés, S. Mean first passage time and the Kramers turnover theory in activated atom-surface diffusion, Phys. Chem. Chem. Phys. 2002, 4, 4985.
94. Sancho, J.M., Lacasta, A.M., Lindenberg, K., Sokolov, I.M., Romero, A.H. Diffusion on a solid surface: anomalous is normal, Phys. Rev. Lett. 2004, 92, 250601.
95. Lacasta, A.M., Sancho, J.M., Romero, A.H., Sokolov, I.M., Lindenberg, K. From subdiffusion to superdiffusion of particles on solid surfaces, Phys. Rev. E 2004, 70, 051104.
96. Hershkovitz, E., Talkner, P., Pollak, E., Georgievski, Y. Multiple hops in multidimensional activated surface diffusion, Surf. Sci. 1999, 421, 73.
97. Braun, O.M., Ferrando, R. Role of long jumps in surface diffusion, Phys. Rev. E 2002, 65, 061107.
98. Ferrando, R., Spadacini, R., Tommei, G.E. Kramers problem in periodic potentials: Jump rate and jump lengths, Phys. Rev. E 1993, 48, 2437.

99. Montalenti, F., Ferrando, R. Jumps and concerted moves in Cu, Ag, and Au(110) adatom self-diffusion, Phys. Rev. B 1999, 59, 5881.
100. Mattsson, T.R., Wahnström, G. Isotope effect in hydrogen surface diffusion, Phys. Rev. B 1997, 56, 14944.
101. Hernandez, R. The projection of a mechanical system onto the irreversible generalized Langevin equation (iGLE), J. Chem. Phys. 1999, 111, 7701.
102. Popov, A.V., Melvin, J., Hernandez, R. Dynamics of swelling hard spheres surmised by an irreversible Langevin equation, J. Phys. Chem. A 2006, 110, 1635.
103. Doering, C.R., Gadoua, J.C. Resonant activation over a fluctuating barrier, Phys. Rev. Lett. 1992, 69, 2318.
104. Gammaitoni, L., Hänggi, P., Jung, P., Marchesoni, F. Stochastic resonance, Rev. Mod. Phys. 1998, 70, 223.
105. Shepherd, T.D., Hernandez, R. Chemical reaction dynamics with stochastic potentials beyond the high-friction limit, J. Chem. Phys. 2001, 115, 2430.
106. Maddox, J. Surmounting fluctuating barriers, Nature 1992, 359, 771.
107. Pechukas, P., Hänggi, P. Rates of activated processes with fluctuating barriers, Phys. Rev. Lett. 1994, 73, 2772.
108. Reimann, P. Thermally driven escape with fluctuating potentials: A new type of resonant activation, Phys. Rev. Lett. 1995, 74, 4576.

Section 5
Quantum Chemistry

Section Editor: T. Daniel Crawford

Department of Chemistry
Virginia Tech
Blacksburg
VA 24061
USA

CHAPTER 9

An Active Database Approach to Complete Rotational–Vibrational Spectra of Small Molecules

Attila G. Császár[*], **Gábor Czakó**[*], **Tibor Furtenbacher**[*] and **Edit Mátyus**[*]

Contents		
	1. Introduction	155
	2. Nonadiabatic Computations—Where Theory Delivers	158
	3. MARVEL—An Active Database Approach	158
	4. Electronic structure computations	160
	4.1 The focal-point approach (FPA)	160
	4.2 Ab initio force fields	162
	4.3 Ab initio (semi)global PESs	163
	4.4 Empirical PESs	164
	4.5 Dipole moment surfaces (DMSs)	165
	5. Variational Nuclear Motion Computations	165
	5.1 Computations in internal coordinates	166
	5.2 Computations in normal coordinates	167
	6. Outlook	169
	Acknowledgement	169
	References	169

1. INTRODUCTION

Spectroscopy has traditionally been considered as the branch of sciences offering the perhaps most precise measurement results. As a consequence, molecular spectroscopic results are usually extremely hard to match even by the most sophisticated nonadiabatic computational approaches based on quantum electrodynamics (QED). Nevertheless, experimental molecular spectroscopy, when the aim is the determination of complete spectra, has several important limitations,

[*] Laboratory of Molecular Spectroscopy, Institute of Chemistry, Eötvös University, P.O. Box 32, H-1518 Budapest 112, Hungary

as follows: (1) while line positions can be measured with outstanding accuracy that is almost impossible to match by computations, line intensities and shapes usually have much larger relative uncertainties; (2) experiments measure transitions but in many application of spectroscopic results, e.g., for the determination of temperature-dependent partition functions through direct summation [1], one needs accurate energy levels; (3) since even for small systems the number of allowed transitions is huge, it is in the billions for each isotopologue of a triatomic species, the complete line-by-line experimental determination of a spectrum is clearly impossible; (4) many important species and many important spectroscopic regions are hardly amenable to experimental scrutiny or require expensive instrumentation, for example, even the stretching fundamentals of the triplet ground electronic state of the CH_2 radical have not been measured [2]; and (5) measurement of transitions without detailed assignment is hardly useful for most practical purposes and as the energy grows the level density increases drastically while the clear description of energy levels using traditional simple schemes starts to fail.

As suggested herein, the best quantum mechanical computations are able to solve or at least remedy all of the above problems. While highly specialized techniques exist for few-electron systems [3], the canonical process of obtaining accurate computational predictions for rotational–vibrational spectra of many-electron systems is normally divided into two steps. First, one or more potential energy surfaces (PESs) [4,5], and possibly property surfaces (like the dipole moment surface, DMS) are obtained, based on solving the electronic part of the Schrödinger equation on a grid including a large number of nuclear configurations. PESs are defined as the total energy of the quantum system as a function of its geometric variables. Property surfaces are defined similarly to PESs. Second, the PESs, usually after proper fitting, are used to solve the nuclear motion problem resulting in a large number of eigenpairs, while the appropriate property surfaces are then used to obtain the full spectrum.

Many approaches have been developed in electronic-structure theory for determining accurate energy and property hypersurfaces [4,5]. Ideally, one would do complete basis set (CBS) full configuration interaction (FCI) computations at a very large number of structures employing an appropriately chosen relativistic Hamiltonian. Of course, this is not practical and introduction of several approximations is mandated. This is a field of electronic-structure theory where sufficiently large experience has been acquired to allow meaningful choices to be made. If the aim is the accurate determination of complete rotational–vibrational spectra, even non-relativistic CBS FCI computations are not sufficient, effects usually considered to be small must also be taken into account, most importantly effects resulting from the theory of special relativity [6,7], even for molecules containing only light atoms, and the (partial) breakdown of the Born–Oppenheimer (BO) [8,9] approximation, especially for H-containing species. Furthermore, at least close to dissociation limit(s) or intersections, interactions between PESs might need to be considered, leading to additional difficulties.

For decades high-resolution rotational–vibrational spectroscopy treated nuclear motion in terms of near-rigid rotations and small-amplitude vibrations, relying heavily on perturbation theory (PT) [10–16]. While the formulas [11,14,15] resulting from PT, even at second order, are often rather complex, they are easy to program, running them is almost cost free, and they reproduce many experimental data though only at low to medium excitation. These low-order PT approaches are unable to yield complete molecular spectra. From the very beginning there have been attempts to compute rovibronic spectra of polyatomic molecules by computationally more intensive variational techniques. Variational nuclear motion computations can be made, at least in principle and within the BO approximation, arbitrarily accurate and in principle allow the determination of complete spectra. Nevertheless, for the first-principles approach to complete rotational–vibrational spectra to be really successful one has to utilize sophisticated procedures. This means that one needs not only highly accurate electronic-structure techniques to compute energy and property surfaces but also involved protocols to represent them, and numerically efficient ways for the (nearly) variational nuclear motion treatments. Recent developments suggest that, in favorable cases, the rovibrational eigenvalues obtained can approach what quantum chemists call spectroscopic accuracy, which is 1 cm^{-1} on average [17].

Neither experiments nor first-principles computations can determine the complete rotational–vibrational spectra of even small molecules with the required accuracy. It seems to us that the most practical approach to overcome most of the difficulties is through an active database approach. This requires building two databases linked together through a unique assignment scheme, one containing energy levels and the other the related transitions. This way one can take advantage of the strengths of the two main sources of spectroscopic information. Variational computations can yield all the possible energy levels, with various assignment possibilities, and thus all the possible labeled transitions, though with limited accuracy deteriorating as the level of excitation increases [17]. Experimental transitions, and the energy levels obtained through an appropriate inversion procedure, have a much higher accuracy but are limited in number even in the spectroscopically most easily accessible regions. We do not see any other possible route to the determination of complete molecular spectra and thus strongly advocate the active database approach what we call MARVEL, standing for Measured Active Rotational–Vibrational Energy Levels [18,19]. MARVEL requires not only complex tools for handling information in the databases but also experimental efforts to obtain and analyze high-resolution spectra of important small species and theoretical developments that allow efficient and accurate computation of complete spectra.

The fields of electronic-structure theory and variational nuclear motion computations are diverse and involve a huge number of papers. Consequently, it is impossible to review the advances in these fields. Only efforts in our group related to the computation of complete rotational–vibrational spectra of small molecules is overviewed and references from other groups are given only when directly relevant to our own efforts.

2. NONADIABATIC COMPUTATIONS—WHERE THEORY DELIVERS

For the smallest quantum systems, comprising perhaps up to five particles, one can afford not introducing the separation of the electronic and nuclear degrees of freedom, i.e., not introducing the BO approximation. For historical reasons such computations of energy levels are usually referred to as nonadiabatic though strictly speaking they should be called diabatic.

For three- and four-particle systems, like H_2^+ and H_2, sophisticated nonadiabatic computations have been performed with specialized techniques [3,20–27]. These computations can yield rovibronic energy levels whose accuracy is limited only by the Hamiltonian used for their evaluation. Unlike a BO treatment, nonadiabatic computations can distinguish between certain spectroscopic characteristics of the different isotopologues. Nevertheless, while nonadiabatic computations yield energy levels in a quantitative way, the qualitative characterization of them is somewhat difficult.

In a recent paper [23] we made an attempt to retain the notion of a PES in a nonadiabatic treatment. This was achieved by fixing the internuclear separation in H_2^+-like systems, a straightforward procedure in Jacobi coordinates. The resulting energy correction to the BO energies was termed adiabatic Jacobi correction (AJC). The AJC numerical values are considerably smaller than the well-established diagonal Born–Oppenheimer corrections (DBOC) [28–33], suggesting that the DBOC might correct for more than simply the translational motion. More work needs to be done to understand better the deviations between the AJC and DBOC corrections and to see which one stands closer to the fully nonadiabatic limit.

The fully nonadiabatic treatment of few-body systems have yielded very accurate energy levels and transition energies. At the limit of these calculations, when even QED effects are considered, the energies have not only internal consistency but are in almost full accord with the relevant results of measurements.

As to many-electron systems, corrections to the BO approximation can be obtained by means of a second-order contact transformation method [28]. This introduces two terms: (a) the simple DBOC, which gives rise to a mass-dependent correction to the PES; and (b) the considerably more difficult second-order (also called non-adiabatic) correction, which introduces coupling between electronic states and primarily results in corrections to the kinetic energy operator. In the most sophisticated first-principles treatments [17,34,35] allowance is made for non-adiabatic effects though further work is required to explore the best possible strategies for computation and utilization of this information.

3. MARVEL—AN ACTIVE DATABASE APPROACH

There are several areas in the sciences where experimentally measured quantities, with well defined uncertainties, and quantities preferred on some theoretical ground, again with appropriate uncertainties, are decidedly distinct but relations can be worked out between the two sets of data. Such areas include thermochemistry [36,37], reaction kinetics [38,39], and, of course, spectroscopy.

In spectroscopy the relation between transitions and energy levels is linear and exceedingly simple. To the best of our knowledge Flaud and co-workers [40] were the first to suggest a useful procedure for inverting the information contained in measured transitions to energy levels. Their method has been extended [18,19] to treat all measured rotational–vibrational transitions available and obtain the related energy levels in one grand inversion and refinement process. The active database protocol and program developed is called MARVEL [18]. The energy levels so obtained are considered measured as they are obtained from experiment. The set of measured energy levels is called active in the sense of the Active Thermochemical Tables approach of Ruscic [36], and implies that if new experimental transitions become available the refinement process must be repeated resulting in a new set of improved rotational–vibrational energy levels.

Determination of a set of energy levels and an improved set of transitions by MARVEL is based on the following steps:

(1) Collect, critically evaluate, and compile all transitions, including their assignments and uncertainties, into a database.
(2) Determine those energy levels which belong to a particular spectroscopic network (SN).
(3) Within a given SN, set up a vector containing all the transitions, another one comprising the requested measured levels, and a sparse inversion matrix describing the relation between transitions and levels.
(4) During solution of the resulting set of linear equations uncertainties in the measured transitions can be incorporated which result in uncertainties for the energy levels. The absolute energy levels of a given SN can only be obtained if the value of the lowest energy level within the SN, with zero uncertainty, is set up correctly.

The MARVEL procedure and code developed has been tested for $H_2^{17}O$ (Table 9.1). $H_2^{17}O$ was chosen as it contains a relatively small number of accurately measured transitions (on the order of 7000) [41–46], including a large number of transitions on the ground vibrational state, and water is probably the single most important polyatomic molecule whose spectroscopy on the ground electronic state is especially relevant in a number of applications, including understanding of the greenhouse effect on Earth. In the case of $H_2^{17}O$, and indeed for all other symmetrically substituted isotopologues of water, the transitions can be divided unequivocally into two main SNs, para and ortho [47].

A good model must be available prior to using MARVEL in order to give unique labels for the upper and lower states participating in the transitions. Approximate Hamiltonians, variational computations based on PESs, and even perturbation-resonance approaches [48] are able to provide these labels. In the MARVEL program the normal mode labeling is used for the states, e.g., $(n_1 n_2 n_3 J K_a K_c)$ in the case of water, where n_1, n_2, and n_3 stand for the symmetric stretching, angle bending, and antisymmetric stretching quantum numbers, respectively, and the standard asymmetric top notation, $J_{K_a K_c}$, applies for the rotational states.

The uncertainties of the MARVEL vibrational levels of $H_2^{17}O$ are on the order of 10^{-6} cm^{-1} (Table 9.1). Since the complete list of vibrational states is available from computations, it is clear that even in the experimentally most accessible low-energy region several vibrational levels are not available from experiment.

MARVEL supplies important information both for spectroscopists and quantum chemists. Running MARVEL for a transitions dataset collected from several publications will determine whether there are any outliers in the transition set assembled and whether the experimental uncertainties are realistic. This contributes to the validation of the experimental results. The resulting energy levels can be used for the empirical improvement of PESs and for checking existing assignments or suggesting new ones. Execution of MARVEL for the most important isotopologue of water, $H_2^{16}O$, is in progress.

4. ELECTRONIC STRUCTURE COMPUTATIONS

It is useful if the energy-level database within MARVEL contains a (complete) set of accurate rotation–vibration levels. This information also helps the assignment of measured transitions. For many-electron systems the levels can only be determined using PESs obtained from sophisticated, though approximate first-principles techniques, like the focal-point approach (FPA) [49,50].

4.1 The focal-point approach (FPA)

A fundamental characteristic of the FPA is the dual extrapolation to the one- and n-particle electronic-structure limits. The process leading to these limits can be described as follows: (a) use families of basis sets, such as the correlation-consistent (aug-)cc-p(wC)VnZ sets [51,52], which systematically approach completeness through an increase in the cardinal number n; (b) apply lower levels of theory with extended [53] basis sets (typically direct Hartree–Fock (HF) [54] and second-order Møller–Plesset (MP2) [55] computations); (c) use higher-order valence correlation treatments [CCSD(T), CCSDTQ(P), even FCI] [5,56] with the largest possible basis sets; and (d) lay out a two-dimensional extrapolation grid based on the assumed additivity of correlation *increments* followed by suitable extrapolations. FPA assumes that the higher-order correlation increments show diminishing basis set dependence. Focal-point [2,49,50,57–62] and numerous other theoretical studies have shown that even in systems without particularly heavy elements, account must also be taken for core correlation and relativistic phenomena, as well as for (partial) breakdown of the BO approximation, i.e., inclusion of the DBOC correction [28–33].

Note that the FPA can be used for more than spectroscopic applications. In fact it has helped to redefine first-principles thermochemistry, see the HEAT (High-accuracy Extrapolated *Ab initio* Thermochemistry) [63,64] and Wn (Weizmann-n) [65] protocols and Refs. [37,66,67], for example.

TABLE 9.1 Ab initio (CVRQD), empirical (FIS3), and experimental (MARVEL) vibrational energy levels (in cm^{-1}) of H$_2$^{17}O up to 14300 cm^{-1}

$n_1n_2n_3$	CVRQD	FIS3	MARVEL[a,b]	Levels[c]
000	4630.367	4630.168	0.000000	279
010	1591.644	1591.346	1591.325655(43)	184
020	3145.535	3144.925	3144.980456(44)	93
100	3653.138	3653.157	3653.142275(96)	113
001	3748.115	3748.200	3748.318070(93)	150
030	4657.888	4657.007		
110	5228.214	5227.816	5227.705615(386)	150
011	5320.482	5320.211	5320.250932(66)	122
040	6122.609	6121.559		
120	6765.588	6764.799	6764.725615(966)	28
021	6857.887	6857.267	6857.272719(105)	78
200	7192.896	7193.186	7193.246625(193)	76
101	7238.116	7238.557	7238.713600(185)	100
002	7430.918	7431.136	7431.076115(1450)	26
050	7528.957	7527.914		1
130	8261.877	8260.780		3
031	8357.439	8356.514		7
210	8750.235	8750.000		34
111	8792.566	8792.525	8792.544310(926)	106
060	8855.488	8854.751		1
012	8983.119	8982.897	8982.869215(966)	51
140	9709.883	9708.660		
041	9814.524	9813.391		
070	10070.793	10070.693		
220	10270.377	10269.703		12
121	10311.631	10311.179	10311.202510(926)	73
022	10502.067	10501.444		
300	10584.956	10585.923		65
201	10597.244	10598.352	10598.475610(926)	100
102	10853.042	10853.561	10853.505315(966)	53
003	11011.416	11011.933		45
150	11082.442	11081.789		
051	11221.331	11220.134		
080	11235.025	11234.647		
230	11750.948	11749.946		
131	11793.487	11792.703	11792.827010(6019)	31
032	11985.407	11984.450		
310	12121.917	12122.290		26
211	12132.557	12133.072	12132.992610(926)	87
160	12360.562	12360.454		
112	12389.311	12389.173		17

TABLE 9.1 *(Continued)*

$n_1n_2n_3$	CVRQD	FIS3	MARVEL[a,b]	Levels[c]
090	12511.995	12510.737		
013	12541.086	12541.118	12541.225510(926)	13
061	12562.935	12561.916		
240	13186.424	13185.271		
141	13234.029	13233.034		1
042	13428.499	13427.338		1
320	13620.792	13620.773		3
221	13631.364	13631.483	13631.499810(1019)	48
170	13640.751	13640.414		
071	13807.707	13809.693	13808.273310(926)	1
400	13809.863	13810.343		25
301	13811.086	13812.195	13812.158110(926)	68
0100	13829.143	13827.954		
122	13890.137	13889.483		
023	14039.711	14039.317		
202	14202.553	14203.479		11
103	14295.039	14296.298	14296.279510(370)	26
			Total:	2308

a   Values in parentheses correspond to 2σ uncertainties, in units of 10^{-6} cm^{-1}. The lowest level was set exactly to zero with zero uncertainty.
b   The ranges (cm^{-1}) of measured transitions: 0–177 [41], 177–600 [42], 500–8000 [43], 8000–9400 [44], 9711–11335 [45], and 11365–14377 [46].
c   Number of rotational energy levels corresponding to the given vibrational energy level.

4.2 Ab initio force fields

One old difficulty of nuclear motion computations for larger systems, namely the representation of PESs, plagues applications of even the most sophisticated procedures. While low-order force fields [68,69] may not provide a good representation of the PES for systems undergoing large-amplitude motions, for many systems of practical interest an anharmonic force field representation of the PES should provide at least the first important stepping stone to understand the complex internal dynamics of the system at low energies.

Internal coordinate quartic force fields have been computed for relatively large systems, e.g., for the 17-atom amino acid L-proline [70]. Nevertheless, despite the fact that electronic-structure programs to compute analytic geometric first and second derivatives of the energy have become available at almost any level [71–74], to the best of our knowledge [68], complete sextic force fields in internal coordinates are available only for a handful of triatomic systems, N_2O [75–78], CO_2 [79–82], and H_2O [48,59]. This is due to several factors. First, it is exceedingly difficult to determine accurate higher-order force constants strictly from experimental information. Second, force fields computed from most electronic-structure codes are given in rectilinear Cartesian or normal coordinates and their nonlinear transformation to more meaningful representations involving curvilinear internal coordinates is nontrivial [83]. Third, polynomial expansions are subject to rather limited ranges of applicability. Fourth, quartic normal coordinate force fields give excellent frequencies when used with VPT2 formulas, a precision of 1–2 cm^{-1} is not uncommon [84], but when used in variational procedures the computed frequencies show much larger deviations from experiment. This was discouraging as variational procedures render the use of somewhat complex and tedious procedures [48,85] treating resonances present in PT treatments unnecessary. Nevertheless, as shown in section 5.2, one can use internal coordinate force fields in an exact and completely general way not only within internal coordinate Hamiltonians but also within Hamiltonians [86,87] expressed in normal coordinates. This should result in a renewed interest in force fields for lower-energy (ro)vibrational studies of systems having more than three atoms.

4.3 Ab initio (semi)global PESs

Since one cannot compute truly high quality PESs and DMSs in a single step, one needs to build them piecewise. It is advantageous to utilize the focal-point approach [49,50] detailed in subsection 4.1 for this purpose. In fact, it has been employed successfully to obtain highly accurate semiglobal PESs for a number of triatomic systems, including H_2O [17,59], [H,C,N] [88], and H_2S [60]. The so far most elaborate and most successful application of FPA yielded the adiabatic CVRQD PESs of the water isotopologues. CVRQD means that the final *ab initio* ground electronic state surface includes corrections due to core (C) and valence (V) correlation, as well as relativistic (R) and QED (Q) contributions, and it is an adiabatic surface utilizing the DBOC correction (D) [28–30]. For purposes of illustration, it is insightful to repeat the steps resulting in the presently most accurate *ab initio* semiglobal surfaces of the water isotopologues, which can reproduce all the measured transitions of all isotopologues with an average accuracy better than 1 cm^{-1} [17].

The CVRQD PESs of the water isotopologues are based upon valence-only aug-cc-pVnZ [51,52], $n = 4,5,6$, internally contracted multi-reference configuration interaction (ICMRCI) [89] calculations including the size-extensive Davidson correction [90], which were extrapolated to the CBS limit. The largest correction to the valence-only surface comes from core correlation, which should be determined using a size-extensive technique. Nevertheless, the core correction surface of the CVRQD PESs was determined at the averaged coupled pair functional (ACPF) [91] level lacking strict size-extensivity. The relativistic surfaces were obtained by first-order perturbation theory as applied to the one-electron mass-velocity (MV) and one- and two-electron Darwin terms (MVD2) [92,93], supplemented by a correction obtained from the inclusion of the higher-order Breit term in the electronic Hamiltonian [93]. A correction surface due to the one-electron Lamb shift has also been determined. Consideration of the Lamb shift was shown to have contributions as much as 1 cm^{-1} for some levels beyond 20000 cm^{-1} [94]. Finally, a DBOC correction surface was obtained, at the cc-pVTZ MRCI level [17]. The un-

TABLE 9.2 Approximate per quanta contributions (in cm^{-1}) of so called small corrections to the low-lying VBOs of H$_2$^{16}O[a]

Correction surface	STRE	BEND
MVD1	$-2.8(n_1 + n_3)$	$+1.4n_2$
D2	$-0.04(n_1 + n_3)$	$+0.12n_2$
Breit	$-0.6(n_1 + n_3)$	$-0.02n_2$
Lamb-shift	$+0.18(n_1 + n_3)$	$-0.11n_2$
Core correction	$+7.3(n_1 + n_3)$	$-0.5n_2^2$
DBOC	$+0.4(n_1 + n_3)$	$-0.45n_2$

[a] MVD1 = one-electron mass-velocity plus Darwin; D2 = two-electron Darwin; DBOC = diagonal Born–Oppenheimer correction. Please see text for details. n_1 and n_3 are the stretching, n_2 is the bending quantum number.

precedented precision of the CVRQD PESs in determining the vibrational levels of water can be judged from the relevant entries in Table 9.1. The importance of the correction surfaces can be judged from entries in Table 9.2, showing the approximate per quanta changes in the vibrational band origins (VBOs) of H$_2$^{16}O.

To use the *ab initio* energies computed over a grid most efficiently in nuclear motion computations we need to fit them to analytical surfaces. Fitting the surfaces involves several delicate choices if the high quality of the underlying electronic-structure calculations is not to be lost. Notwithstanding the importance of this step the fitting process is not discussed here; for important details please consult, for example, Refs. [59,95,96].

4.4 Empirical PESs

Whatever complicated procedures are employed for their determination, *ab initio* PESs can hardly produce transitions matching the accuracy of experimentally determined transitions. A partial remedy to this problem is offered by the empirical adjustment of the surface to best match the available experimental data in a least-squares sense.

Ab initio PESs, like CVRQD for water, provide an excellent starting point for the refinement of empirical PESs. The rule of thumb seems to be that the higher the quality of the initial surface the better the resulting empirical PES. In fact the best empirical PES for water, termed FIS3 [96], as it is a joint fitted surface for three isotopologues, H$_2$^{16}O, H$_2$^{17}O, and H$_2$^{18}O, utilized the CVRQD PESs as a starting point. To assemble a reliable set of experimental rotational–vibrational energy levels for the refinement process is far from trivial. As detailed in section 3, it is possible to invert the directly measured transitions to energy levels and obtain a partial set of high-accuracy levels. As to the functional form of the fit, several choices are possible and these mainly depend on the accuracy of the starting PES. The methods that can be used to fit PESs are basically the same as the nuclear motion methods described in section 5, thus they need no further discussion here.

The energy levels determined with the help of empirical PESs cannot match the extreme accuracy of the MARVEL levels but they also provide a complete set.

Empirical PESs interpolate very well but their extrapolation potential is inferior to those of the *ab initio* PESs. Therefore, even if a highly accurate empirical PES is available, the *ab initio* surface must be retained as it might prove to be a better choice for finding new transitions in a new region of the spectrum and a better starting point for further refinement of the surface if more detailed experimental information became available.

4.5 Dipole moment surfaces (DMSs)

Determination of first-principles transition intensities of rotational–vibrational levels relies on knowledge of the DMS and the nuclear motion wavefunctions of the states involved in the transition. The former can be obtained from electronic-structure computations while the latter can be determined from a variational solution of the nuclear motion problem. The DMS is a two- or three-component vector function. While a lot of work has been devoted to obtaining high-accuracy PESs and the corresponding rotation-vibration energy levels and wavefunctions for small molecules, there is only limited experience accumulated about the determination of high-accuracy DMSs. Furthermore, while empirical adjustment of PESs is common practice, empirical adjustment of DMSs does not seem to be viable, partly due to the inferior quality of the available experimental data. Accurate measurement of the intensities of rotational–vibrational transitions in the laboratory is a technically demanding task even at room and especially at elevated temperatures. The range of intensities and their observational uncertainties are much larger for transition intensities than those for line positions. New, high-precision experiments have started to appear but this changes the present-day scenario rather slowly [97,98].

Ab initio studies of the PESs of triatomic molecules [17,59,60,88] have shown the importance of appending so-called small corrections to standard non-relativistic valence-only *ab initio* predictions. So far these have not been considered for the DMSs of polyatomic molecules. It is up to future high-accuracy computation of DMSs and the utilization of new measurements to decide whether such corrections have a significant effect on computed rotational–vibrational intensities making their computation worth pursuing.

Obtaining a high-quality analytical fit to *ab initio* dipole data is a challenging problem [99]. This is connected to the fact that the resulting DMS must be able to reproduce transition intensities which vary by many orders of magnitude. Fits using procedures proven adequate for PESs may suffer from small unphysical oscillations. Construction of a new DMS, including relativistic effects, is underway for water [100].

5. VARIATIONAL NUCLEAR MOTION COMPUTATIONS

Breaking away from the traditional treatment of molecular spectra using perturbative approaches, variational computation of rovibronic energy levels was introduced in the early 70s [101,102], following the prior derivation of simplified and exact normal coordinate Hamiltonians for nonlinear [86] and linear [87] molecules. Two routes can be followed in variational-type nuclear motion computations. One employs Hamiltonians in curvilinear, preferably orthogonal internal coordinates [103–106] offering the advantage that such Hamiltonians, with appropriately chosen basis sets, matrix element computations, diagonalization techniques, and PESs, can yield the complete eigenspectrum. Due to obvious dimensionality problems, this technique could only be pursued for small species, most notably for triatomics. Recognizing the difficulties associated with the development and use of tailor-made internal coordinate Hamiltonians, the other direction prefers to have a unique Hamiltonian which would be the same for almost all molecular systems. This is offered by the Hamiltonians derived by Watson [86,87] applying an Eckart frame. Perhaps the so far most elaborate use of the Eckart–Watson Hamiltonians has been allowed by the MULTIMODE set of programs [107].

For tri- and tetratomic systems solution of the rovibrational problem was made particularly tractable by the introduction of the discrete variable representation (DVR) [108–115] of the Hamiltonian. Initially, the DVR was developed with standard orthogonal polynomial bases and the associated Gaussian quadratures, employing the same number of basis functions and quadrature points. DVRs based on such basis sets, quadrature points, and weights possess remarkable properties. The most relevant is the diagonality of the potential energy matrix $\mathbf{V}$ making DVR a nearly ideal technique for nuclear motion computations eventhough the simplifications introduced in the computation of $\mathbf{V}$ make the eigenvalues not strictly variational. Nowadays solution strategies have started to appear to not only the four- [116–125] but also the five- [126–128] and six-atomic [129] problems.

5.1 Computations in internal coordinates

As Refs. [116–134] testify, there are several strategies to set up matrix representations of multidimensional rotational–vibrational Hamiltonians. One of the simplest ones is the following. The rotational–vibrational Hamiltonian is expanded in orthogonal (O) coordinates [103,135] so that there are no cross-derivative terms in the kinetic energy operator, its matrix is represented by the discrete variable representation (D) [108–112] coupled with a direct product (P) basis for the vibrational modes multiplied by a rotation function formed by combining the normalized Wigner rotation functions, and advantage is taken of the sparsity of the resulting Hamiltonian matrix whose selected eigenvalues can thus be determined extremely efficiently by variants of the iterative (I) Lanczos technique [136]. The resulting procedure has been termed DOPI [2,137].

A particularly important feature of internal coordinate rovibrational Hamiltonians is that singularities will always be present in them when expressed in the moving body-fixed frame [138]. Protocols that do not treat the singularities in these rovibrational Hamiltonians may result in sizeable errors for some of the rovibrational wave functions which depend on coordinates characterizing the singularity, thereby preventing their use for the computation of the complete rovibrational eigenspectrum.

Apart from approaches which avoid the introduction of certain singularities during construction of the Hamiltonian [139–141], it seems that there are only a few *a posteriori* strategies to cope with singular terms in rovibrational Hamiltonians when solving the related time-independent Schrödinger equation by means of (nearly) variational techniques. Building partially on previous efforts [142–146], Czakó and co-workers [147–149] developed a generalized finite basis representation (GFBR) strategy based on the use of the Bessel-DVR functions of Littlejohn and Cargo [150], and several resulting implementations for coping with the radial singularities present, for example, in the Sutcliffe–Tennyson triatomic rovibrational Hamiltonian expressed in orthogonal internal coordinates. In this strategy a non-polynomial nondirect-product basis is employed. An efficient GFBR has been developed with nondirect-product basis functions having structure similar to that of spherical harmonics [148]. It was shown there that the use of an FBR which couples different grid points to each basis function can be useful even if it results in a non-symmetric representation of the Hamiltonian.

5.2 Computations in normal coordinates

The Eckart–Watson Hamiltonians [86,87] expressed in normal coordinates are universal and thus make the introduction and programming of tailor-made Hamiltonians for each new system exhibiting unique bonding arrangements unnecessary. While the use of these Hamiltonians for systems having more than three atoms has a long and successful history [151–154], their application is not without difficulties. In particular, due to the numerical integration schemes employed for the potential, in general it has proved to be impossible to use PESs expressed in arbitrary coordinates with this Hamiltonian without resorting to some kind of an expansion of the PES in normal coordinates, thus separating, to a certain extent, otherwise non-separable functions. One of the best approximate techniques developed so far for computing the matrix representation of the potential is due to Gerber [152] and Carter et al. [154] and is called the *n*-mode representation.

This shortcoming has so far excluded the possibility of the exact inclusion of general high-quality PESs in vibrational computations for systems having more than three atoms even if they were available. Nevertheless, as shown here and in Ref. [155] in more detail, this problem can be eliminated. To achieve this, one needs to (a) represent the Hamiltonian using the DVR technique; and (b) apply a formalism allowing the exact expression of arbitrary internal coordinates in terms of normal coordinates.

To express curvilinear internal coordinates in terms of normal coordinates, bond vectors in terms of normal coordinates are needed. A bond vector pointing from nucleus p to i ($i, p = 1, 2, \ldots, N$ and $i \neq p$) in a molecule with N nuclei is given as

$$\mathbf{r}_{pi} = \mathbf{C}\left[\mathbf{a}_i - \mathbf{a}_p + \sum_{k=1}^{3N-F}\left(\frac{1}{\sqrt{m_i}}\mathbf{l}_{ik} - \frac{1}{\sqrt{m_p}}\mathbf{l}_{pk}\right)Q_k\right],$$

where the orthogonal matrix $\mathbf{C}$ describes spatial orientation, $\mathbf{a}_i$ ($i = 1, 2, \ldots, N$) are the Cartesian coordinates of the chosen reference structure, and elements of $\mathbf{l}_{ik}$

TABLE 9.3 Variational vibrational band origins (VBOs, in cm^{-1}) with $l = 0$ up to the highest fundamental of ^{12}C^{16}O$_2$ obtained with Chédin's [80] sextic empirical force field[a]

$(n_1, n_2^{|l|}, n_3)$[b]	Internal[c]	Normal[d]	Expt.[e]
$(0, 0^0, 0)$	2535.4	2535.4	–
$(1, 0^0, 0)$	1285.0	1285.0	1285.4
$(0, 2^0, 0)$	1387.5	1387.5	1388.2
$(0, 0^0, 1)$	2347.3	2347.3	2349.2

[a] A potential energy cutoff of 20000 cm^{-1} was applied, as described in detail in Ref. [137].
[b] Standard normal coordinate notation of the VBOs for a triatomic linear molecule.
[c] The variational results based on a triatomic internal coordinate Hamiltonian were obtained with the DOPI algorithm [137], the results are the same as in Table 9.3 of Ref. [137].
[d] Variational results obtained with the DEWE algorithm [155].
[e] Experimental vibrational frequencies taken from Ref. [80].

($i = 1, 2, \ldots, N$, $k = 1, 2, \ldots, 3N - F$, where $F = 5/6$ if the molecule is linear/nonlinear) are the transformation coefficients between normal coordinates and the instantaneous displacement coordinates in the Eckart frame. Bond vectors are thus expressed in terms of Q_k ($k = 1, 2, \ldots, 3N - F$) and the Euler angles (through **C**). Curvilinear internal coordinates, expressed as scalar and triple products of bond vectors, are functions of only the normal coordinates [156]. Due to this transformation, arbitrary potentials given in curvilinear internal coordinates can be called in a program working in the Q_k ($k = 1, 2, \ldots, 3N - F$) normal coordinates.

Along these lines an efficient protocol, called DEWE has been developed [155] which is based on the DVR of the Eckart–Watson Hamiltonians involving an exact inclusion of potentials expressed in an arbitrary set of coordinates. The DEWE procedure has been tested both for nonlinear (H$_2$O, H$_3^+$, and CH$_4$) and linear (CO$_2$, HCN, and HNC) molecules.

For H$_2$^{16}O, employing the high-accuracy CVRQD PES [17,59], the lower vibrational energy levels obtained with DEWE were the same, within numerical precision, as those determined with the DOPI procedure. However, unlike in the case of DOPI, the higher bending levels, with the bending quantum number $n_2 \geq 4$, could not be converged tightly. This convergence problem corresponds to the singularity present in the Eckart–Watson Hamiltonian.

VBOs applying Chédin's sextic empirical force field [80] are presented in Table 9.3 for ^{12}C^{16}O$_2$ using exactly the same potential with the internal and normal coordinate Hamiltonians. When comparison can be made, the two approaches result in the same eigenenergies. It is also worth mentioning that for CO$_2$ no convergence (singularity) problems appeared, in clear contrast to the case of the nonlinear H$_2$O molecule.

6. OUTLOOK

Understanding the complete rotational–vibrational spectra of small molecules is an almost formidable task. This is partly due to the fact that complete spectra contain information about billions of lines even for a triatomic species. Understanding these spectra requires sophisticated instrumentation and experiments, involving measurement and assignment of high-resolution molecular spectra, high-accuracy first-principles computations, involving electronic-structure and nuclear-motion determinations, empirical adjustments of *ab initio* PESs, and allowance for nonadiabatic effects. Only by interplay of all these experimental and computational elements can one expect that for polyatomic species the intricacies of complete molecular spectra will be unraveled some day. It seems most advantageous to us to combine results from experiment and theory by centering on a database approach sketched in this report. Work along these lines is underway for the isotopologues of water, arguably the most important polyatomic molecule.

ACKNOWLEDGEMENT

Over the years the research described received support from several granting agencies, the ongoing work has been supported by OTKA, the NKTH, the EU, INTAS, and IUPAC. The authors are grateful to the many discussions with Professors Jonathan Tennyson (London, U.K.), Viktor Szalay (Budapest, Hungary), and Wesley D. Allen (Athens, GA, U.S.A.) related to the topic of this report and to all the co-authors for their work in the cited joint papers.

REFERENCES

1. Neale, L., Tennyson, J. A high-temperature partition function for H_3^+, Astrophys. J. 1995, 454, L169–73.
2. Furtenbacher, T., Czakó, G., Sutcliffe, B.T., Császár, A.G., Szalay, V. The methylene saga continues: stretching fundamentals and zero-point energy of $\tilde{X}\,^3B_1$ CH_2, J. Mol. Struct. 2006, 780–81, 283–94.
3. Harris, F.E. Current methods for Coulomb few-body problems, Adv. Quant. Chem. 2004, 47, 129–55;
   Armour, E.A.G., Richard, J.-M., Varga, K. Stability of few-charge systems in quantum mechanics, Phys. Rep. 2005, 413, 1–90.
4. Császár, A.G., Tarczay, G., Leininger, M.L., Polyansky, O.L., Tennyson, J., Allen, W.D. Dream or reality: complete basis set full configuration interaction potential energy hypersurfaces, in Demaison, J., Sarka, K., Cohen, E.A., editors. Spectroscopy from Space. Dordrecht: Kluwer; 2001, p. 317–39.
5. Császár, A.G., Allen, W.D., Yamaguchi, Y., Schaefer III, H.F. *Ab initio* determination of accurate ground electronic state potential energy hypersurfaces for small molecules, in Jensen, P., Bunker, P.R., editors. Computational Molecular Spectroscopy. New York: Wiley; 2000, p. 15–68.
6. Tarczay, G., Császár, A.G., Klopper, W., Quiney, H.M. Anatomy of relativistic energy corrections in light molecular systems, Mol. Phys. 2001, 99, 1769–94.
7. Balasubramanian, K. Relativistic Effects in Chemistry. Part A. Theory and Techniques, New York: Wiley; 1997.
8. Born, M., Oppenheimer, J.R. Zur Quantentheorie der Molekeln, Ann. Phys. 1927, 84, 457–84;

Born, M. Kopplung der Elektronen- und Kernbewegung in Molekeln und Kristallen, Göttinger Nachr. Acad. Wiss. Math. Nat. Kl. 1951, 6, 1–3;
Born, M., Huang, K. Dynamical Theory of Crystal Lattices, London: Oxford University Press; 1956.
9. Bunker, P.R., Jensen, P. The Born–Oppenheimer approximation, in Jensen, P., Bunker, P.R., editors. Computational Molecular Spectroscopy. New York: Wiley; 2000, p. 1–12.
10. Nielsen, H.H. The vibration-rotation energies of molecules, Rev. Mod. Phys. 1951, 23, 90–136.
11. Papoušek, D., Aliev, M.R. Molecular Vibration-Rotation Spectra, Amsterdam: Elsevier; 1982.
12. Mills, I.M., in Rao, K.N., Mathews, C.W., editors. Molecular Spectroscopy: Modern Research. New York: Academic Press; 1972.
13. Gaw, J.F., Willetts, A., Green, W.H., Handy, N.C. in Advances in Molecular Vibrations and Collision Dynamics. Greenwich, CT: JAI Press; 1990.
14. Clabo, D.A., Allen, W.D., Remington, R.B., Yamaguchi, Y., Schaefer III, H.F. A systematic study of molecular vibrational anharmonicity and vibration-rotation interaction by self-consistent field higher derivative methods—asymmetric-top molecules, Chem. Phys. 1988, 123, 187–239.
15. Allen, W.D., Yamaguchi, Y., Császár, A.G., Clabo, D.A., Remington, R.B., Schaefer III, H.F. A systematic study of molecular vibrational anharmonicity and vibration–rotation interaction by self-consistent field higher derivative methods—linear polyatomic molecules, Chem. Phys. 1990, 145, 427–66.
16. Aliev, M.R., Watson, J.K.G., in Narahari Rao, K., editor. Molecular Spectroscopy: Modern Research, vol. III. San Diego, CA: Academic Press; 1985, p. 1–67.
17. Polyansky, O.L., Császár, A.G., Shirin, S.V., Zobov, N.F., Barletta, P., Tennyson, J., Schwenke, D.W., Knowles, P.J. High accuracy *ab initio* rotation-vibration transitions for water, Science 2003, 299, 539–42.
18. Furtenbacher, T., Császár, A.G., Tennyson, J. MARVEL: Measured active rotational-vibrational energy levels, J. Mol. Spectry., 2007, submitted.
19. Császár, A.G., Furtenbacher, T., Czakó, G. The greenhouse effect on Earth and the complete spectroscopy of water, Magy. Kém. Foly. 2006, 112, 123–8 (in Hungarian).
20. Bubin, S., Bednarz, E., Adamowicz, L. Charge asymmetry in HD^+, J. Chem. Phys. 2005, 122, 041102.
21. Kolos, W., Szalewicz, K., Monkhorst, H.J. New Born–Oppenheimer potential energy curve and vibrational energies for the electronic ground state of the hydrogen molecule, J. Chem. Phys. 1986, 84, 3278–83.
22. Hilico, L., Billy, N., Grémaud, B., Delande, D. *Ab initio* calculation of $J = 0$ and $J = 1$ states of the H_2^+, D_2^+, and HD^+ molecular ions, Eur. Phys. J. D 2000, 12, 449–66.
23. Czakó, G., Császár, A.G., Szalay, V., Sutcliffe, B.T. Adiabatic Jacobi corrections for H_2+-like systems, J. Chem. Phys. 2007, 126, 024102.
24. Moss, R.E. Energies of low-lying vibration-rotation levels of H_2+ and its isotopomers, J. Phys. B 1999, 32, L89–91.
25. Cencek, W., Komasa, J., Rychlewski, J. Configuration interaction and Hylleraas configuration interaction methods in valence bond theory. Diatomic two-electron systems, J. Chem. Phys. 1991, 95, 2572–6.
26. Wolniewicz, L. Nonadiabatic energies of the ground state of the hydrogen molecule, J. Chem. Phys. 1995, 103, 1792–9.
27. Frolov, A.M., Smith, V.H. Scaled-time dynamics of ionization of Rydberg Stark states by half-cycle pulses, J. Phys. B 1995, 28, L449–56.
28. Bunker, P.R., Moss, R.E. Breakdown of Born–Oppenheimer approximation—effective vibration-rotation Hamiltonian for a diatomic molecule, Mol. Phys. 1977, 33, 417–24;
Bunker, P.R., Moss, R.E. Effect of the breakdown of the Born–Oppenheimer approximation on the rotation-vibration Hamiltonian of a triatomic molecule, J. Mol. Spectry. 1980, 80, 217–28.
29. Sellers, H., Pulay, P. The adiabatic correction to molecular potential surfaces in the SCF approximation, Chem. Phys. Lett. 1984, 103, 463–5.
30. Handy, N.C., Yamaguchi, Y., Schaefer III, H.F. The diagonal correction to the Born–Oppenheimer approximation—its effect on the singlet–triplet splitting of CH_2 and other molecular effects, J. Chem. Phys. 1986, 84, 4481–4.
31. Kutzelnigg, W. The adiabatic approximation I. The physical background of the Born–Handy ansatz, Mol. Phys. 1997, 90, 909–16.

32. Valeev, E.F., Sherrill, C.D. The adiabatic correction to molecular potential surfaces in the SCF approximation, J. Chem. Phys. 2003, 118, 3921–7.
33. Gauss, J., Tajti, A., Kállay, M., Stanton, J.F., Szalay, P.G. Analytic calculation of the diagonal Born–Oppenheimer correction within configuration-interaction and coupled-cluster theory, J. Chem. Phys. 2006, 125, 144111.
34. Schwenke, D.W. Beyond the potential energy surface: *Ab initio* corrections to the Born–Oppenheimer approximation for H_2O, J. Phys. Chem. A 2001, 105, 2352–60.
35. Schwenke, D.W. A first principle effective Hamiltonian for including nonadiabatic effects for H_2+ and HD^+, J. Chem. Phys. 2001, 114, 1693–9.
36. Ruscic, B., Pinzon, R.E., Morton, M.L., Von Laszevski, G., Bittner, S.J., Nijsure, S.G., Amin, K.A., Minkoff, M., Wagner, A.F. Introduction to active thermochemical tables: several "key" enthalpies of formation revisited, J. Phys. Chem. A 2004, 108, 9979–97.
37. Ruscic, B., Boggs, J.E., Burcat, A., Császár, A.G., Demaison, J., Janoschek, R., Martin, J.M.L., Morton, M., Rossi, M.J., Stanton, J.F., Szalay, P.G., Westmoreland, P.R., Zabel, F., Bérces, T. IUPAC critical evaluation of thermochemical properties of selected radicals, Part I, J. Phys. Chem. Ref. Data 2005, 34, 573–656.
38. Feeley, R., Seiler, P., Packard, A., Frenklach, M. Consistency of a reaction dataset, J. Phys. Chem. A 2004, 108, 9573–83.
39. Frenklach, M., Packard, A., Seiler, P., Feeley, R. Collaborative data processing in developing predictive models of complex reaction systems, Int. J. Chem. Kinet. 2004, 36, 57–66.
40. Flaud, J.-M., Camy-Peyret, C., Maillard, J.P. Higher ro-vibrational levels of H_2O deduced from high resolution oxygen-hydrogen flame spectra between 2800–6200 cm^{-1}, Mol. Phys. 1976, 32, 499–521.
41. De Lucia, F.C. Microwave spectrum and ground state energy levels of $H_2^{17}O$, J. Mol. Spectry. 1975, 56, 138–45;
Matsushima, F., Nagase, H., Nakauchi, T., Odashima, H., Takagi, K. Frequency measurement of pure rotational transitions of $H_2^{17}O$ and $H_2^{18}O$ from 0.5 to 5 THz, J. Mol. Spectry. 1999, 193, 217–23.
42. Kauppinen, J., Kyro, E. High resolution pure rotational spectrum of water vapor enriched by $H_2^{17}O$ and $H_2^{18}O$, J. Mol. Spectry. 1980, 84, 405–23.
43. SISAM database http://mark4sun.jpl.nasa.gov/.
44. Liu, A.-W., Hu, S.-M., Camy-Peyret, C., Mandin, J.-Y., Naumenko, O., Voronin, B. Fourier-transform absorption spectra of $H_2^{17}O$ and $H_2^{18}O$ in the 8000–9400 cm^{-1} spectral region, J. Mol. Spectry. 2006, 237, 53–62.
45. Camy-Peyret, C., Flaud, J.-M., Mandin, J.-Y., Bykov, A., Naumenko, O., Sinitsa, L., Voronin, B. Fourier-transform absorption spectrum of the $H_2^{17}O$ molecule in the 9711–11335 cm^{-1} spectral region: the first decade of resonating states, J. Quant. Spectrosc. Rad. Transfer 1999, 61, 795–812.
46. Tanaka, M., Naumenko, O., Brault, J.W., Tennyson, J. Fourier-transform absorption spectra of $H_2^{18}O$ and $H_2^{17}O$ in the $3\nu + \delta$ and 4ν polyad region, J. Mol. Spectry. 2005, 234, 1–9.
47. Tennyson, J., Zobov, N.F., Williamson, R., Polyansky, O.L., Bernath, P.F. Experimental energy levels of the water molecule, J. Phys. Chem. Ref. Data 2001, 30, 735–831.
48. Császár, A.G., Mills, I.M. Vibrational energy levels of water, Spectrochim. Acta 1997, 53A, 1102–22.
49. Allen, W.D., East, A.L.L., Császár, A.G. *Ab initio* anharmonic vibrational analysis of non-rigid molecules, in Laane, J., Dakkouri, M., van der Veken, B., Oberhammer, H., editors. Structures and Conformations of Non-Rigid Molecules. Dordrecht: Kluwer; 1993, p. 343–73.
50. Császár, A.G., Allen, W.D., Schaefer III, H.F. In pursuit of the *ab initio* limit for conformational energy prototypes, J. Chem. Phys. 1998, 108, 9751–64.
51. Dunning, T.H. Jr. Gaussian basis sets for use in correlated molecular calculations. I. The atoms boron through neon and hydrogen, J. Chem. Phys. 1989, 90, 1007–23.
52. The (aug-)cc-p(wC)VnZ basis sets can be obtained from the Extensible Computational Chemistry Environment Basis Set Database, Version 1.0, as developed and distributed by the Molecular Science Computing Facility, Environmental and Molecular Sciences Laboratory, which is part of the Pacific Northwest Laboratory, P.O. Box 999, Richland, Washington 99352, USA and funded by the U.S. Department of Energy. The Pacific Northwest Laboratory is a multi-program laboratory operated by Battelle Memorial Institute for the U.S. Department of Energy under contract DE-AC06-76RLO 1830.

53. Note that the Hartree–Fock limit (HFL) can be achieved with just a few dozen Gaussian functions in a fully variational computation through optimization of the positions and the exponents of the Gaussians, as discussed in Tasi, G., Császár, A.G., Hartree–Fock-limit energies and structures with a few dozen distributed Gaussians, Chem. Phys. Lett., 2007, 438, 139–43.
54. Hartree, D.R. The wave mechanics of an atom with a non-Coulomb central field. Part I. Theory and methods, Proc. Cambridge Philos. Soc. 1928, 24, 89–110;
   Fock, V. Naherungsmethode zur Lösung des quantenmechanischen Mehrkörperproblems, Z. Phys. 1930, 61, 126–48.
55. Møller, C., Plesset, M.S. Note on an approximation treatment for many-electron systems, Phys. Rev. 1934, 46, 618–22.
56. Kállay, M., Surján, P.R. Computing coupled-cluster wave functions with arbitrary excitations, J. Chem. Phys. 2000, 113, 1359–65;
   Kállay, M., Gauss, J., Szalay, P.G. Analytic first derivatives for general coupled-cluster and configuration interaction models, J. Chem. Phys. 2003, 119, 2991–3004;
   Kállay, M., Gauss, J. Analytic second derivatives for general coupled-cluster and configuration-interaction models, J. Chem. Phys. 2004, 120, 6841–8.
57. Nielsen, I.M.B., Allen, W.D., Császár, A.G., Schaefer III, H.F. Toward resolution of the silicon dicarbide (SiC$_2$) saga: *Ab initio* excursions in the web of polytopism, J. Chem. Phys. 1997, 107, 1195–211.
58. Tarczay, G., Császár, A.G., Klopper, W., Szalay, V., Allen, W.D., Schaefer III, H.F. The barrier to linearity of water, J. Chem. Phys. 1999, 110, 11971–81.
59. Barletta, P., Shirin, S.V., Zobov, N.F., Polyansky, O.L., Tennyson, J., Valev, E.F., Császár, A.G. CVRQD *ab initio* ground-state adiabatic potential energy surfaces for the water molecule, J. Chem. Phys. 2006, 125, 204307.
60. Tarczay, G., Császár, A.G., Polyansky, O.L., Tennyson, J. *Ab initio* rovibrational spectroscopy of hydrogen sulfide, J. Chem. Phys. 2001, 115, 1229–42.
61. Aarset, K., Császár, A.G., Sibert III, E.L., Allen, W.D., Schaefer III, H.F., Klopper, W., Noga, J. Anharmonic force field, vibrational energies, and barrier to inversion of SiH$_3^-$, J. Chem. Phys. 2000, 112, 4053–63.
62. Tarczay, G., Császár, A.G., Leininger, M.L., Klopper, W. The barrier to linearity of hydrogen sulphide, Chem. Phys. Lett. 2000, 322, 119–28.
63. Tajti, A., Szalay, P.G., Császár, A.G., Kállay, M., Gauss, J., Valeev, E.F., Flowers, B.A., Vázquez, J., Stanton, J.F. HEAT: High accuracy extrapolated *ab initio* thermochemistry, J. Chem. Phys. 2004, 121, 11599–613.
64. Bomble, Y.J., Vázquez, J., Kállay, M., Michauk, C., Szalay, P.G., Császár, A.G., Gauss, J., Stanton, J.F. High accuracy extrapolated *ab initio* thermochemistry. II. Minor improvements to the protocol and a vital simplification, J. Chem. Phys. 2006, 125, 064108.
65. Boese, A.D., Oren, M., Atasoylu, O., Martin, J.M.L., Kállay, M., Gauss, J. W3 theory: robust computational thermochemistry in the kJ/mol accuracy range, J. Chem. Phys. 2004, 120, 4129–41;
   Karton, A., Rabinovich, E., Martin, J.M.L., Ruscic, B. W4 theory for computational thermochemistry: in pursuit of confident sub-kJ-mol predictions, J. Chem. Phys. 2006, 125, 144108.
66. Tasi, G., Izsák, R., Matisz, G., Császár, A.G., Kállay, M., Ruscic, B., Stanton, J.F. The origin of systematic error in the standard enthalpies of formation of hydrocarbons computed via atomization schemes, ChemPhysChem 2006, 7, 1664–7.
67. Császár, A.G., Szalay, P.G., Leininger, M.L. The enthalpy of formation of $^2\Pi$ CH, Mol. Phys. 2002, 100, 3879–83;
   Császár, A.G., Leininger, M.L., Szalay, V. The enthalpy of formation of CH$_2$, J. Chem. Phys. 2003, 118, 10631–42.
68. Császár, A.G. Anharmonic molecular force fields, in Schleyer, P.v.R., Allinger, N.L., Clark, T., Gasteiger, J., Kollman, P.A., Schaefer, III H.F., Schreiner, P.R., editors. The Encyclopedia of Computational Chemistry, vol. 1. Chichester: Wiley; 2001, p. 13–30.
69. Sarka, K., Demaison, J. Perturbation theory, effective Hamiltonians and force constants, in Jensen, P., Bunker, P.R., editors. Computational Molecular Spectroscopy. Chichester: Wiley; 2000, p. 255–303.
70. Allen, W.D., Czinki, E., Császár, A.G. Molecular structure of proline, Chem. Eur. J. 2004, 10, 4512.

71. Maslen, P.E., Jayatilaka, D., Colwell, S.M., Amos, R.D., Handy, N.C. Higher analytical derivatives. 2. The 4th derivative of self-consistent-field energy, J. Chem. Phys. 1991, 95, 7409–17.
72. Yamaguchi, Y., Osamura, Y., Goddard, J.D., Schaefer III, H.F. A New Dimension to Quantum Chemistry: Analytic Derivative Methods in *Ab Initio* Molecular Electronic Structure Theory, New York: Oxford University Press; 1994.
73. Kállay, M., Gauss, J., Szalay, P.G. Analytic first derivatives for general coupled-cluster and configuration interaction models, J. Chem. Phys. 2003, 119, 2991–3004.
74. Kállay, M., Gauss, J. Analytic second derivatives for general coupled-cluster and configuration-interaction models, J. Chem. Phys. 2004, 120, 6841–8.
75. Lacy, M., Whiffen, D.H. The anharmonic force field of nitrous-oxide, Mol. Phys. 1982, 45, 241–52.
76. Császár, A.G. The anharmonic force field of N_2O, J. Phys. Chem. 1994, 98, 8823–6.
77. Kobayashi, M., Suzuki, I. Sextic force-field of nitrous-oxide, J. Mol. Spectry. 1987, 125, 24–42.
78. Teffo, J.-L., Chédin, A. Internuclear potential and equilibrium structure of the nitrous-oxide molecule from rovibrational data, J. Mol. Spectry. 1989, 135, 389–409.
79. Császár, A.G. The anharmonic force field of CO_2, J. Phys. Chem. 1992, 96, 7898–904.
80. Chédin, A. Carbon-dioxide molecule—potential, spectroscopic, and molecular constants from its infrared spectrum, J. Mol. Spectry. 1979, 76, 430–91.
81. Lacy, M. The anharmonic force field of carbon-dioxide, Mol. Phys. 1982, 45, 253–8.
82. Suzuki, I. General anharmonic force constants of carbon dioxide, J. Mol. Spectry. 1968, 25, 479–500.
83. Allen, W.D., Császár, A.G. On the *ab initio* determination of higher-order force-constants at nonstationary reference geometries, J. Chem. Phys. 1993, 98, 2983–3015.
84. Bailleux, S., Bogey, M., Demuynck, C., Destombes, J.L., Dujardin, N., Liu, Y., Császár, A.G. *Ab initio* study and millimeter-wave spectroscopy of P_2O, J. Chem. Phys. 1997, 107, 8317.
85. Martin, J.M.L., Lee, T.J., Taylor, P.R., Francois, J.-P. The anharmonic force-field of ethylene, C_2H_4, by means of accurate *ab initio* calculations, J. Chem. Phys. 1995, 103, 2589–602;
Martin, J.M.L., Taylor, P.R. Accurate *ab initio* quartic force field for *trans*-HNNH and treatment of resonance polyads, Spectrochim. Acta 1997, 53A, 1039–50.
86. Watson, J.K.G. Simplification of the molecular vibration-rotation Hamiltonian, Mol. Phys. 1968, 5, 479–90.
87. Watson, J.K.G. The vibration-rotation hamiltonian of linear molecules, Mol. Phys. 1970, 19, 465–87.
88. van Mourik, T., Harris, G.J., Polyansky, O.L., Tennyson, J., Császár, A.G., Knowles, P.J. *Ab initio* global potential, dipole, adiabatic, and relativistic correction surfaces for the HCN-HNC system, J. Chem. Phys. 2001, 115, 3706.
89. Werner, H.-J., Knowles, P.J. An efficient internally contracted multiconfiguration reference configuration-interaction method, J. Chem. Phys. 1988, 89, 5803–14.
90. Langhoff, S., Davidson, E.R. Configuration-interaction calculations on the nitrogen molecule, Int. J. Quant. Chem. 1974, 8, 61–72.
91. Gdanitz, R.J., Ahlrichs, R. The averaged coupled-pair functional (ACPF)—a size-extensive modification of MR-CISD, Chem. Phys. Lett. 1988, 143, 413–20.
92. Császár, A.G., Kain, J.S., Polyansky, O.L., Zobov, N.F., Tennyson, J. Relativistic correction to the potential energy surface and vibration-rotation levels of water, Chem. Phys. Lett. 1998, 293, 317–23;
Császár, A.G., Kain, J.S., Polyansky, O.L., Zobov, N.F., Tennyson, J. Relativistic correction to the potential energy surface and vibration-rotation levels of water, Chem. Phys. Lett. 1999, 312, 613–6.
93. Quiney, H.M., Barletta, P., Tarczay, G., Császár, A.G., Polyansky, O.L., Tennyson, J. Two-electron relativistic corrections to the potential energy surface and vibration-rotation levels of water, Chem. Phys. Lett. 2001, 344, 413–20.
94. Pyykkö, P., Dyal, K.G., Császár, A.G., Tarczay, G., Polyansky, O.L., Tennyson, J. Estimation of Lamb-shift effects for molecules: Application to the rotation-vibration spectra of water, Phys. Rev. A 2001, 63, 024502.
95. Partridge, H., Schwenke, D.W. The determination of an accurate isotope dependent potential energy surface for water from extensive *ab initio* calculations and experimental data, J. Chem. Phys. 1997, 106, 4618–39.
96. Shirin, S.V., Polyansky, O.L., Zobov, N.F., Ovsyannikov, R.I., Császár, A.G., Tennyson, J. Spectroscopically determined potential energy surfaces of the $H_2^{16}O$, $H_2^{17}O$, and $H_2^{18}O$ isotopologues of water, J. Mol. Spectry. 2006, 236, 216–23.

97. Callegari, A., Theule, P., Tolchenov, R.N., Zobov, N.F., Polyansky, O.L., Tennyson, J., Muenter, J.S., Rizzo, T.R. Dipole moments of highly vibrationally excited water, Science 2002, 297, 993–5.
98. Lisak, D., Hodges, J.-T., Cyurilo, R. Comparison of semiclassical line-shape models to rovibrational H_2O spectra measured by frequency-stabilized cavity ring-down spectroscopy, Phys. Rev. A 2006, 73, 012507.
99. Schwenke, D.W., Partridge, H. Convergence testing of the analytic representation of an *ab initio* dipole moment function for water: Improved fitting yields improved intensities, J. Chem. Phys. 2000, 113, 6592–7.
100. Lodi, L., Tolchenov, R.N., Tennyson, J., Lynas-Gray, A.-E., Shirin, S.V., Zobov, N.F., Polyansky, O.L., Császár, A.G., van Stralen, J.N.P., Visscher, L. A new *ab initio* ground state dipole moment surface for the water molecule, J. Chem. Phys. 2007, submitted.
101. Bucknell, M.G., Handy, N.C., Boys, S.F. Vibration-rotation wavefunctions and energies for any molecule obtained by a variational method, Mol. Phys. 1974, 28, 759–76.
102. Whitehead, R.J., Handy, N.C. Variational calculation of vibration-rotation energy levels for triatomic molecules, J. Mol. Spectry. 1975, 55, 356–73.
103. Sutcliffe, B.T., Tennyson, J. A general treatment of vibration-rotation coordinates for triatomic-molecules, Int. J. Quant. Chem. 1991, 39, 183–96.
104. Handy, N.C. The derivation of vibration-rotation kinetic energy operators in internal coordinates, Mol. Phys. 1987, 61, 207–23.
105. Császár, A.G., Handy, N.C. Exact quantum-mechanical vibrational kinetic-energy operator of sequentially bonded molecules in valence internal coordinates, J. Chem. Phys. 1995, 102, 3962–7.
106. Lukka, T.J. A simple method for the derivation of exact quantum-mechanical vibration-rotation Hamiltonians in terms of internal coordinates, J. Chem. Phys. 1995, 102, 3945–55.
107. Bowman, J.M., Carter, S., Huang, X. MULTIMODE: a code to calculate rovibrational energies of polyatomic molecules, Int. Rev. Phys. Chem. 2003, 22, 533–49.
108. Harris, D.O., Engerholm, G.G., Gwinn, W.D. Calculation of matrix elements for one-dimensional quantum-mechanical problems and the application to anharmonic oscillators, J. Chem. Phys. 1965, 43, 1515–7.
109. Dickinson, A.S., Certain, P.R. Calculation of matrix elements for one-dimensional quantum-mechanical problems, J. Chem. Phys. 1968, 49, 4209–11.
110. Light, J.C., Hamilton, I.P., Lill, J.V. Generalized discrete variable approximation in quantum-mechanics, J. Chem. Phys. 1985, 82, 1400–9.
111. Bačić, Z., Light, J.C. Theoretical methods for rovibrational states of floppy molecules, Annu. Rev. Phys. Chem. 1989, 40, 469–98.
112. Szalay, V. Discrete variable representations of differential-operators, J. Chem. Phys. 1993, 99, 1978–84.
113. Szalay, V. The generalized discrete variable representation. An optimal design, J. Chem. Phys. 1996, 105, 6940–56.
114. Light, J.C., Carrington, T. Jr. Discrete-variable representations and their utilization, Adv. Chem. Phys. 2000, 114, 263–310.
115. Szalay, V., Czakó, G., Nagy, Á., Furtenbacher, T., Császár, A.G. On one-dimensional discrete variable representations with general basis functions, J. Chem. Phys. 2003, 119, 10512–8.
116. Bramley, M.J., Carrington, T. Jr. A general discrete variable method to calculate vibrational-energy levels of 3-atom and 4-atom molecules, J. Chem. Phys. 1993, 99, 8519–41.
117. Mladenović, M. Rovibrational Hamiltonians for general polyatomic molecules in spherical polar parametrization. I. Orthogonal representations, J. Chem. Phys. 2000, 112, 1070–81.
118. Mladenović, M. Discrete variable approaches to tetratomic molecules Part I: DVR(6) and DVR(3)+DGB methods, Spectrochim. Acta 2002, 58A, 795–807.
119. Antikainen, J., Friesner, R., Leforestier, C. Adiabatic pseudospectral calculation of vibrational-states of 4 atom molecules—application to hydrogen-peroxide, J. Chem. Phys. 1995, 102, 1270–9.
120. Schwenke, D.W. Variational calculations of rovibrational energy levels and transition intensities for tetratomic molecules, J. Chem. Phys. 1996, 100, 2867–84.
121. Carter, S., Pinnavaia, N., Handy, N.C. The vibrations of formaldehyde, Chem. Phys. Lett. 1995, 240, 400–8.
122. Luckhaus, D. 6D vibrational quantum dynamics: Generalized coordinate discrete variable representation and (a)diabatic contraction, J. Chem. Phys. 2000, 113, 1329–47.

123. Handy, N.C., Carter, S., Colwell, S.M. The vibrational energy levels of ammonia, Mol. Phys. 1999, 96, 477–91.
124. Koput, J., Carter, S., Handy, N.C. Potential energy surface and vibrational-rotational energy levels of hydrogen peroxide, J. Phys. Chem. A 1998, 102, 6325–30.
125. Kozin, I.N., Law, M.M., Hutson, J.M., Tennyson, J. Calculating energy levels of isomerizing tetraatomic molecules. I. The rovibrational bound states of Ar_2HF, J. Chem. Phys. 2003, 118, 4896–904.
126. Schwenke, D.W., Partridge, H. Vibrational energy levels for CH_4 from an *ab initio* potential, Spectrochim. Acta 2001, 57A, 887–95;
Schwenke, D.W. Towards accurate *ab initio* predictions of the vibrational spectrum of methane, Spectrochim. Acta 2002, 58A, 849–61.
127. Wang, X.-G., Carrington, T. Jr. A contracted basis-Lanczos calculation of vibrational levels of methane: Solving the Schrodinger equation in nine dimensions, J. Chem. Phys. 2003, 119, 101–17.
128. Yu, H.-G. An exact variational method to calculate vibrational energies of five atom molecules beyond the normal mode approach, J. Chem. Phys. 2002, 117, 2030–7;
Yu, H.-G. Two-layer Lanczos iteration approach to molecular spectroscopic calculation, J. Chem. Phys. 2002, 117, 8190–6.
129. Yu, H.-G. Full-dimensional quantum calculations of vibrational spectra of six-atom molecules. I. Theory and numerical results, J. Chem. Phys. 2004, 120, 2270–84.
130. Carter, S., Handy, N.C. The variational method for the calculation of ro-vibrational energy levels, Comp. Phys. Rep. 1986, 5, 115–72.
131. Carney, G.D., Porter, R.N. H_3^+—*Ab initio* calculation of vibration spectrum, J. Chem. Phys. 1976, 65, 3547–65.
132. Tennyson, J., Kostin, M.A., Barletta, P., Harris, G.J., Polyansky, O.L., Ramanlal, J., Zobov, N.F. DVR3D: a program suite for the calculation of rotation-vibration spectra of triatomic molecules, Comp. Phys. Comm. 2004, 163, 85–116.
133. Schwenke, D.W. On the computation of ro-vibrational energy levels of triatomic molecules, Comp. Phys. Comm. 1992, 70, 1–14;
Klepeis, N.E., East, A.L.L., Császár, A.G., Allen, W.D., Lee, T.J., Schwenke, D.W. The $[FHCl]^-$ molecular anion: structural aspects, global surface, and vibrational eigenspectrum, J. Chem. Phys. 1993, 99, 3865–97.
134. Wang, X.-G., Carrington, T. Jr. New ideas for using contracted basis functions with a Lanczos eigensolver for computing vibrational spectra of molecules with four or more atoms, J. Chem. Phys. 2003, 117, 6923–34.
135. Makarewicz, J. Rovibrational Hamiltonian of a triatomic molecule in local and collective internal coordinates, J. Phys. B 1988, 21, 1803–19.
136. Lanczos, C. An iteration method for the solution of the eigenvalue problem of linear differential and integral operators, J. Res. Natl. Bur. Stand. 1950, 45, 255–82.
137. Czakó, G., Furtenbacher, T., Császár, A.G., Szalay, V. Variational vibrational calculations using high-order anharmonic force fields, Mol. Phys. 2004, 102, 2411–23.
138. Sutcliffe, B.T. The coupling of nuclear and electronic motions in molecules, J. Chem. Soc. Faraday Trans. 1993, 89, 2321–35.
139. Meremianin, A.V., Briggs, J.S. The irreducible tensor approach in the separation of collective angles in the quantum N-body problem, Physics Rep. 2003, 384, 121–95.
140. For example, for triatomic vibrational calculations in hyperspherical coordinates, used by Carter, S., Meyer, W., A variational method for the calculation of vibrational-energy levels of triatomic molecules using a Hamiltonian in hyperspherical coordinates, J. Chem. Phys. 1990, **93**, 8902–14 and others, no radial singularity is present in the Hamiltonian. Recently the approach using hyperspherical coordinates has been extended to obtain eigenstates beyond the barrier to linearity by Schiffels, P., Alijah, A., Hinze, J., Rovibrational states of H_3^+. Part 2: The energy region between 9000 cm^{-1} and 13000 cm^{-1} including empirical corrections for the non-adiabatic effects, Mol. Phys. 2003, **101**, 189–209.
141. Spirko, V., Cejhan, A., Jensen, P. A new Morse-oscillator based Hamiltonian for H_3^+—Explicit expressions for some vibrational matrix elements, J. Mol. Spectry. 1987, 134, 430–6.
142. Henderson, J.R., Tennyson, J., Sutcliffe, B.T. All the bound vibrational states of H_3^+—A reappraisal, J. Chem. Phys. 1993, 98, 7191–203.

143. Watson, J.K.G. Vibration-rotation calculations for H_3^+ using a Morse-based discrete variable representation, Can. J. Phys. 1994, 72, 702–13;
Watson, J.K.G. The vibration-rotation spectrum and anharmonic potential of H_3^+, Chem. Phys. 1995, 190, 291–300.
144. Bramley, M.J., Tromp, J.W., Carrington, T. Jr, Corey, G.C. Efficient calculation of highly excited vibrational energy levels of floppy molecules—The band origins of H_3^+ up to 35000 cm^{-1}, J. Chem. Phys. 1994, 100, 6175–94.
145. Mandelshtam, V.A., Taylor, H.S. The quantum resonance spectrum of the H_3^+ molecular ion for $J = 0$. An accurate calculation using filter diagonalization, J. Chem. Soc. Faraday Trans. 1997, 93, 847–60.
146. Vincke, M., Malegat, L., Baye, D. Regularization of singularities in Lagrange-mesh calculations, J. Phys. B: At. Mol. Opt. Phys. 1993, 26, 811–26.
147. Czakó, G., Szalay, V., Császár, A.G., Furtenbacher, T. Treating singularities present in the Sutcliffe–Tennyson vibrational Hamiltonian in orthogonal internal coordinates, J. Chem. Phys. 2005, 122, 024101.
148. Czakó, G., Szalay, V., Császár, A.G. Finite basis representations with nondirect product basis functions having structure similar to that of spherical harmonics, J. Chem. Phys. 2006, 124, 014110.
149. Czakó, G., Furtenbacher, T., Barletta, P., Császár, A.G., Szalay, V., Sutcliffe, B.T. Use of a nondirect-product basis for treating singularities in triatomic rotational-vibrational calculations, Phys. Chem. Chem. Phys. 2007, 9, 3407–15.
150. Littlejohn, R.G., Cargo, M. Bessel discrete variable representation bases, J. Chem. Phys. 2002, 117, 27–36.
151. Dunn, K., Boggs, J.E., Pulay, P. Vibrational energy levels of methyl-fluoride, J. Chem. Phys. 1987, 86, 5088–93.
152. Leonard, C., Handy, N.C., Carter, S., Bowman, J.M. The vibrational levels of ammonia, Spectrochim. Acta 2002, 58A, 825–38.
153. Jung, J.O., Gerber, R.B. Vibrational wave functions and spectroscopy of $(H_2O_n, n = 2, 3, 4, 5)$: Vibrational self-consistent field with correlation corrections, J. Chem. Phys. 1996, 105, 10332–48.
154. Carter, S., Culik, S., Bowman, J.M. Vibrational self-consistent field method for many-mode systems: A new approach and application to the vibration of CO absorbed on Cu(100), J. Chem. Phys. 1997, 107, 10458–69.
155. Mátyus, E., Czakó, G., Sutcliffe, B.T. Császár, A.G. Vibrational energy levels with arbitrary potentials using the Eckart–Watson Hamiltonians and the discrete variable representation, J. Chem. Phys. 2007, in press.
156. Sutcliffe, B.T. Molecular Hamiltonians, in Wilson, S., editor. Handbook of Molecular Physics and Quantum Chemistry. Chichester: Wiley; 2003.

CHAPTER 10

The Effective Fragment Potential: A General Method for Predicting Intermolecular Interactions

Mark S. Gordon[*], Lyudmilla Slipchenko[*], Hui Li[**]
and Jan H. Jensen[***]

Contents		
	1. Introduction	177
	2. EFP2 Theory	178
	2.1 Contributing interaction terms	178
	2.2 Energy gradients and molecular dynamics	180
	2.3 Interface with continuum	181
	2.4 EFP-QM interactions	182
	2.5 EFP-QM across covalent bonds	182
	3. Example Applications	183
	3.1 Benzene dimer	183
	3.2 Benzene–water	186
	3.3 The prediction of the pK_a value of Lys55 using QM/MM	188
	4. Summary and Future Developments	190
	Acknowledgements	191
	References	191

1. INTRODUCTION

The interactions between molecules and molecular systems play key roles in many important phenomena in chemistry, the biological sciences, materials science and engineering, and chemical and mechanical engineering, among many other disciplines. Important examples include the structure and properties of weakly interacting clusters; the behavior and properties of liquids; solvent effects on ions,

[*] Department of Chemistry, Iowa State University and Ames Laboratory, Ames, IA 50011, USA
[**] Department of Chemistry, University of Nebraska, Lincoln, NE 68588, USA
[***] Department of Chemistry, Universitetsparken 5, University of Copenhagen, 2100 Copenhagen, Denmark

electrolytes, amino acids and other biomolecules, and the mechanisms of chemical reactions; the structures and structure-activity relationships of large molecules like polymers, proteins and enzymes; aggregation of polymers in solution; and the nature of interfacial (e.g., gas–liquid and liquid–solid) phenomena.

Because intermolecular interactions are so important, one needs theoretical methodologies that can accurately account for the broad range of such interactions. The most desirable theoretical approach would be to use a level of quantum mechanics (QM) that can treat both intermolecular and intramolecular (typically covalent) interactions with an acceptable level of accuracy. The range of intermolecular interactions summarized above will involve many different *types* of intermolecular forces, while intramolecular interactions will include the same forces as well as strong covalent bonds. Realistically, the minimal QM levels of theory that can adequately treat all of these phenomena are second order perturbation theory (MP2) [1] and (preferably) coupled cluster (CC) theory with some accounting of triples; i.e., CCSD(T) [2]. Unfortunately, a sufficiently high level of QM comes at a significant computational cost; for example, CCSD(T) scales $\sim N^7$ with problem size, where N is the number of atomic basis functions. This places serious limitations on the sizes of accessible molecular systems.

An alternative approach to the study of intermolecular interactions is to employ a model potential. Such potentials, broadly referred to as molecular mechanics (MM), can generally not account for bond-breaking, but can, in principle, account for the range of intermolecular interactions. If one is concerned with both intermolecular interactions and breaking chemical bonds, a combined QM/MM approach can be used [3]. Ideally, model potentials should be derived from first principles and should contain all of the essential underlying physics.

A particularly promising model potential is the effective fragment potential (EFP) that has been developed by the authors and many co-workers [4,5]. The EFP method was originally developed specifically to describe aqueous solvent effects on biomolecular systems and chemical reaction mechanisms. This EFP1 method contains fitted parameters for the repulsive term and, while very successful for its original purpose [6,7], it is difficult to extend beyond water. Therefore, for the last decade, the authors and co-workers have been developing a more general (EFP2) method that includes all of the essential physics and that has no empirically fitted parameters. The remainder of this work focuses exclusively on the EFP2 method and a few illustrative examples. For simplicity the method will henceforth be referred to as EFP.

2. EFP2 THEORY

2.1 Contributing interaction terms

All of the terms in the EFP method may be thought of as truncated expansions. At present, the EFP interaction energy is a sum of five terms:

$$E = E_{\text{coul}} + E_{\text{ind}} + E_{\text{exrep}} + E_{\text{disp}} + E_{\text{ct}}. \tag{1}$$

Equation (1) specifically refers to EFP-EFP interactions. EFP-QM interactions are discussed later. E_{coul} refers to the Coulomb portion of the electrostatic interaction. This term is obtained using the distributed multipolar expansion introduced by Stone, with the expansion carried out through octopoles. The expansion centers are taken to be the atom centers and the bond midpoints. So, for water, there are five expansion points (three at the atom centers and two at the O–H bond midpoints), while in benzene there are 24 expansion points. E_{ind} is the induction or polarization part of the electrostatic interaction. This term is represented by the interaction of the induced dipole on one fragment with the permanent dipole on another fragment, expressed in terms of the dipole polarizability. Although this is just the first term of the polarizability expansion, it is robust, because the molecular polarizability is expressed as a tensor sum of localized molecular orbital (LMO) polarizabilities. That is, the number of polarizability points is equal to the number of bonds and lone pairs in the molecule. This dipole–induced dipole term is iterated to self-consistency, so some many body effects are included.

The exchange repulsion E_{exrep} is derived as an expansion in the intermolecular overlap. When this overlap expansion is expressed in terms of frozen LMOs on each fragment, the expansion can reliably be truncated at the quadratic term [8]. This term does require that each EFP carries a basis set, and the smallest recommended basis set is 6-31++G(d,p) [9] for acceptable results. Since the basis set is used only to calculate overlap integrals, the computation is very fast and quite large basis sets are realistic. The dispersion interaction can be expressed as the familiar inverse R expansion,

$$E_{disp} = \sum_n C_n R^{-n}. \qquad (2)$$

The coefficients C_n may be derived from the (imaginary) frequency dependent polarizabilities summed over the entire frequency range [10,11]. If one employs only dipole polarizabilities the dispersion expansion is truncated at the leading term, with $n = 6$. In the current EFP code, an estimate is used for the $n = 8$ term, in addition to the explicitly derived $n = 6$ term. Rather than express a molecular C_6 as a sum over atomic interaction terms, the EFP dispersion is expressed in terms of LMO-LMO interactions. In order to ensure that the dispersion interaction goes to zero at short distances, the damping term proposed by Tang and Toennies [12] is employed.

The charge transfer interaction E_{ct} is derived by considering, using a supermolecule approach, the interactions between the occupied valence molecular orbitals on one fragment with the virtual orbitals on another fragment. This leads to significant energy lowering in *ab initio* calculations on ionic or highly polar species when incomplete basis sets are employed. An approximate formula [13] for the charge transfer interaction in the EFP2 method was derived and implemented using a second order perturbative treatment of the intermolecular interactions for a pair of molecules at the Hartree–Fock level of theory. The approximate formula is formulated in canonical orbitals from Hartree–Fock calculations of independent molecules and uses a multipolar expansion (through quadrupoles) of the molecular electrostatic potentials. Orthonormality between the virtual orbitals of the other molecule to all the orbitals of the considered molecule is enforced so the charge transfer is not contaminated with induction. Implemented in the EFP method, the approximate formula gives charge transfer energies comparable to those obtained from Hartree–Fock calculations. The analytic gradients of the charge transfer energy were also derived and implemented, enabling efficient geometry optimization and molecular dynamics simulations [13].

The two electrostatic terms discussed above, E_{coul} and E_{ind} must be modulated by damping, or screening, expressions. The Coulomb point multipole model breaks down when fragments approach too closely, since then the actual electron density on the two fragments is not well approximated by point multipoles. The latter interactions become too repulsive and must be moderated by a screening term [14]. On the other hand, the induction interaction becomes too attractive if fragments approach each other too closely, so a damping term is needed here as well. To avoid this unphysical behavior, the multipole electrostatic potential is augmented by exponential damping functions $f_{damp} = 1 - \exp(-\alpha R)$, with parameters α being determined at each multipole expansion point by fitting the multipole damped potential to the Hartree–Fock one. Damping terms in the electrostatic energy are derived explicitly from the damped potential and the charge density. The damping procedure can be extended to higher electrostatic terms, such as charge–dipole, dipole–dipole, etc. Charge–dipole, dipole–dipole, and dipole–quadrupole damping is applied to the polarization energy.

EFPs currently have internally fixed geometries. Analytic gradients for all terms have been derived and implemented, so full geometry optimizations and Monte Carlo and molecular dynamics simulations [15,16] can be performed. Because the method involves no empirically fitted parameters, an EFP for any system can be generated by a "makefp" run in the GAMESS suite of programs.

2.2 Energy gradients and molecular dynamics

Because the analytic gradients of all the interaction terms in the EFP2 method have been derived and implemented, molecular dynamics simulations can be performed [16]. The EFP electrostatic and dispersion interactions have relatively simple expressions, and the corresponding analytic energy gradients (forces and torques) can be derived in a straightforward manner. The exchange repulsion and charge transfer interactions are modeled with approximate formulas that employ MOs from SCF calculations on independent molecules. The MOs and their energies are the EFP parameters for these two interaction terms. The gradients are obtained by differentiating the corresponding energy expressions with respect to molecular rotational and translational displacements. The timings for the gradient evaluation for exchange repulsion and charge transfer are only 1–2 times those for the corresponding energy evaluation [17].

All of the EFP interaction terms are pairwise additive, except the induction (polarization) energy, which is modeled with asymmetric anisotropic polarizability tensors located at the centroids of the localized molecular orbitals. The analytic energy gradients (both forces and torques) for anisotropic polarizability tensors are derived via a direct differentiation approach in the form of matrix equations.

The forces and torques on the polarizability tensors can be evaluated with the induced dipoles (as if they were permanent) and the total electric fields at the polarizability tensors. Once the induced dipoles have been determined, the exact polarization energy gradients can be evaluated analytically at a very low cost [16].

Periodic boundary conditions (PBC) with the minimum image convention (MIC) [18] have been implemented for the EFP method by using the distances between the centers of masses of the (rigid) molecules as the inter-molecular distances [15]. To ensure energy conservation in molecular dynamics simulations, switching functions are applied to modify the intermolecular potential so that the interaction potential energies and forces for molecular pairs smoothly decrease to zero within the periodic cell [19]. The application of the MIC-PBC and switching functions can be rigorously applied to both the pairwise interaction terms and the EFP polarization energy. Using the MIC-PBC and a fifth-order polynomial switching function, very good energy conservation has been realized in molecular dynamics simulations with the EFP method [16].

2.3 Interface with continuum

Low-cost continuum models are often used to assess bulk solvation effects. The polarizable continuum models (PCM) [20] are continuum solvation models in which the solvent effects are described with induced surface charges. In a PCM calculation, the solutes can be modeled with *ab initio* methods or force fields, or both. In a combined QM/EFP/PCM calculation [21], the EFP induced dipoles and PCM induced charges are iterated to self-consistency as the QM wavefunction converges.

The EFP induced dipoles and PCM induced charges may be described by a supermatrix equation [22]:

$$\mathbf{B} \cdot \mathbf{w} = \mathbf{p}. \qquad (3)$$

The matrix p is a combined set of the external electrostatic fields that represent the effects of the QM field on the EFP polarizability tensors and the PCM potential, while w is a combined set of induced dipoles and surface charges. The physical meaning of the supermatrix equation (3) is that the EFP induced dipoles and PCM induced charges are uniquely determined by the external field and potential; therefore, the right hand side of Eq. (3) involves only the external field/potential, and the left side involves only the induced EFP dipoles and PCM charges. The interactions among the induced dipoles and charges are implicitly described with the matrix B. The supermatrix Eq. (3) can be solved either with direct inversion or various iterative methods.

The gradients (both forces and torques) of the polarization energy in a combined EFP/PCM calculation have been derived and implemented [22]. It is found that all of the energy gradient terms can be formulated as simple electrostatic forces and torques on the induced dipoles and charges as if they were permanent static dipoles and charges, in accordance with the electrostatic nature of these models. Geometry optimizations can be performed efficiently with the analytic gradients of the EFP + PCM polarization energies. Due to the intrinsic discontinuity in the molecular surface tessellation in the PCM method, the gradients are not strictly continuous. Although this non-continuity rarely affects geometry optimizations, it prevents good energy conservation in molecular dynamics simulations.

2.4 EFP-QM interactions

The discussion in the foregoing sections has focused primarily on interactions among fragments. In order to have a general method that is able to treat solvent effects on chemical reactions, the analogous EFP-QM interactions are required. The first two interaction terms in Eq. (1), E_{coul} and E_{ind}, have already been developed for the EFP-QM interface [4,5], including energy gradients. A general expression for the QM-EFP exchange repulsion interaction, E_{exch}, has been derived and coded [23], and the corresponding expressions for the energy gradients are in progress [24]. Once these gradients have been implemented, one will have an EFP-QM interface at a level of theory that is comparable to Hartree–Fock. The derivation of the most important remaining component, the dispersion interaction E_{disp}, and its energy gradient has been completed, and the implementation is in progress [25]. This will provide a correlated EFP-QM interface.

2.5 EFP-QM across covalent bonds

Since the fragments are represented by model potentials (EFPs), the method may be considered to be in the general category of QM/MM (quantum mechanics/molecular mechanics) methods. In other contexts QM/MM methods have also been very useful for describing extended systems in which the QM and MM regions are separated by covalent bonds rather than weak intermolecular forces. To make the link between the *ab initio* and MM portions a covalently bonded *ab initio*/EFP interface has been developed [26] and implemented in GAMESS [27]. The method is similar in spirit to that of Assfeld and Rivail [28]. The essential features of the approach are as follows:

(1) A buffer region consisting of several LMOs, typically surrounding the α-carbon of a given side-chain, is defined as the *ab initio*/EFP boundary. Once the buffer region is defined, these LMOs are obtained by an *ab initio* calculation on all or a subset of the system, projected onto the buffer atom basis functions [29]. These LMOs are subsequently frozen in the EFP calculations by setting select MO Fock matrix elements to zero [30,31]. The *ab initio*/buffer region interactions are calculated by including the exact quantum mechanical Coulomb and exchange operators corresponding to the charge distribution of the buffer region, in the *ab initio* Hamiltonian. This requires calculation of two-electron integrals over basis functions in the buffer region. Since the buffer MOs are frozen, the changes in induction (polarization) contributions from the buffer region are neglected during a geometry optimization of the *ab initio* region. The effect of this approximation on the chemical reaction of interest can be systematically reduced by increasing the size of the *ab initio* region.

(2) Variational collapse of the *ab initio* wavefunction into the buffer region is avoided by keeping the *ab initio* MOs orthogonal to the buffer LMOs by Schmidt orthogonalization. This is an approximation relative to a full *ab initio* calculation because the MOs are allowed to build up "orthogonality tails" only in the buffer region, not in the EFP region. The associated error can again be systematically reduced by increasing the size of the buffer region.

(3) The remaining part of the system (or within a defined radius of the active region) is represented by an EFP. The presence of the buffer region provides sufficient separation between the EFP and the *ab initio* regions to ensure that the remaining interactions can be treated as non-bonded interactions via the EFP terms presented above.

3. EXAMPLE APPLICATIONS

3.1 Benzene dimer

The benzene dimer, a prototype for $\pi-\pi$ interactions, has attracted extensive theoretical and experimental attention [32–39]. $\pi-\pi$ interactions govern structures of proteins and DNA, self-assembly of aromatic macromolecules, and drug-intercalation into DNA. Combined theoretical and experimental studies suggest that there are two minima on the potential energy surface of the benzene dimer: The perpendicular T-shaped and parallel-slipped configurations; the transition state sandwich structure is highest in energy (Figure 10.1). A rotational experiment by Arunan and Gutowsky [35] determined a 4.96 Å separation between the benzene centers of mass in the T-shaped configuration. The binding energy of the dimer was determined to be $D_0 = 1.6 \pm 0.2$ kcal/mol by Krause et al. [37] and as 2.4 ± 0.4 kcal/mol by Grover et al. [36].

Accurate *ab initio* calculations for the benzene dimer require using both an extensive basis set with diffuse functions and a high level of dynamic correlation. Recently, several independent studies were devoted to theoretical investigation of the potential energy surface of the benzene dimer [33,34,38–40]. This work closely follows the analysis presented by Sherrill and co-workers [33,34]. They estimated

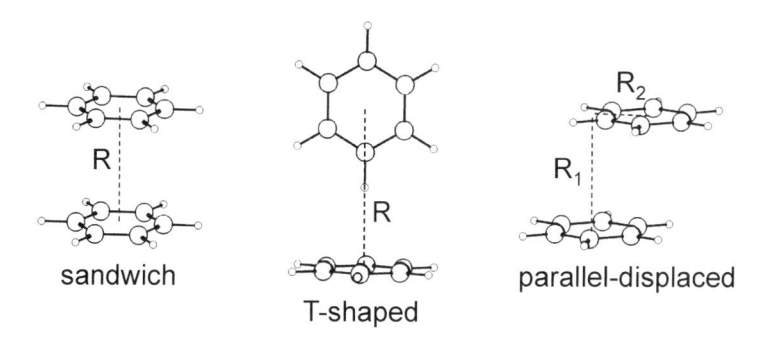

FIGURE 10.1 Sandwich, T-shaped, and parallel-displaced configurations of the benzene dimer.

sandwich

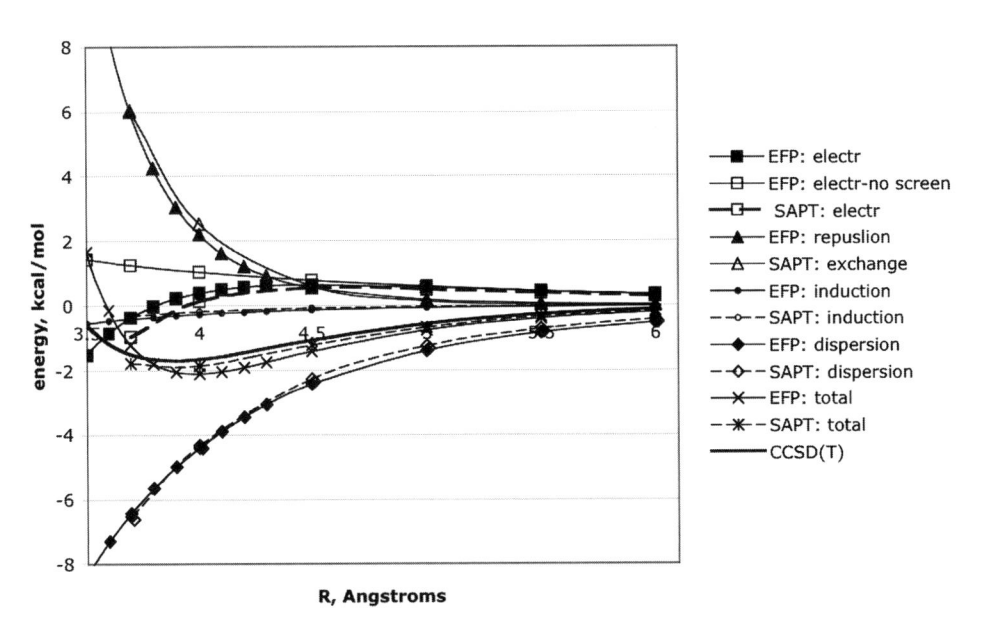

FIGURE 10.2 Comparison of the EFP and SAPT energy components (kcal/mol) in the sandwich benzene dimer. SAPT and CCSD(T) energies from Ref. [33].

potential energy curves for the dimer by the coupled-cluster method including singles and doubles with perturbative triples corrections [CCSD(T)] [2], using the aug-cc-pVQZ basis set. They also analyzed the nature of the $\pi-\pi$ interactions by using symmetry-adapted perturbation theory (SAPT) [41]. To analyze the quality of the EFP results for the benzene dimer, the EFP and SAPT/aug-cc-pVDZ potential curves for each energy term are compared separately, as well as the EFP and CCSD(T) total binding energy curves.

The EFP potential for benzene was constructed using the 6-311++G(3df,2p) basis set, at the MP2/aug-cc-pVTZ [42] monomer geometry from Ref. [34]. Multipoles were generated using a numerical DMA, with high-order electrostatic screening [14].

Figures 10.2 and 10.3 present comparisons of the EFP and SAPT results for electrostatic, exchange-repulsion, polarization, and dispersion terms. The total EFP, SAPT, and CCSD(T) binding energies are also plotted. Figure 10.2 shows the potential energy curves for the sandwich configuration and Figure 10.3 gives the T-shaped curves. The equilibrium inter-monomer distances (defined in Figure 10.1) in the benzene dimer vary from $R = 3.7$ Å to 4.0 Å in the sandwich and from $R = 4.9$ Å to 5.1 Å in the T-shaped configurations depending on the level of theory and basis set used. The CCSD(T)/aug-cc-pVQZ values are $R = 3.9$ Å and $R = 5.0$ Å, respectively. These intermolecular separations are used as the reference values in the following discussion. Note that the sandwich and T-shaped configurations used here are not minima on the potential energy surface of the dimer;

T-shaped

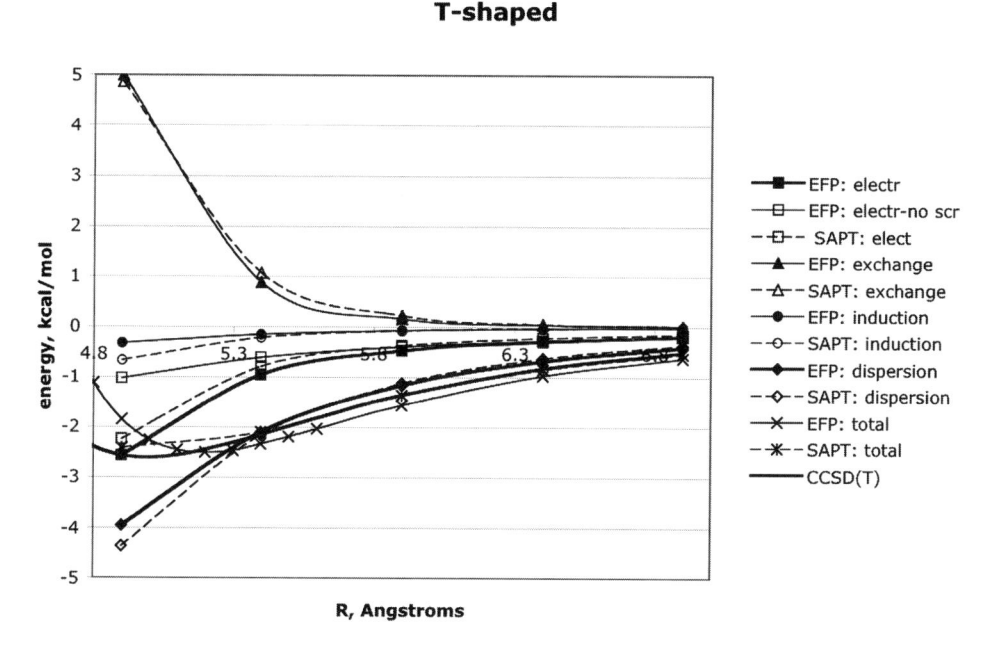

FIGURE 10.3 Comparison of the EFP and SAPT energy components (kcal/mol) in the T-shaped benzene dimer. SAPT and CCSD(T) energies from Ref. [33].

these structures were chosen as representative ones. In the real T-shaped-like minimum, the "upper" benzene is tilted, so that one C–C bond is almost parallel to the plane of the "lower" benzene (see orientation of the benzenes in Figure 10.1). The sandwich structure is a transition state between different parallel-displaced configurations [39].

The electrostatic curves are plotted both with and without damping. The EFP curves without damping underestimate the electrostatic interaction in the equilibrium region by >1 kcal/mol. The damping correction accounts for most of the charge-penetration energy, so that the damped EFP energies differ from the SAPT values by 0.2–0.3 kcal/mol at the equilibrium geometries. Electrostatic damping is very important for a system like the benzene dimer, because both equilibrium geometries and binding energies would be significantly in error without the damping term.

For each term in the interaction energy, EFP is in excellent agreement with SAPT, generally within less than 0.5 kcal/mol and often much better. Overall, EFP over-binds the sandwich dimer by about 0.4 kcal/mol and under-binds the T-shaped structure by 0.1 kcal/mol, as compared to CCSD(T). The equilibrium intermolecular separations calculated by EFP are 0.1–0.2 Å longer than those calculated by CCSD(T).

Table 10.1 summarizes the interaction energies of the three structures of the dimer calculated by MP2, CCSD(T), and EFP. Relative to CCSD(T) with the same basis set, MP2 underestimates the equilibrium distances by 0.1–0.2 Å and over-

TABLE 10.1 Equilibrium geometries (angstroms) and interaction energies (kcal/mol) for different configurations of the benzene dimer

Method	Basis	Sandwich		T-shaped		Parallel-displaced		
		R	Energy	R	Energy	R_1	R_2	Energy
MP2[a]	aug-cc-pVDZ[b]	3.8	−2.83	5.0	−3.00	3.4	1.6	−4.12
	aug-cc-pVTZ	3.7	−3.25	4.9	−3.44	3.4	1.6	−4.65
	aug-cc-pVQZ[b]	3.7	−3.35	4.9	−3.48	3.4	1.6	−4.73
CCSD(T)[a]	aug-cc-pVDZ[b]	4.0	−1.33	5.1	−2.24	3.6	1.8	−2.22
	aug-cc-pVQZ[b]	3.9	−1.70	5.0	−2.61	3.6	1.6	−2.63
EFP	6-311++G(3df,2p)	4.0	−2.11	5.2	−2.50	3.8	1.2	−2.34

[a] Reference [33].
[b] Basis sets as described in Ref. [33].

estimates the binding energies by 0.7–2.1 kcal/mol. The best agreement between MP2 and CCSD(T) is for the T-shaped structure, while the worst is for the parallel-displaced configuration. EFP overestimates the inter-monomer separations by 0.1–0.2 Å, and inaccuracies in the interaction energies are 0.1–0.4 kcal/mol. In general, the agreement between the EFP and CCSD(T) methods is very reasonable, and EFP is in better agreement with CCSD(T) than is MP2. This is striking in view of the orders of magnitude less computer time required by EFP. For example, a single-point energy calculation in the 6-311++G(3df,2p) basis set (660 basis functions) by MP2 requires 142 minutes of CPU time on one IBM Power5 processor, whereas the analogous EFP calculation requires only 0.4 seconds.

3.2 Benzene–water

Interactions of aromatic molecules with solvent are of fundamental interest, since these interactions are common in bio-systems. The simplest systems of this type, small benzene-water complexes, have attracted both experimental and theoretical attention [43–53]. Zwier and coworkers [45–47] have presented accurate IR data on (benzene)$_{1-2}$ (water)$_{1-8}$ complexes. Accurate assignments would provide unambiguous insight on the structures of these clusters. However, accurate theoretical investigation of these clusters is still very challenging. Both an extensive basis set with diffuse functions and a high level of dynamic correlation are required for an accurate treatment of these systems [43]. Moreover, binding in water–benzene complexes is complicated, since both electrostatic interactions, for hydrogen-bonded water and benzene $\pi-\pi$ interactions play a role. Thus, an accurate analysis of these systems requires a balanced description of the different types of intermolecular interactions.

In this example application, the predictions of EFP and accurate *ab initio* methods are compared with experimental values for the water–benzene dimer, with a focus on the question, what is the nature of the water–benzene interaction?

TABLE 10.2 Intermolecular distances (angstroms) and binding energies (kcal/mol) in the water–benzene dimer

Method	Ref.	R_e[a]	D_e[b]	D_0
MP2/aVTZ	Feller [43]	3.21	−4.01 (−3.13)	−3.01 (−2.13)[c]
CCSD(T)/aVTZ	Feller [43]		−3.85	−2.85[c]
Est. MP2/CBS	Feller [43]		−3.9 ± 0.2	−2.9 ± 0.2[c]
EFP	This work	3.38	−3.90	−2.87
Expt.	Gotch, Zwier [45]	3.32		−1.63 to −2.78
	Suzuki et al. [52]	3.35		
	Gutowsky et al. [51]	3.33		
	Cheng et al. [49]			2.25 ± 0.28
	Courty et al. [50]			2.44 ± 0.09

a Distance between the water and benzene centers of mass.
b Values in parentheses correspond to counterpoise-corrected (CP) binding energies.
c Using estimated zero-point vibrational energy from Ref. [43].

EFP parameters for water and benzene were obtained with the 6-311++G(3df,2p) basis set. The benzene EFP is the same as that discussed above. Electrostatic parameters for water were obtained with a numerical DMA, and electrostatic interactions were screened by the high-order electrostatic damping functions. Additionally, for both benzene and water, polarization interactions were screened as well, with damping parameters at all centers being set to 1.5.

Table 10.2 summarizes the experimental and theoretical intermolecular distances and binding energies for the lowest energy structure of the water-benzene dimer. Experimental values of the intermolecular distances (determined as the distance between the centers of mass in water and benzene) are 3.32–3.35 Å, so MP2 underestimates these values by more than 0.1 Å, whereas EFP overestimates them by about 0.05 Å. There is significant disagreement in the measured binding energies in the benzene dimer, e.g., the binding energy was determined to be 1.63–2.78 kcal/mol in Ref. [45], 2.25 ± 0.28 kcal/mol in Ref. [49] and 2.44 ± 0.09 kcal/mol in Ref. [50]. Theoretical binding energies dramatically depend on the method and basis set used.

Even though there is no accurate estimate of the CCSD(T)/complete basis set (CBS) binding energy in the literature, it seems that MP2 generally overestimates the CCSD(T) binding by about 0.2 kcal/mol. This results in an estimated 2.7 kcal/mol CCSD(T) binding energy, in reasonable agreement with the available experimental data. The EFP binding energy in the water-benzene dimer is 3.9 kcal/mol, which, when combined with the EFP ZPE value of 1.03 kcal/mol, results in a net binding of 2.87 kcal/mol. This agrees very well with the MP2/CBS limit and over-binds the experimental [CCSD(T)] values by about 0.5 [0.2] kcal/mol.

Figure 10.4 compares binding in the water dimer and benzene dimers with that in the water-benzene dimer. The energy components were calculated by EFP

	electrost.	exch.-repulsion	induction	dispers.	total binding	total+ZPE
	-8.59	5.33	-0.99	-0.85	-5.10	-2.57
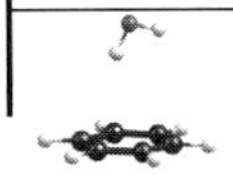	-3.85	2.32	-0.57	-1.81	-3.90	-2.87
	-3.91	3.21	-0.43	-1.44	-2.58	-1.59
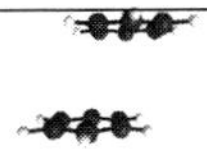	-0.10	2.86	-0.27	-4.85	-2.38	-2.02
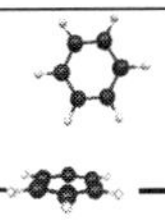	-1.82	2.39	-0.21	-3.21	-2.86	-2.44

FIGURE 10.4 Binding in water dimer, benzene dimer, and water–benzene dimer by EFP. All values are in kcal/mol.

at the EFP equilibrium geometries for each dimer. It is well known that the dominant contribution to binding in the water dimer is the electrostatic interaction (−8.6 kcal/mol), whereas the polarization and dispersion interactions are almost 10 times weaker. Contrarily, binding in the parallel-displaced benzene dimer is dominated by dispersion forces (−4.9 kcal/mol). The T-shaped benzene dimer has significant contributions from both dispersion and electrostatic forces, and, not surprisingly, this is also true for the two structures of the benzene-water dimer. It is also educational to compare the total binding energies in the dimers shown in Figure 10.4. The water dimer is the most strongly bound, the benzene dimers have the weakest interaction energies, but the water–benzene dimer is in between. However, including ZPEs makes the situation less obvious. The ZPE values are largest in the water dimer and the smallest in the benzene dimers, benzene–water dimers again being in the middle. This results in a striking observation that the ZPE-corrected binding energies of the dimers are much less spread energetically. Thus, the immiscibility of benzene in water is due to an unfavorable entropy, rather than enthalpy, contribution.

3.3 The prediction of the pK_a value of Lys55 using QM/MM

The approach taken in this work to predict the pK_a of an amine group in the protein, HA^+, is related to the standard free energy change, ΔG, due to proton transfer to a reference compound (using Lys residues as an example),

$$pK_a = 10.60 + \{[G(A) - G(HA^+)] - [G(CH_3NH_2) - G(CH_3NH_3^+)]\}/1.36. \quad (4)$$

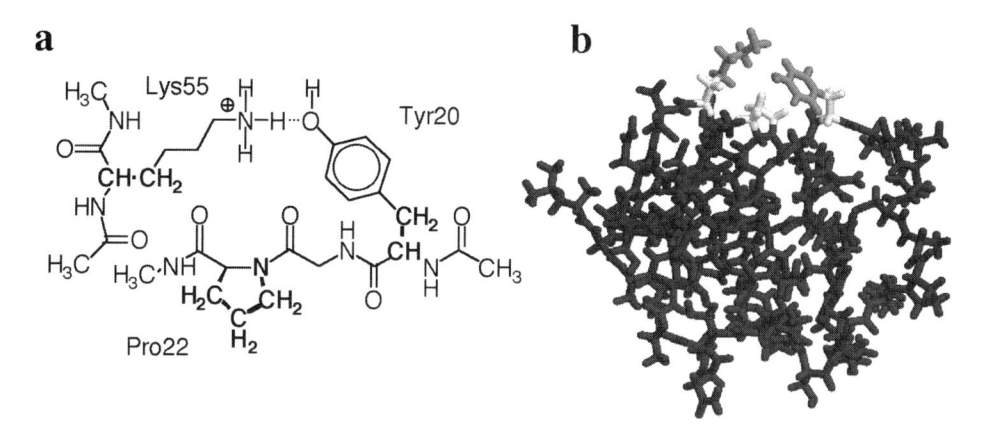

FIGURE 10.5 (a) Subsystem of OMTKY3 used to obtain the buffer region (bold) used for (b) ab initio/buffer/EFP regions (red/blue/green) used for the computation of the pK_a of Lys55.

Here, 10.60 is the experimentally determined pK_a of methylamine at 298 K [54] and 1.36 is $RT \ln 10$ for $T = 298$ K in kcal/mol. $G(X)$ is the total free energy (in kcal/mol) of molecule X, which is the sum of the ground state electronic energy (E_{ele}), thermochemical energy (G_{trv}), and solvation energy (G_{sol}):

$$G = E_{ele} + G_{trv} + G_{sol}. \tag{5}$$

The solution structure of OMTKY3 has been determined using NMR by Hoogstraten et al. [55], and was obtained from the Protein Data Bank (entry 1OMU). The first of the 50 conformers is used without further refinement of the overall structure.

The electronic and geometric structures of the Lys55 and Tyr20 side chains are treated quantum mechanically at the MP2/6-31+G(2d,p)//RHF/6-31G(d) level of theory (Figure 10.5), while the rest of the protein is treated with an EFP, described in more detail below. The use of the diffuse functions on atoms near the buffer region causes SCF convergence problems due to couplings with the induced dipoles in the EFP region, so the 6-31+G(2d,p) basis set was used only for the $C^\delta H_2 C^\varepsilon H_2 NH_3 \cdots HO-C^\xi (C^{\varepsilon 1,2} H)_2$ atoms in the MP2 calculation.

The *ab initio* region is separated from the protein EFP by a buffer region [26] comprised of frozen LMOs corresponding to all the bond LMOs connecting the **bold** atoms in Figure 10.5, as well as the core and lone pair LMOs belonging to those atoms. The Pro22 buffer is needed to describe its short-range interactions with Tyr20 [56]. The buffer LMOs are generated by an RHF/6-31G(d) calculation on a subset of the system (shown in Figure 10.5), projected onto the buffer atom basis functions, and subsequently frozen in the EFP calculations by setting select off-diagonal MO Fock matrix elements to zero. The *ab initio*/buffer region interactions are calculated *ab initio*, and thus include short-range interactions.

The EFP describing the rest of the protein is generated by nine separate *ab initio* calculations on overlapping pieces of the protein truncated by methyl groups. Two different regions of overlap are used depending on whether it occurs on the protein backbone or on a disulfide bridge, as described in Ref. [56]. The electrostatic potential of each protein piece is expanded in terms of multipoles through octupoles centered at all atomic and bond midpoint centers using Stone's Distributed Multipole Analysis [57]. The monopoles of the entire EFP are scaled to ensure a net integer charge and the dipole polarizability tensor due to each LMO in the EFP region is calculated by a perturbation expression, as described in Ref. [56].

The vibrational free energy (G^{vib}) of the optimized part of the *ab initio* region is calculated by the Partial Hessian Vibrational Analysis (PHVA) method [58]. The solvation energy (ΔG_s) is calculated with the EFP/PCM interface developed by Bandyopadyay et al. [59] for small molecules and extended to macromolecules by Li, Pomelli, and Jensen [60,61].

The pK_a of Lys55 computed using this approach is 11.4 pH units, in good agreement with the experimental value [62] of 11.1 considering the uncertainty in the experimental values is roughly ±0.1 pH units. A more thorough summary of this work can be found in reference [63], and further application of this general approach to other residue types and proteins can be found in [64–67].

4. SUMMARY AND FUTURE DEVELOPMENTS

As illustrated in this work, the effective fragment potential is an accurate method for treating the broad range of intermolecular interactions, at a small fraction of the cost of *ab initio* calculations that produce comparable accuracy. Because no empirically fitted parameters are required, an EFP can easily be generated automatically for any closed shell species simply by running the appropriate GAMESS calculation on the isolated molecule. The long-term goal of the method is to use it both as a stand-alone method to study intermolecular interactions and as an interface with electronic structure methods to provide a sophisticated QM/MM approach to such phenomena as solvent effects on chemical reactions and solvent-induced spectroscopic shifts. To attain this goal, several new features are in progress.

While the leading R^{-6} term in the dispersion expansion generally accounts for the largest part of the dispersion interaction, higher order terms are clearly needed when polar or ionic species are present in the system of interest. The current method has an estimate for the R^{-8} term[11], but a more rigorous accounting for this and other terms is needed. At present, an EFP can only be generated for closed shell species, but one can imagine many instances in which an EFP for an open shell compound (e.g., radical) would be useful. A preliminary open shell EFP version is nearly completed and will be made available shortly [24]. At present, the geometry of an EFP is internally frozen. This is not unreasonable for simple molecules like water, but if one wishes to use EFPs to study biomolecules or polymers, it is desirable to at least allow the molecule to relax along torsional coordinates.

As discussed in section 2.4, several new developments are in progress in order to make the EFP-QM interface fully viable. In addition, EFP interfaces are being built with methods that can treat excited electronic states. In addition to the existing MCSCF interface, these include CI singles and time-dependent density functional theory in the short term and more sophisticated CI and coupled cluster methods in the longer term.

ACKNOWLEDGEMENTS

The development of the effective fragment potential method has been supported by both the US Air Force Office of Scientific Research and by the US Department of Energy through its Scientific Discovery through Advanced Computing (SciDAC) initiative. JHJ acknowledges past and present support from the US National Science Foundation (MCB 209941) and a Skou Fellowship to JHJ from the Danish Research Agency (Forskningsrådet for Natur og Univers).

REFERENCES

1. Møller, C., Plesset, M.S. Phys. Rev. 1934, 46, 618;
   Pople, J.A., Binkley, J.S., Seeger, R. Int. J. Quantum Chem. 1976, S10, 1.
2. Paldus, J., in Wilson, S., editor. Handbook of Molecular Physics and Quantum Chemistry, vol. 2. Chichester: Wiley; 2003, p. 272–313.
3. Friesner, R.A., Guallar, V. Ann. Rev. Phys. Chem. 2005, 56, 389.
4. Day, P.N., Jensen, J.H., Gordon, M.S., Webb, S.P., Stevens, W.J., Krauss, M., Garmer, D., Basch, H., Cohen, D. J. Chem. Phys. 1996, 105, 1968.
5. Gordon, M.S., Freitag, M.A., Bandyopadhyay, P., Kairys, V., Jensen, J.H., Stevens, W.J. J. Phys. Chem. 2001, 105, 293.
6. Webb, S.P., Gordon, M.S. J. Phys. Chem. 1999, A103, 1265.
7. Adamovic, I., Gordon, M.S. J. Phys. Chem. 2005, 109, 1629.
8. Jensen, J.H., Gordon, M.S. Mol. Phys. 1996, 89, 1313.
9. Hehre, W.J., Ditchfield, R., Pople, J.A. J. Chem. Phys. 1972, 56, 2257.
10. Amos, R.D., Handy, N.C., Knowles, P.J., Rice, J.E., Stone, A.J. J. Phys. Chem. 1985, 89, 2186.
11. Adamovic, I., Gordon, M.S. Mol. Phys. 2005, 103, 379.
12. Tang, K.T., Toennies, J.P. J. Chem. Phys. 1984, 80, 3726.
13. Li, H., Gordon, M.S., Jensen, J.H. J. Chem. Phys. 2006, 124, 214107.
14. Slipchenko, L., Gordon, M.S. J. Comp. Chem. 2006, 28, 276.
15. Netzloff, H., Gordon, M.S. J. Chem. Phys. 2004, 121, 2711.
16. Li, H., Netzloff, H.M., Gordon, M.S. J. Chem. Phys. 2006, 125, 194103.
17. Li, H., Gordon, M.S. Theor. Chem. Accts. 2006, 115, 385.
18. Metropolis, N., Rosenbluth, A.W., Rosenbluth, M.N., Teller, A.H., Teller, E. J. Chem. Phys. 1953, 21, 1087.
19. Leach, A.R. Molecular Modelling—Principles and Applications, 2nd edn. Harlow: Prentice Hall; 2001.
20. Cances, E., Mennucci, B., Tomasi, J. J. Chem. Phys. 1997, 107, 3032;
    Barone, V., Cossi, M. J. Phys. Chem. A 1998, 102, 1995.
21. Bandyopadhyay, P., Mennucci, B., Tomasi, J., Gordon, M.S. J. Chem. Phys. 2002, 116, 5023.
22. Li, H., Gordon, M.S. J. Chem. Phys. 2007, 126, 124112.
23. Jensen, J.H. J. Chem. Phys. 2001, 114, 8775.
24. Kemp, D., Gordon, M.S., in preparation.
25. Smith, T., Gordon, M.S., in preparation.
26. Kairys, V., Jensen, J.H. J. Phys. Chem. A 2000, 104, 6656.
27. Schmidt, M.W., Baldridge, K.K., Boatz, J.A., Elbert, S.T., Gordon, M.S., Jensen, J.H., Koseki, S., Matsunaga, N., Nguyen, K.A., Su, S., Windus, T.L., Dupuis, M., Montgomery Jr., J.A. J. Comput. Chem. 1993, 14, 1347;
    Gordon, M.S., Schmidt, M.W., in Dykstra, C.E., Frenking, G., Kim, K.S., Scuseria, G.E., editors. Theory and Applications of Computational Chemistry. Elsevier, 2005. Ch. 41.
28. Assfeld, X., Rivail, J.L. Chem. Phys. Lett. 1996, 263, 100.
29. King, H.F., Stanton, R.E., Kim, H., Wyatt, R.E., Parr, R.G. J. Chem. Phys. 1967, 47, 1936.
30. Bagus, P.S., Hermann, K., Bauschlicher, C.W. J. Chem. Phys. 1984, 80, 4378.

31. Stevens, W.J., Fink, W.H. Chem. Phys. Lett. 1987, 139, 15.
32. Hobza, P., Selzle, H.L., Schlag, E.W. J. Am. Chem. Soc. 1994, 116(8), 3500;
    Hobza, P., Selzle, H.L., Schlag, E.W. J. Phys. Chem. 1996, 100(48), 18790;
    Jaffe, R.L., Smith, G.D. J. Chem. Phys. 1996, 105(7), 2780;
    Park, Y.C., Lee, J.S. J. Phys. Chem. A 2006, 110, 5091;
    Spirko, V., Engvist, O., Soldan, P., Selzle, H.L., Schlag, E.W., Hobza, P. J. Chem. Phys. 1999, 111(2), 572;
    Tsuzuki, S., Honda, K., Uchimaru, T., Mikami, M., Tanabe, K. J. Am. Chem. Soc. 2002, 124(1), 104;
    Tsuzuki, S., Uchimaru, T., Sugawara, K., Mikami, M. J. Chem. Phys. 2002, 117(24), 11216;
    Bornsen, K.O., Selzle, H.L., Schlag, E.W. J. Chem. Phys. 1986, 85(4), 1726;
    Felker, P.M., Maxton, P.M., Schaeffer, M.W. Chem. Rev. 1994, 94(7), 1787;
    Janda, K.C., Hemminger, J.C., Winn, J.S., Novick, S.E., Harris, S.J., Klemperer, W. J. Chem. Phys. 1975, 63(4), 1419;
    Law, K.S., Schauer, M., Bernstein, E.R. J. Chem. Phys. 1984, 81(11), 4871;
    Scherzer, W., Kratzschmar, O., Selzle, H.L., Schlag, E.W. Zeitschrift Fur Naturforschung Section a-a Journal of Physical Sciences 1992, 47(12), 1248;
    Steed, J.M., Dixon, T.A., Klemperer, W. Journal of Chemical Physics 1979, 70(11), 4940;
    Puzder, A., Dion, M., Langreth, D.C. J. Chem. Phys. 2006, 124(16);
    Sato, T., Tsuneda, T., Hirao, K. J. of Chem. Phys. 2005, 123(10).
33. Sinnokrot, M.O., Sherrill, C.D. J. Phys. Chem. A 2004, 108(46), 10200.
34. Sinnokrot, M.O., Valeev, E.F., Sherrill, C.D. J. Am. Chem. Soc. 2002, 124(36), 10887.
35. Arunan, E., Gutowsky, H.S. J. Chem. Phys. 1993, 98(5), 4294.
36. Grover, J.R., Walters, E.A., Hui, E.T. J. Phys. Chem. 1987, 91(12), 3233.
37. Krause, H., Ernstberger, B., Neusser, H.J. Chem. Phys. Lett. 1991, 184(5–6), 411.
38. Jung, Y., Head-Gordon, M. Phys. Chem. Chem. Phys. 2006, 8(24), 2831.
39. Podeszwa, R., Bukowski, R., Szalewicz, K. J. Phys. Chem. A 2006, 110(34), 10345.
40. Park, Y.C., Lee, J.S. J. Phys. Chem. A 2006, 110(15), 5091;
    Piecuch, P., Kucharski, S.A., Bartlett, R.J. J. Chem. Phys. 1999, 110(13), 6103.
41. Jeziorski, B., Moszynski, R., Szalewicz, K. Chem. Rev. 1994, 94(7), 1887.
42. Dunning, T.H. J. Chem. Phys. 1989, 90(2), 1007;
    Kendall, R.A., Dunning, T.H., Harrison, R.J. J. Chem. Phys. 1992, 96(9), 6796.
43. Feller, D. J. Phys. Chem. A 1999, 103(38), 7558.
44. Fredericks, S.Y., Pedulla, J.M., Jordan, K.D., Zwier, T.S. Theor. Chem. Accts. 1997, 96(1), 51.
45. Gotch, A.J., Zwier, T.S. J. Chem. Phys. 1992, 96(5), 3388.
46. Gruenloh, C.J., Carney, J.R., Arrington, C.A., Zwier, T.S., Fredericks, S.Y., Jordan, K.D. Science 1997, 276(5319), 1678;
    Gruenloh, C.J., Carney, J.R., Hagemeister, F.C., Arrington, C.A., Zwier, T.S., Fredericks, S.Y., Wood, J.T., Jordan, K.D. J. Chem. Phys. 1998, 109(16), 6601;
    Pribble, R.N., Garrett, A.W., Haber, K., Zwier, T.S. J. Chem. Phys. 1995, 103(2), 531;
    Pribble, R.N., Zwier, T.S. Faraday Discussions 1994, 97, 229.
47. Pribble, R.N., Zwier, T.S. Science 1994, 265(5168), 75.
48. Tarakeshwar, P., Choi, H.S., Lee, S.J., Lee, J.Y., Kim, K.S., Ha, T.K., Jang, J.H., Lee, J.G., Lee, H. J. Chem. Phys. 1999, 111(13), 5838;
    Tsuzuki, S., Honda, K., Uchimaru, T., Mikami, M., Tanabe, K. J. Am. Chem. Soc. 2000, 122(46), 11450;
    Baron, M., Kowalewski, V.J. J. Phys. Chem. A 2006, 110(22), 7122;
    Jedlovszky, P., Keresztúri, A., Horvai, G. Faraday Discussions 2005, 129, 35.
49. Cheng, B.M., Grover, J.R., Walters, E.A. Chem. Phys. Lett. 1995, 232(4), 364.
50. Courty, A., Mons, M., Dimicoli, N., Piuzzi, F., Gaigeot, M.P., Brenner, V., de Pujo, P., Millie, P. J. Phys. Chem. A 1998, 102(33), 6590.
51. Gutowsky, H.S., Emilsson, T., Arunan, E. J. Chem. Phys. 1993, 99(7), 4883.
52. Suzuki, S., Green, P.G., Bumgarner, R.E., Dasgupta, S., Goddard, W.A., Blake, G.A. Science 1992, 257(5072), 942.
53. Day, P.N., Pachter, R., Gordon, M.S., Merrill, G.N. J. Chem. Phys. 2000, 112(5), 2063.
54. CRC Handbook of Chemistry and Physics, 83 edn. Cleveland, OH: CRC Press; 2003.
55. Hoogstraten, C.G., Choe, S., Westler, W.M., Markley, J.L. Protein Science 1995, 4, 2289.

56. Minikis, R.M., Kairys, V., Jensen, J.H. J. Phys. Chem. A 2001, 105, 3829.
57. Stone, A.J., Price, S.L. J. Phys. Chem. 1988, 92, 3325.
58. Li, H., Jensen, J.H. Theor. Chem. Accts. 2002, 107, 211.
59. Bandyopadhyay, P., Gordon, M.S., Mennucci, B., Tomasi, J. J. Chem. Phys. 2002, 116, 5023.
60. Li, H., Pomelli, C.S., Jensen, J.H. Theor. Chem. Accts. 2003, 109, 71.
61. Li, H., Jensen, J.H. J. Comp. Chem. 2004, 25, 1449.
62. Forsyth, W.R., Gilson, M.L., Antosiewicz, J., Jaren, O.R., Robertson, A.D. Biochem. 1998, 37, 8643.
63. Jensen, J.H., Li, H., Robertson, A.D., Molina, P.A. J. Phys. Chem. A 2005, 109, 6634.
64. Li, H., Robertson, A.D., Jensen, J.K.H. Proteins-Structure Function and Bioinformatics 2004, 55, 689.
65. Naor, M.M., Jensen, J.H. Proteins-Structure Function and Bioinformatics 2004, 57, 799.
66. Porter, M.A., Hall, J.R., Locke, J.C., Jensen, J.H., Molina, P.A. Proteins-Structure Function and Bioinformatics 2006, 63, 621.
67. Wang, P.F., Flynn, A.J., Naor, M.M., Jensen, J.H., Cui, G.L., Merz, K.M., Kenyon, G.L., McLeish, M.J. Biochem. 2006, 45, 11464.

CHAPTER 11

Gaussian Basis Sets Exhibiting Systematic Convergence to the Complete Basis Set Limit

Kirk A. Peterson[*]

Contents
1. Introduction 195
2. Correlation Consistent Basis Sets: A Review 196
   2.1 Background 196
   2.2 Diffuse-augmented basis sets 198
   2.3 Core-valence basis sets 198
   2.4 The alkali and alkaline earth metals 199
3. Recent Advances in Correlation Consistent Basis Sets 200
   3.1 Heavy main group elements 200
   3.2 Transition metal basis sets 201
   3.3 Core correlation basis sets for the post-3d elements 203
4. Conclusions 203
Acknowledgements 203
References 203

1. INTRODUCTION

In the solution of the electronic Schrödinger equation via wavefunction-based methods, there are two major sources of error that must be considered. One is the expansion of the many-electron wavefunction in terms of Slater determinants (the "method") and the other is the representation of the 1-particle orbitals by a suitable basis set, typically consisting of Gaussian-type functions, from which these determinants are constructed. In general, similar considerations also apply in common implementations of density functional theory (DFT), however the first approximation then involves the chosen form of the correlation and exchange functionals. In any event, each of these two expansions, except in very special cases, are necessarily incomplete and they separately impact the final accuracy of an electronic

[*] Department of Chemistry, Washington State University, Pullman, WA 99164-4630, USA

structure calculation. Their respective errors can also be strongly coupled, which can lead to fortuitous error cancellations. Hence in certain cases, low level methods, when combined with small basis sets, can produce better agreement with experimental properties than more sophisticated and computationally expensive methods and basis sets. Thus, *apparent* errors arising from a given method with a finite basis set can often be much different than their *intrinsic* errors, which are obtained at the complete basis set (CBS) limit, i.e., the exact solution for a given wavefunction- or density-based method. The latter is indicative of the true accuracy of the chosen method and is generally much more transferable between related chemical systems.

Therefore the route to understanding the accuracy of a given method for a particular atomic or molecular property is by carrying out the calculation at the CBS limit, which removes all errors due to the basis set approximation. Unfortunately, single, finite basis sets are generally far from the CBS limit. The solution, however, is to extrapolate the results obtained from a sequence of calculations that utilize a series of finite basis sets that systematically approach the CBS limit. A desirable requirement of these basis sets is that they should lead to regular convergence of not only the total energy of the system but also other properties depending on the accuracy of the wavefunction. The most commonly-used family of basis sets that generally satisfies these requirements is the *correlation consistent* family of basis sets, which were originally developed by Dunning for the first-row atoms [1]. Systematically-convergent basis sets explicitly designed for HF and DFT calculations, denoted *polarization consistent*, have also been developed by Jensen [2,3]. However since correlation consistent basis sets can be accurately used for both DFT and accurate wavefunction-based approaches, these will be the emphasis of the present work. Because this paper only focuses on basis sets that can lead to accurate estimates of the CBS limit, it will also necessarily neglect the discussion of basis sets that may be of high quality but are not members of a well-defined series leading to the CBS limit.

2. CORRELATION CONSISTENT BASIS SETS: A REVIEW

2.1 Background

Following the seminal work of Almlöf and Taylor [4] introducing atomic natural orbital (ANO) basis sets, as well the work of Ahlrichs and co-workers [5], Dunning reported the family of correlation consistent Gaussian basis sets for the first row elements and hydrogen, (1) denoted cc-pVnZ with n = D, T, Q, which allowed for the systematic extension of the 1-particle basis set towards the CBS limit in correlated calculations. These sets were optimized for the atoms (or H_2) at the singles and doubles configuration interaction (CISD) level of theory. The construction of these sets followed a very simple prescription—starting with a set of contracted functions describing the Hartree–Fock (HF) wavefunction, additional functions for correlation were added in shells, where each function within the shell individually recovered very similar amounts of correlation energy. For the first-row atoms, inspection of the resulting incremental correlation energy lowerings indicated that a single d-type function, which accounted for approximately 60 mE_h of correlation energy in the oxygen atom, should be used for the smallest cc-pVDZ basis set, while a 2d1f set was appropriate for cc-pVTZ (the 2nd d-type and 1st f-type functions *each* recovered about 15 mE_h of correlation energy for the O atom), and the cc-pVQZ basis set should include a 3d2f1g group of correlating functions. In addition, sets of s and p functions appropriate for correlation in the atoms, i.e., (1s1p), (2s2p), and (3s3p), were also included in the cc-pVDZ, cc-pVTZ, and cc-pVQZ sets, respectively, by uncontracting from the underlying, generally contracted [6] HF basis set.

It should be noted that an increase in the cardinal number of a correlation consistent basis set, i.e., $n = 2$ (DZ), $n = 3$ (TZ), etc., results in an increase in the maximum angular momentum of the basis, as well as the addition of multiple functions of each existing angular momenta. Hence this prescription systematically expands both the angular and radial parts of the basis set. After the introduction of the correlation consistent basis set family, it was empirically observed that the convergence of various properties to the CBS limit followed a sufficiently regular pattern that a variety of simple extrapolation formulas could be used to predict the limit (see, for instance, Refs. [7–9]). It should also be recognized that the underlying set of s and p functions describing the HF wavefunction also systematically increases in size with each member of the series of correlation consistent basis sets. Hence accurate estimates of the HF limit can also be achieved with sequences of cc-pVnZ basis sets.

The above correlation consistent prescription has since been used essentially unchanged for all the 2nd-row main group elements Al–Ar and 3rd-row elements Ga–Kr [10–14]. While the sizes of these basis sets generally range from $n = $ D to $n = 5$ or 6, selected elements have been covered up to as large as cc-pV10Z [7]. In the case of the post-3d elements, the HF set also included d-type functions, however these cc-pVnZ basis sets defined the 3d electrons to lie within the frozen core approximation. Hence the pattern of valence correlating functions is identical in these cases to the 1st and 2nd row p-block atoms.

After the initial development of cc-pVnZ basis sets for the 2nd-row elements, it was observed that these basis sets lacked sufficiently tight d-type functions that were necessary at the HF level of theory to describe molecular inner shell polarization effects [15–19]. Subsequently, a systematic study by Dunning et al. [15] resulted in the cc-pV($n + d$)Z series of basis sets, where the number of d-type exponents was increased by one compared to the cc-pVnZ sets and were completely reoptimized. CBS extrapolations of dissociation energies and bond lengths were found to be much more accurate when the new cc-pV($n + d$)Z basis sets were used compared to those obtained with the original sets. Thus, it was strongly recommended that all frozen-core calculations involving 2nd-row atoms use these newer basis sets that include an additional tight d function at each level. Recently a requirement for tight f-type functions to polarize the inner d shells of the post-d p-block elements has been noted by Weigend and Ahlrichs [20] in their QZVPP basis sets.

2.2 Diffuse-augmented basis sets

The standard cc-pVnZ basis sets are optimized for the ground states of the neutral atoms and while they can accurately describe neutral and cationic systems, they generally are not adequate for calculating properties dependent on the long-range part of the wavefunction, e.g., electron affinities of atoms and molecules or weak interactions. To address this deficiency, additional diffuse functions are added to each basis set to form the aug-cc-pVnZ series of basis sets [21]. These functions are optimized in correlated calculations on the atomic anions, and one diffuse function is added for each angular momentum symmetry present in the cc-pVnZ basis set. Diffuse functions with angular momentum symmetries present in the HF basis are optimized at the HF level of theory. For very weakly-bound van der Waals systems or in the calculation of hyperpolarizabilities, additional diffuse functions are often necessary. These are added to the aug-cc-pVnZ basis sets in an even-tempered fashion and lead to doubly- and triply-augmented basis sets denoted by d-aug-cc-pVnZ and t-aug-cc-pVnZ [22]. Because of the highly diffuse nature of these latter basis sets, extra care should be taken in molecular calculations to remove basis set superposition error (BSSE), e.g., via the counterpoise correction [23].

2.3 Core-valence basis sets

Standard cc-pVnZ or cc-pV$(n + d)$Z, including their diffuse-augmented counterparts, are designed to be used within the frozen-core approximation, i.e., 1s frozen for the 1st-row, 1s2s2p for 2nd-row, and 1s2s2p3s3p3d for 3rd-row main group elements. In each case these basis sets are only single-zeta in the core regions, i.e., one contracted function is present to describe each atomic core orbital. It is thus not surprising that these basis sets do not accurately describe correlation of the low-lying core electrons. For most applications, invoking the frozen-core approximation is well justified—the core electrons are low-lying and contribute little to chemical bonding. However, in the prediction of accurate thermochemical or spectroscopic properties, effects due to correlation of the core electrons, particularly the $(n - 1)$ shell, can be non-negligible. Additional tight functions appropriate for core correlation *must* then be added to the valence basis sets.

There are two families of correlation consistent basis sets designed to recover core correlation effects, denoted cc-pCVnZ and cc-pwCVnZ [24,25]. Both add shells of functions to the standard cc-pVnZ sets in the usual correlation consistent prescription and both systematically lead to identical CBS limits. The number and type of added functions is dependent on the core definition. First-row atoms have a [He] core, so the DZ core correlation set adds a (1s1p) set to the cc-pVDZ, while the TZ basis adds a (2s2p1d) set, etc. In the case of the 2nd-row atoms, only 2s2p correlation is treated, i.e., the 1s electrons are not correlated and the [Ne] core results in additional (1s1p1d), (2s2p2d1f), etc. added to the cc-pVDZ, cc-pVTZ, etc. sets to construct the core correlation basis sets.

The cc-pCVnZ and cc-pwCVnZ basis sets differ only in how the additional exponents are optimized. The new functions that are added to obtain the cc-pCVnZ basis sets are optimized in atomic CISD calculations for the *total* core correlation energy, e.g., for a 1st-row atom, $E_{opt} = E_{1s2s2p} - E_{2s2p}$. The *weighted* core-valence basis sets [24], cc-pwCVnZ, differentiate between intrashell (core–core) and intershell (core–valence) correlation effects. Anticipating that core–valence correlation has the largest effect on molecular properties, this contribution to the total core correlation energy is strongly weighted compared to the core-core correlation energy in the optimization procedure, i.e., $E_{opt} = E_{core-val} + 0.01 \times E_{core-core}$. The resulting tight exponents in the cc-pwCVnZ sets are generally more diffuse than the corresponding ones in the cc-pCVnZ basis sets, and the total energies are higher, especially for the smaller members of the series. More importantly however, in regards to the calculated core–valence correlation effects on molecular properties, e.g., $\Delta r_e = r_e$(valence + core correlation) $- r_e$(frozen-core), the cc-pwCVnZ basis sets often yield results close to cc-pCV$(n+1)$Z quality. It should also be stressed at this point that in the calculation of the effects of core correlation, a core-valence basis set should be used for *both* the core correlation and frozen-core calculations since the additional tight core functions in the core-valence set generally will yield non-negligible contributions to the frozen-core property. It should also be noted that these core correlation basis sets naturally include the tight d function present in the cc-pV$(n+d)$Z basis sets, so their use for the valence calculation does not lead to a loss in accuracy.

2.4 The alkali and alkaline earth metals

Correlation consistent basis sets for Li, Be, Na, and Mg have been previously developed by Woon et al. [26], but while these sets have been widely distributed, they have never been published. In terms of basis set optimizations, these elements are similar in a sense to both the hydrogen atom and transition metals. In the cases of Li and Na, the correlating functions for the cc-pVnZ basis sets cannot be optimized in atomic calculations since the valence correlation energy is zero. Thus, similar to the earlier work on the hydrogen atom where basis functions were optimized for the H_2 molecule [1], valence correlating functions for Li and Na were optimized in calculations on Li_2 and Na_2. The number and type of correlating functions were identical to those used in the p-block elements. In the optimization of core correlation basis sets, however, both cc-pCVnZ and cc-pwCVnZ sets were optimized for the atoms in the manner described in the previous section. It should be noted that Martin and co-workers [27] have recently published core correlation basis sets for the alkali and alkaline earth metals by adding correlation consistent shells to the cc-pVnZ valence sets, yielding cc-pCVnZ sets. These are essentially equivalent to the unpublished sets of Woon et al. [26].

These elements are similar to the transition metals in that the valence p orbitals are unoccupied in the ground state HF wavefunction. In the correlation consistent sets, the valence p HF functions are then taken from the ^{2}P (Li, Na) or ^{3}P (Be, Mg) excited states. In the cases of Na and Mg the HF 2p functions were taken from ground state calculations while exponents describing the 3p electrons were obtained from the excited states.

Recently correlation consistent basis sets for the Ca atom have been reported by Koput and Peterson [28]. Analogous to adding p-type functions to the Mg sets at the HF level of theory, additional HF d-type functions were included and optimized for the $4s^13d^1$ excited state of Ca. After contraction, these functions played a very similar role in molecular calculations as the tight d functions in the cc-pV$(n + d)$Z basis sets described above for the 3p main group atoms.

3. RECENT ADVANCES IN CORRELATION CONSISTENT BASIS SETS

The previous section outlined the development of correlation consistent basis sets involving mostly light, p-block elements. In the extension of these ideas to heavier elements, the effects of relativity on the basis set should be introduced. In addition to relativistic effects, the influence of low-lying electronic states must also be considered in the cases of the transition metals. For most cases only scalar relativistic effects will be considered, i.e., even if spin-orbit coupling is included in the calculations, each j component will be described by the same contracted basis set. The exceptions are the correlation consistent basis sets of Dyall [29–33], which were developed in fully-relativistic, 4-component Dirac–Hartree–Fock calculations. These basis sets, which are of DZ–QZ quality, are currently available for the heavier p-block elements, as well as the 4d and 5d transition metals.

3.1 Heavy main group elements

Other than the systematic work of Dyall, two general approaches have been used for introducing scalar relativity into correlation consistent basis sets. One is to retain the same prescription as the previous section, but include relativity via the Douglas–Kroll–Hess (DKH) Hamiltonian [34,35]. The simplest scheme in this approach is to actually retain the nonrelativistic exponents and just recontract the HF portions of the cc-pVnZ basis sets in DKH atomic calculations. This strategy is sufficient for the lighter elements where the values of the exponents are not strongly affected by relativity. This has been carried out previously by de Jong et al. [36] for H–He, B–Ne, Al–Ar, and Ga–Kr with resulting basis sets denoted by appending "DK" to the usual notation, e.g., cc-pVTZ-DK. Use of the unmodified contracted basis sets in DKH calculations were shown in that work to lead to very large errors in atomic and molecular properties. For a more complete treatment, exponent optimizations can be carried out with inclusion of the DKH Hamiltonian throughout. This approach was used, for example, in the recent development of correlation consistent basis sets for the 3d transition metal atoms by Balabanov and Peterson [37] (see also below).

As an alternative to all-electron calculations using either 4-component or DKH calculations, relativistic pseudopotentials (PPs) or effective core potentials (ECPs) provide a very convenient route to accurately including relativistic effects into electronic structure calculations [38,39]. Since PPs replace the low-lying core electrons, their use also results in smaller basis sets. Correlation consistent basis sets from DZ through 5Z quality were first developed with relativistic PPs for the post-d elements Ga–Xe by Martin and Sundermann [40], denoted SDB-cc-pVnZ. These sets employed quasi-relativistic, large-core PPs, i.e., only the valence ns and np electrons were explicitly treated. Shortly thereafter, Peterson et al. [41–43] reported correlation consistent basis sets (denoted cc-pVnZ-PP and aug-cc-pVnZ-PP) for the entire set of post-d elements Ga–Rn using newly developed Stuttgart–Köln PPs that were adjusted in multiconfiguration Dirac–Hartree–Fock calculations [42–45]. Since the motivation was to use PPs only for the accurate recovery of relativistic effects and not to decrease the computational expense, only small-core PPs were used in that work, i.e., only the $(n-2)$ rare gas core was replaced by the effective core potentials (e.g., 1s2s2p for Ga–Kr).

While the procedure for optimization of correlating functions for PP-based correlation consistent basis sets is essentially identical to the all-electron cases, special considerations do apply, particularly for the low angular momentum functions [46,47]. Specifically, the resulting pseudo-orbitals, which are nodeless or nearly so, go smoothly to zero in the core region. Describing this behavior requires modification of the basis set when it is used in correlated calculations. In the cc-pVnZ-PP basis sets of Peterson et al. for the post-d main group elements [41–43], this was addressed by addition of extra s and p correlating functions at each zeta level, e.g., a 2s2p1d correlating set is included in the cc-pVDZ-PP basis set instead of the typical 1s1p1d set included in cc-pVDZ. In addition, the optimization of the HF s and p functions would sometimes lead to near coalescence of two exponents. This undesirable behavior was avoided in the work of Peterson et al. by carrying out constrained optimizations with a minimum ratio of 1.6 between any two consecutive exponents. The resulting basis sets exhibited systematic convergence to both the HF and correlated basis set limits, and the errors due to the pseudopotential approximation was shown to be nearly negligible for both dissociation energies and spectroscopic properties [42,43].

3.2 Transition metal basis sets

One of the challenges of carrying out accurate electronic structure calculations of transition metal-containing species is having to address the plethora of low-lying electronic states that are often present in these systems. While this certainly must be addressed for the chosen correlation method, it is also an issue in developing systematically convergent transition metal basis sets since they should not be strongly biased for any given electronic state. This was previously accounted for in the framework of ANO basis sets by Bauschlicher and co-workers [48,49] and Pou-Amérigo et al. [50] In these studies the correlating functions were optimized for an average of several electronic states of the atoms, and the spd HF functions were contracted based on the average density matrices. Subsequently, Bauschlicher and co-workers [51,52] also reported correlation consistent basis sets for both Ti and Fe. These sets employed fixed-size HF sets constructed from the spd primitive functions of Partridge [53]. In the cases of the transition metals, the appropriate correlation consistent shells were 1f for DZ, 2f1g for TZ, 3f2g1h for QZ, and 4f3g2h1i for 5Z.

Recently, Balabanov and Peterson [37] reported nonrelativistic and DKH relativistic correlation consistent basis sets, cc-pVnZ and cc-pVnZ-DK with $n =$ T, Q, and 5, for the 3d transition metals Sc–Zn. Similar to the previous studies, the majority of the correlating functions were uncontracted and optimized for an average of 2–3 electronic states in CISD calculations, but the correlating functions of spd symmetries corresponded to ANO contractions. In regards to the primitive HF spd functions, the sp sets for TZ and QZ were taken from the work of Partridge, while large (28s18p) sets appropriate for the 5Z basis sets were optimized using the Legendre expansion method of Petersson et al. [54]. In order to provide a description of the 4p orbitals, the outermost p exponent in each set was replaced by 3 that were optimized for the $4s^2 3d^{m-3} 4p^1$ excited states (m is the number of valence electrons). Since flexibility in the d sets is essential for an unbiased description of the low-lying electronic states even at the HF level of theory, the primitive d sets were optimized for the average HF energy of up to 3 electronic states.

In addition to basis sets for frozen core calculations, Balabanov and Peterson also developed core-valence, cc-pwCVnZ(-DK) sets for the 3d metals, as well as diffuse-augmented sets. Since the optimal core correlation functions for the former overlapped with the existing valence functions in the cc-pVnZ(-DK) sets, simply adding extra functions for core correlation, i.e., 1f1g for TZ, 2f2g1h for QZ, and 3f3g2h1i for 5Z, was not carried out. Instead, only the most diffuse valence functions from the cc-pVnZ(-DK) sets were retained and the remainder were reoptimized with the additional core correlation functions for the total weighted-core-valence (3s3p) and valence (3d4s) correlation energy.

In the category of PP-based basis sets for transition metals, Peterson and co-workers have reported correlation consistent basis sets (cc-pVnZ-PP, aug-cc-pVnZ-PP, and cc-pwCVnZ-PP, $n =$ D – 5) paired with (small-core) Stuttgart-Köln relativistic pseudopotentials for the group 11 and 12 metals (Cu, Ag, Au, Zn, Cd, Hg) [55,56], as well as the remaining 4d metals, Y–Pd [57]. Analogous PP-based basis sets have also been developed for the 3d transition metals [58]. In each of these cases, the optimization procedure was analogous to that used by Balabanov and Peterson [37] for the all-electron 3d transition metal basis sets, although new primitive spd sets were optimized throughout and the basis set sizes were considerably smaller due to the use of PPs in the calculations. In order to accurately evaluate the errors due to the PP approximation, all-electron DKH basis sets at the TZ level (standard, aug-, and wCVTZ) were also developed. Atomic benchmark calculations [55,57–59] at the coupled cluster level of theory demonstrated that the additional errors due to the pseudopotential approximation were negligible in frozen core calculations, but rose to more than 1 kcal/mol for certain electronic excitation energies of the late metals when the outer core electrons (3s3p or 4s4p) were correlated. The recommended scheme was then to first estimate the CBS limit using sequences of cc-pVnZ-PP or cc-pwCVnZ-PP basis sets and then calibrate the PP approximation via an all-electron calculation with a cc-pwCVTZ-DK basis.

3.3 Core correlation basis sets for the post-3d elements

Among the available correlation consistent basis sets for the post-d elements discussed above, only valence basis sets have generally been defined (see, however, Refs. [40,60] for Ga, Ge, and In). Outer core correlation, however, can be substantial for the early elements of these series, e.g., groups 13 and 14, due to strong interactions with the $(n-1)$d shell [20,60–62]. Recently DeYonker et al. [63] have reported cc-pCVnZ and cc-pwCVnZ ($n =$ D, T, Q, 5; nonrelativistic and DKH relativistic) core-valence basis sets for the 3d elements that are appropriate for correlating the 3s3p3d electrons of the post-3d elements.

4. CONCLUSIONS

This contribution has attempted to give an overview of the current state of development of correlation consistent basis sets. It is probably safe to say that no other family of basis sets has spanned the number of elements and levels of accuracy as these basis sets. As emphasized in this work, however, their main advantage is in providing a systematic route towards estimates of the complete basis set limit for correlated methods. This has revolutionized the field of computational thermochemistry, as well as in the prediction of spectroscopic properties and calculation of potential energy surfaces. Since the transition metal elements are now becoming covered by these basis sets, the next challenge is in the lanthanide and actinide series. Even for main group elements there are still significant advances being made, such as the development of correlation consistent basis sets for explicitly correlated methods [64].

All of the basis sets discussed in this paper can be downloaded in various formats from either the Pacific Northwest National Laboratory basis set website or the author's website.

ACKNOWLEDGEMENTS

The author would like to thank Dr. David Feller for his careful reading of the manuscript prior to submission.

REFERENCES

1. Dunning, T.H. Jr. Gaussian basis sets for use in correlated molecular calculations. I. The atoms boron through neon and hydrogen, J. Chem. Phys. 1989, 90, 1007.
2. Jensen, F. Polarization consistent basis sets: principles, J. Chem. Phys. 2001, 115, 9113.
3. Jensen, F., Helgaker, T. Polarization consistent basis sets. V. The elements Si–Cl, J. Chem. Phys. 2004, 121, 3463.
4. Almlöf, J., Taylor, P.R. General contraction of Gaussian basis sets. I. Atomic natural orbitals for first- and second-row atoms, J. Chem. Phys. 1987, 86, 4070.
5. Jankowski, K., Becherer, R., Scharf, P., Schiffer, H., Ahlrichs, R. The impact of higher polarization basis functions on molecular ab initio results. I. The ground state of F_2, J. Chem. Phys. 1985, 82, 1413.

6. Raffenetti, R.C. General contraction of Gaussian atomic orbitals: core, valence, polarization, and diffuse basis sets; Molecular integral evaluation, J. Chem. Phys. 1973, 58, 4452.
7. Feller, D., Peterson, K.A., Crawford, T.D. Sources of error in electronic structure calculations on small chemical systems, J. Chem. Phys. 2006, 124, 054107.
8. Klopper, W., Bak, K.L., Jørgensen, P., Olsen, J., Helgaker, T. Highly accurate calculations of molecular electronic structure, J. Phys. B 1999, 32, R103.
9. Schwenke, D.W. The extrapolation of one-electron basis sets in electronic structure calculations: how it should work and how it can be made to work, J. Chem. Phys. 2005, 122, 014107.
10. Woon, D.E., Dunning, T.H. Jr. Gaussian basis sets for use in correlated molecular calculations. III. The atoms aluminum through argon, J. Chem. Phys. 1993, 98, 1358.
11. van Mourik, T., Wilson, A.K., Dunning, T.H. Jr. Benchmark calculations with correlated molecular wavefunctions. XIII. Potential energy curves for He_2, Ne_2 and Ar_2 using correlation consistent basis sets through augmented sextuple zeta, Mol. Phys. 1999, 99, 529.
12. van Mourik, T., Dunning, T.H. Jr. Gaussian basis sets for use in correlated molecular calculations. VIII. Standard and augmented sextuple zeta correlation consistent basis sets for aluminum through argon, Int. J. Quantum Chem. 2000, 76, 205.
13. Wilson, A.K., van Mourik, T., Dunning, T.H. Jr. Gaussian basis sets for use in correlated molecular calculations. VI. Sextuple-zeta correlation-consistent sets for boron through neon, J. Mol. Struct. (Theochem) 1996, 388, 339.
14. Wilson, A.K., Peterson, K.A., Woon, D.E., Dunning, T.H. Jr. Gaussian basis sets for use in correlated molecular calculations. IX. Correlation consistent basis sets for the atoms gallium through krypton, J. Chem. Phys. 1999, 110, 7667.
15. Dunning, T.H. Jr., Peterson, K.A., Wilson, A.K. Gaussian basis sets for use in correlated molecular calculations. X. The atoms aluminum through argon revisited, J. Chem. Phys. 2001, 114, 9244.
16. Bauschlicher, C.W. Jr., Ricca, A. Atomization energies of SO and SO_2: basis set extrapolation revisited, J. Phys. Chem. A 1998, 102, 8044.
17. Martin, J.M.L., Uzan, O. Basis set convergence in second-row compounds. The importance of core polarization functions, Chem. Phys. Lett. 1998, 282, 16.
18. Martin, J.M.L. Basis set convergence study of the atomization energy, geometry, and anharmonic force field of SO_2: the importance of inner polarization functions, J. Chem. Phys. 1998, 108, 2791.
19. Bauschlicher, C.W. Jr., Partridge, H. The sensitivity of B3LYP atomization energies to the basis-set and a comparison of basis-set requirements for CCSD(T) and B3LYP, Chem. Phys. Lett. 1995, 240, 533.
20. Weigend, F., Ahlrichs, R. Balanced basis sets of split valence, triple zeta valence and quadruple zeta valence quality for H to Rn: design and assessment of accuracy, Phys. Chem. Chem. Phys. 2005, 7, 3297.
21. Kendall, R.A., Dunning, T.H. Jr., Harrison, R.J. Electron affinities of the first-row atoms revisited. Systematic basis sets and wave functions, J. Chem. Phys. 1992, 96, 6796.
22. Woon, D.E., Dunning, T.H. Jr. Gaussian basis sets for use in correlated molecular calculations. IV. Calculation of static electrical response properties, J. Chem. Phys. 1994, 100, 2975.
23. Boys, S.F., Bernardi, F. The calculation of small molecular interactions by the differences of separate total energies. Some procedures with reduced errors, Mol. Phys. 1970, 19, 553.
24. Peterson, K.A., Dunning, T.H. Jr. Accurate correlation consistent basis sets for molecular core-valence correlation effects. The second row atoms Al–Ar, and the first row atoms B–Ne revisited, J. Chem. Phys. 2002, 117, 10548.
25. Woon, D.E., Dunning, T.H. Jr. Gaussian basis sets for use in correlated molecular calculations. V. Core-valence basis sets for boron through neon, J. Chem. Phys. 1995, 103, 4572.
26. Woon, D.E., Peterson, K.A., Dunning, T.H. Jr. Gaussian basis sets for use in correlated molecular calculations. VII. Valence and core–valence basis sets for Li, Na, Be, and Mg, in preparation.
27. Iron, M.A., Oren, M., Martin, J.M.L. Alkali and alkaline earth metal compounds: core-valence basis sets and importance of subvalence correlation, Mol. Phys. 2003, 101, 1345.
28. Koput, J., Peterson, K.A. The ab initio potential energy surface and vibrational-rotational energy levels of $X^2\Sigma^+$ CaOH, J. Phys. Chem. A 2002, 106, 9595.
29. Dyall, K.G. Relativistic and nonrelativistic finite nucleus optimized double zeta basis sets for the 4p, 5p and 6p elements, Theor. Chem. Acc. 1998, 99, 366.

30. Dyall, K.G. Relativistic and nonrelativistic finite nucleus optimized triple-zeta basis sets for the 4p, 5p and 6p elements, Theor. Chem. Acc. 2002, 108, 335.
31. Dyall, K.G. Relativistic double-zeta, triple-zeta, and quadruple-zeta basis sets for the 5d elements Hf–Hg, Theor. Chem. Acc. 2004, 112, 403.
32. Dyall, K.G. Relativistic double-zeta, triple-zeta, and quadruple-zeta basis sets for the 4d elements Y–Cd, Theor. Chem. Acc. 2007, 117, 483.
33. Dyall, K.G. Relativistic quadruple-zeta and revised triple-zeta and double-zeta basis sets for the 4p, 5p, and 6p elements, Theor. Chem. Acc. 2006, 115, 441.
34. Douglas, M., Kroll, N.M. Quantum electrodynamical corrections to the fine structure of helium, Ann. Phys. New York 1974, 82, 89.
35. Jansen, G., Hess, B.A. Revision of the Douglas–Kroll transformation, Phys. Rev. A 1989, 39, 6016.
36. de Jong, W.A., Harrison, R.J., Dixon, D.A. Parallel Douglas–Kroll energy and gradients in NWChem: Estimating scalar relativistic effects using Douglas–Kroll contracted basis sets, J. Chem. Phys. 2001, 114, 48.
37. Balabanov, N.B., Peterson, K.A. Systematically convergent basis sets for transition metals. I. All-electron correlation consistent basis sets for the 3d elements Sc–Zn, J. Chem. Phys. 2005, 123, 064107.
38. Dolg, M., in Schwerdtfeger, P., editor. Relativistic Electronic Structure Theory, Part 1: Fundamentals, Theoretical and Computational Chemistry. Amsterdam: Elsevier; 2002, p. 793.
39. Stoll, H., Metz, B., Dolg, M. Relativistic energy-consistent pseudopotentials—recent developments, J. Comp. Chem. 2002, 23, 767.
40. Martin, J.M.L., Sundermann, A. Correlation consistent valence basis sets for use with the Stuttgart–Dresden–Bonn relativistic effective core potentials: The atoms Ga–Kr and In–Xe, J. Chem. Phys. 2001, 114, 3408.
41. Peterson, K.A. Systematically convergent basis sets with relativistic pseudopotentials. I. Correlation consistent basis sets for the post-d group 13–15 elements, J. Chem. Phys. 2003, 119, 11099.
42. Peterson, K.A., Figgen, D., Goll, E., Stoll, H., Dolg, M. Systematically convergent basis sets with relativistic pseudopotentials. II. Small-core pseudopotentials and correlation consistent basis sets for the post-d group 16–18 elements, J. Chem. Phys. 2003, 119, 11113.
43. Peterson, K.A., Shepler, B.C., Figgen, D., Stoll, H. On the spectroscopic and thermochemical properties of ClO, BrO, IO, and their anions, J. Phys. Chem. A 2006, 110, 13877.
44. Metz, B., Schweizer, M., Stoll, H., Dolg, M., Liu, W. A small-core multiconfiguration Dirac–Hartree–Fock-adjusted pseudopotential for Tl—application to TlX (X = F, Cl, Br, I), Theor. Chem. Acc. 2000, 104, 22.
45. Metz, B., Stoll, H., Dolg, M. Small-core multiconfiguration Dirac–Hartree–Fock-adjusted pseudopotentials for post-d main group elements: Application to PbH and PbO, J. Chem. Phys. 2000, 113, 2563.
46. Blaudeau, J.-P., Brozell, S.R., Matsika, S., Zhang, Z., Pitzer, R.M. Atomic orbital basis sets for use with effective core potentials, Int. J. Quantum Chem. 2000, 77, 516.
47. Christiansen, P.A. Basis sets in correlated effective potential calculations, J. Chem. Phys. 2000, 112, 10070.
48. Bauschlicher, C.W. Jr. Large atomic natural orbital basis sets for the first transition row atoms, Theor. Chim. Acta 1995, 92, 183.
49. Bauschlicher, C.W. Jr., Taylor, P.R. Atomic natural orbital basis sets for transition metals. Theor. Chim. Acta 1993, 86, 13.
50. Pou-Amérigo, R., Merchan, M., Nebot-Gil, I., Widmark, P.-O., Roos, B.O. Density matrix averaged atomic natural orbital (ANO) basis sets for correlated molecular wave functions, Theor. Chim. Acta 1995, 92, 149.
51. Bauschlicher, C.W. Jr. TiCl, TiH, and TiH$^+$ bond energies: a test of a correlation-consistent Ti basis set, Theor. Chem. Acc. 1999, 103, 141.
52. Ricca, A., Bauschlicher, C.W. Jr. A correlation consistent basis set for Fe, Theor. Chem. Acc. 2001, 106, 314.
53. Partridge, H. Near Hartree–Fock quality GTO basis sets for the first- and third-row atoms, J. Chem. Phys. 1989, 90, 1043.
54. Petersson, G.A., Zhong, S., Montgomery, J.A. Jr., Frisch, M.J. On the optimization of Gaussian basis sets, J. Chem. Phys. 2003, 118, 1101.

55. Peterson, K.A., Figgen, D., Dolg, M., Stoll, H. Energy-consistent relativistic pseudopotentials and correlation consistent basis sets for the 4d elements Y–Pd, J. Chem. Phys. 2007, 126, 124101.
56. Figgen, D., Rauhut, G., Dolg, M., Stoll, H. Energy-consistent pseudopotentials for group 11 and 12 atoms: adjustment to multi-configuration Dirac–Hartree–Fock data, Chem. Phys. 2005, 311, 227.
57. Peterson, K.A., Puzzarini, C. Systematically convergent basis sets for transition metals. II. Pseudopotential-based correlation consistent basis sets for the group 11 (Cu, Ag, Au) and 12 (Zn, Cd, Hg) elements, Theor. Chem. Acc. 2005, 114, 283.
58. Peterson, K.A., Dolg, M., Stoll, H. Energy-consistent relativistic pseudopotentials and correlation consistent basis sets for the 3d elements Sc–Ni, in preparation.
59. Balabanov, N.B., Peterson, K.A. Basis set limit electronic excitation energies, ionization potentials, and electron affinities for the 3d transition metal atoms: Coupled cluster and multireference methods, J. Chem. Phys. 2006, 125, 074110.
60. Bauschlicher, C.W. Jr. Correlation consistent basis sets for indium, Chem. Phys. Lett. 1999, 305, 446.
61. Bauschlicher, C.W. Jr. Heats of formation of $GaCl_3$ and its fragments, J. Phys. Chem. A 1998, 102, 10424.
62. Bauschlicher, C.W. Jr. Accurate indium bond energies, J. Phys. Chem. A 1999, 103, 6429.
63. DeYonker, N., Peterson, K.A., Wilson, A.K. Systematically convergent correlation consistent basis sets for molecular core-valence correlation effects: the third row atoms gallium through krypton, J. Phys. Chem. A, submitted.
64. Peterson, K.A., Adler, T.B., Werner, H.-J. Correlation consistent-type basis sets for explicitly correlated F12 methods. The elements B–Ne and Al–Ar, in preparation.

Section 6
Emerging Technologies

Section Editor: Wendy Cornell

Merck Research Laboratories
P.O. Box 2000
Rahway, NJ 07065
USA

CHAPTER 12

Principles of G-Protein Coupled Receptor Modeling for Drug Discovery

Irache Visiers[*]

Contents		
	1. Introduction	209
	2. Homology Models of Rhodopsin-Like GPCRs	211
	2.1 Multiple sequence alignment: structural motifs as anchor points	211
	2.2 Multiple sequence alignment: other considerations	212
	2.3 Model building	213
	2.4 Modeling the loops	214
	2.5 Model refinement	216
	3. Ab Initio Methods	219
	3.1 PREDICT	219
	3.2 TASSER	220
	3.3 MEMBSTRUK	220
	4. Modeling the Activated State	220
	5. Conclusions	222
	Acknowledgement	222
	References	222

1. INTRODUCTION

G-Protein Coupled Receptors (GPCRs) constitute an important family of membrane proteins responsible for the transmission of a large number of signals from the exterior to the interior of the cell. They are involved in most physiological processes and therefore, their regulation through the discovery of modulators (agonists, antagonists, inverse agonists or even through allosteric mechanisms) is the objective of numerous discovery programs in pharmaceutical companies. This

---

[*] Millennium Pharmaceuticals, Department of Computational Chemistry, 40 Landsdowne Street, Cambridge, MA 02139, USA. E-mail: visiers@mpi.com

family has about 950 members [1] that can be classified according to sequence homology into 5 subfamilies with no sequence similarity between them: Glutamate, Rhodopsin, Adhesion, Frizzled/taste2, and Secretin families. This is also known as the GRAFS classification system [2].

GPCRs all share a common scaffold of seven helical transmembrane segments (TM1-TM7) forming a tight bundle, connected by intracellular (IL) and extracellular loops (EL). To date, only the structure of bovine rhodopsin has been solved by X-ray crystallography. Currently there are 8 crystal structures for bovine rhodopsin in the Protein Data Bank: 1F88 (2.8 Å), 1GZM (2.65 Å) [3], 1HZX (2.8 Å) [4], 1L9H (2.6 Å) [5], 1U19 (2.2 Å) [6], 2I35 (3.8 Å) [7], 2I36 (4.1 Å) [7], 2I37 (4.15 Å) [7]. These structures differ mainly on the resolution, the crystal symmetry and the degree of mobility observed for the loops, which in some cases are incomplete (1F88 and 1HZX). Overlap of each of these crystal structures reveals that the overall conformation of the eight structures is quite similar. The differences can be found mainly in the conformation of the loops with a large conformational change observed in the third intracellular loop (IL3) linking TM5 and TM6. The conformation of the proline 267 kink in TM6 of 2I36 is also different from the kink induced at that position in the other crystals, being less pronounced in 2I36 (bend angle $= 30°$) than in the rest (about $38°$) [8].

Given the scarcity of crystal structures for GPCRs, the study of this family of proteins has been greatly advanced through the use of structural models, which combined with experimental testing have proven to be successful tools to obtain a structural understanding of their function [9–14] and the shape and characteristics of the ligand binding site [15–18] as well as for virtual high throughput screening [19–22].

In a drug discovery context, building reliable models of GPCRs that can be used as predictive tools for lead discovery and optimization is not an easy task. The difficulty comes from the scarcity of templates to choose from, and from the low sequence identity that most GPCRs show to bovine rhodopsin, the only template available to date. To complicate matters further, GPCRs have evolved to bind a large variety of ligands, in contrast to other proteins of high pharmaceutical interest, like kinases, which have evolved to bind ATP, maintaining their binding site relatively constant. This divergent evolution in GPCRs makes the reliable prediction of the shape and properties of the binding site more challenging. It has become evident that straight homology modeling using rhodopsin as a template is not a correct approach, since it does not take into consideration the likely structural differences among GPCRs, that must account for the large variety of ligands they bind. This chapter will cover the main issues involved in building reliable GPCR homology models and will discuss some ab initio methods that can be extremely useful to probe the structure and function of receptors that do not belong to the rhodopsin-like family. To take full advantage of GPCR models in drug discovery it is advisable to test experimentally any hypothesis derived from the model in order to reinforce the understanding of the binding determinants for a particular set of ligands. This information often comes in the form of site-directed mutagenesis experiments which can help determine whether or not the ligand is in close proximity to a particular residue, and can also give an idea of the overall integrity of the construct. Structure based drug design projects for GPCRs should start with a model consistent with the published experimental information regarding the binding mode of a set of known ligands, or with an internal experimentally validated set of ligand-receptor contacts. In the absence of such information, great caution should be used to employ models of GPCRs in structure based drug discovery programs.

Throughout the paper the numbering scheme proposed by Ballesteros and Weinstein [23] will be used to refer to residues in the transmembrane domains. This numbering scheme allows for easy comparison of residues among GPCRs. For residues outside the seven transmembrane domains, the actual residue number in the sequence of rhodopsin will be used.

2. HOMOLOGY MODELS OF RHODOPSIN-LIKE GPCRs

2.1 Multiple sequence alignment: structural motifs as anchor points

Multiple sequence alignment of most GPCRs with rhodopsin reveals a low sequence identity, on average 19% [24]. Given such a low homology, the creation of a reliable sequence alignment can be daunting. Luckily, all rhodopsin-like GPCRs share a common fingerprint formed by conserved structural motifs that act as functional microdomains [10,13,14] localized mostly in the transmembrane segments. These structural motifs are groups of residues with well-established function that are highly conserved and often in close spatial proximity with each other. Given the high degree of conservation of those residues in the rhodopsin-like family, they can and should be used as anchor points to define the location of the transmembrane domains and to align them with those of rhodopsin. Those structural motifs are:

2.1.1 N1.50, D2.50, N7.49

These residues are conserved in the rhodopsin-like family and form a hydrogen bond network, directly or mediated by water molecules, as revealed by the crystal structure of rhodopsin [25]. This interaction has also been confirmed by site-directed mutagenesis experiments showing that D2.50 and N7.49 have a coordinated substitution pattern in a number of GPCRs, so that when D2.50 changes to N2.50, N7.49 changes to D7.49 [11,26–28]. These three residues contribute to the stabilization of the three dimensional structure of TM1, TM2 and TM7 and their mutation produces serious effects in the function of the receptor. This is the case for the β2 adrenoreceptor [11], TRHR [26], NK2, [27,28], and CB1 [29] among others. N7.49 is also part of the well know NPXXY motif that will be discussed in detail later.

2.1.2 The D/ERY motif in TM3 and E6.30 in TM6

The central arginine (R3.50) of the D/ERY motif makes a double hydrogen bond with the adjacent D3.49 and a highly conserved glutamic acid in TM6, E6.30. For a number of GPCRs, non-conservative mutations in this motif induce a constitutive active (CA) phenotype, and therefore it is assumed that the D/ERY motif holds the receptor constrained in the ground state. This behavior has been identified in the β2 adrenergic receptor [30], 5HT2A receptor [13], and lutropin receptor [31] among others. When E6.30 is missing, it is possible that adjacent residues take on a similar role stabilizing the inactive conformation. This is the case of the μ-opiod receptor, where T6.34 replaces the missing glutamic acid at position 6.30 [32,33]. In contrast, there is a group of GPCRs where mutation of residues in the DRY motif in TM3 impairs receptor activation, indicating that for this subset of receptors, R3.50 might be directly involved in receptor activation, possibly mediating G-protein coupling. This is the case for the GnRH receptor [9], the tromboxan A2 receptor [34], the α2 adrenergic receptor [35], the CB2 receptor [36] and the CCR5 receptor [37] among others.

2.1.3 The aromatic cluster in TM6 including the conserved W4.48, F6.44 and the less conserved F6.51 and F6.52, together with P6.50

Lin et al. [38] using tryptophan UV-absorbance changes and site directed mutagenesis have demonstrated a conformational change of W6.48 upon rhodopsin activation, consistent with a change in the relative orientation of helices TM3 and TM6. Computational studies have shown that the conformation of W6.48, and F6.52 are highly correlated, constituting a rotamer "toggle switch" that may modulate the TM6 proline kink [8,10,39]. W6.48 and P6.50 are highly conserved in the rhodopsin-like family of GPCRs where they are likely to share a common structural/functional role enabling the activation of the receptor.

2.1.4 The NPXXY motif in TM7

The crystal structure of rhodopsin reveals that this motif induces a large distortion in TM7. This distortion is stabilized by a number of hydrogen bonds between the polar N7.49 and a network of waters that connects it with D2.50. In addition to this hydrogen bond network, a hydrophobic interaction between Y7.53 (306 in rhodopsin) and F313 in helix 8 [40,41] also stabilizes the conformation of TM7 and links the NPXXY motif with helix 8. Moreover, changes in the environment of helix 8 that occur upon activation in rhodopsin have been shown to be coupled to changes in the hydrogen bonding network that connects D2.50 to the NPXXY motif preceding helix 8 [42]. A number of additional studies [43–45] have demonstrated that the NPXXY motif and helix 8 play an important role in the activation process, the binding to the G-protein and may act as a conformational switch domain involved in activation [11,41,46,47].

2.2 Multiple sequence alignment: other considerations

After a careful alignment of the conserved structural motifs, there are other criteria that should be applied to obtain a meaningful alignment [14]. Briefly, these criteria can be summarized as: (i) exclusion of insertions or deletions in the transmembrane domains and relocation of these gaps to the loops (preferably close to non-conserved glycine or proline), (ii) proper identification of the limits of the transmembrane domains through the analysis of the periodicity of conservation and hydrophobicity and the identification of non-conserved arginines and lysines at the end of the TM segments where they interact with the phospholipid head groups, (iii) and application of the helix initiator properties of proline, which is favored at position i to $i + 4$ of a helix. It is possible to use the same transmembrane ends as appear in the crystal structure of rhodopsin (TM1: P34–Q64; TM2: P71–H100; TM3: G106–V139; TM4: E150–L172; TM5: N200–Q225 (most of the rhodopsin crystal structures have one more helical turn at the end of TM5, with the end at V230); TM6: Q244–T277; TM7: P285–M309), but there is evidence showing that each receptor type may have different transmembrane ends [48]. Thus, a more detailed sequence alignment analysis might capture some of the specific peculiarities of each receptor.

Once the transmembrane domains have been properly aligned using the information outlined above, the alignment should focus on the loops. Generally the sequence homology is higher for the transmembrane segments and lower for the loop regions. For the chemokine subfamily, for example, the overall sequence identity with rhodopsin is about 15% with a 30% homology. When only the transmembrane domains are considered, the sequence identity rises to 20% with about 45% in sequence homology. Given the extremely low sequence identity between the loops it can be challenging or near to impossible to obtain a meaningful alignment. The main loop that sometimes maintains some similarity with rhodopsin is the second extracellular loop (EL2). This loop contains a highly conserved cysteine (C187 in rhodopsin) making a disulfide bridge with another cysteine at the top of TM3 (C3.25). The conserved cystein in EL2 should be aligned with rhodopsin's C187 even when the overall conservation of EL2 is low. After aligning the conserved cysteine in EL2, it is important to manually refine the output of the automatic alignment such that the homology modeling program can create a loop without introducing gross undesired distortions when trying to close a gap between two residues too far apart in space.

In many other cases, there is no significant homology between the loops of the GPCR we are modeling and the loops in rhodopsin. In those cases ab initio methods should be used to predict the most likely conformations of the loops; some of those methods will be discussed below.

2.3 Model building

There are number of software programs that can be used to build a reasonable initial homology model, and most of them will yield a good model of the transmembrane domains:

2.3.1 MODELLER

MODELLER [49] builds models of three-dimensional structures of proteins by satisfaction of spatial restraints: distances and dihedral angles in the target sequence, stereochemical restraints such as bond length and bond angle obtained from the CHARMM forcefield, and statistical preferences of dihedral angles and non-bonded atomic distances obtained from a representative set of all known protein structures. The model is then calculated by an optimization method relying on conjugate gradients and molecular dynamics with simulated annealing, which seeks to minimize violations of spatial restraints.

2.3.2 MOE

MOE (Molecular Operating Environment) [50] builds an initial model using the full coordinates of aligned identical residues, and only backbone coordinates otherwise. Backbone fragments to span insertions are identified by searching backbone segments from high resolution chains from the PDB, and selection of those which superpose onto anchor residues on either side of the insertion area. Alternative sidechain conformations for non-conserved residues (uncopied sidechains) are assembled from an extensive rotamer library which was generated by systematic clustering of high resolution PDB data. Loops are modeled first in random order and using a contact energy function (best scored loop is placed in the model). The contact energy function takes into account atoms already modeled, as well as any atoms specified by the user as belonging to the model environment (e.g., bound ligands). Missing sidechains are modeled in random order and sorted using a contact energy function (best scored sidechain is placed). If the residue backbone was available from the initial partial geometry, these residues' sidechains are modeled first; sidechains of modeled loops are next, and finally sidechains of outgaps are modeled. The final step involves some energy minimization of either the Cartesian average structure or the best intermediate model as defined by the user.

2.3.3 PRIME

PRIME [51] consists of tools for sequence alignment, homology model building, model refinement, and fold recognition. Sequence and secondary structure matching is used to produce an alignment between the query and one or more templates. Homology models are built by means of a coordinate transfer of all aligned residues that match between query and template. For residues that are aligned, but not matching, the backbone atoms are placed at the same position as that of the template, and the side chain is predicted using a rotamer library and the OPLS_2005 force field and SGB implicit solvent model [2]. Residues in the query that do not exist in the template are built ab initio using a simplified version of Prime's loop sampling algorithm (see below for a more detailed description). Gaps in the query (template insertions) are closed by attempting to connect the two ends of the gap; if that is not possible, the terminal tether residues are allowed to move. Following model building, residues in loops can be further refined using PRIME's loop sampling algorithm.

2.4 Modeling the loops

Modeling the loops of GPCRs and other proteins has been a matter of extensive research over the years, and there are many new methodologies that attempt to predict them in a reliable way. Among all the loops connecting the TM segments in GPCRs we should make a distinction between extracellular and intracellular loops. The first are likely to have a stronger influence in ligand binding, whereas the intracellular loops are not likely to have a direct effect on ligand binding.

Therefore, for drug discovery purposes, prediction of the extracellular loops is more relevant than the intracellular loops. Among the extracellular loops, the second extracellular loop connecting TM4 and TM5 is particularly important. This is in general a long loop (26 residues in rhodopsin) that contains two β-strands forming a β-sheet (β1 and β4) diving deep into the binding pocket, with some residues of β4 forming contacts with retinal. The structure of this loop is maintained by a disulfide bridge between a conserved cysteine in the middle of the loop (C187 in rhodopsin) and C3.25 at the top of TM3. A number of studies suggest that this loop limits the size of the binding site and forms contacts with the ligand, and is likely conserved among aminergic receptors. Using cysteine scanning of the dopamine D2 second extracellular loop Javitch at al. [52] showed that reaction of five of these mutants with sulfhydryl reagents inhibited antagonist binding, whereas bound antagonist protected two of them from reaction. The authors show how the pattern of accessibility in EL2 is consistent with a loop conformation similar to that in rhodopsin [53]. Soderhall et al. [54] showed how HMM secondary structure prediction of the EL2 in GNRHR predicts a beta strand corresponding to the β4 strand in rhodopsin. Indeed, GNRH contains the conserved disulfide bridge between EL2 and TM3 characteristic of rhodopsin-like GPCRs, and the authors postulate that the conformation of EL2 in rhodopsin is preserved in GNRHR. Chimeric receptor as well as site-directed mutagenesis in the EL2 of adenosine receptors have shown that the C-terminal residues in the adenosine receptor are important for both agonist and antagonist binding [55,56].

Given this information we infer that when the length of the EL2 loop of the particular target is close to that of rhodopsin, and when the disulfide bridge is conserved, an initial homology model based on the rhodopsin template can be built. Other GPCRs, including the MC4 receptor or CB1R, do not contain the conserved cysteines at the top of TM3 and middle of EL2, and the length of EL2 is shorter than in rhodopsin. The structure of the extracellular loops in these cases is very likely to differ from that of rhodopsin and therefore ab initio modeling is better suited. For most of the cases where homology modeling of the loops is not possible, two main techniques can be used to build the loops, ab initio loop building and threading or database based loop prediction.

Ab initio techniques seek to predict loop conformation by performing a conformational search using Monte Carlo, dihedral angle search, molecular dynamics or any other conformational sampling technique, followed by a filtering method and conformational scoring based on an energy function (or pseudo energy function). There are many ab initio methods for loop prediction, in this article we will review just a few that are available in different modeling programs.

Threading techniques consist on searching a database of known proteins for a segment whose sequence fits the sequence to be modeled. The matches are ranked according to the ability to meet the geometry of the start and end amino acids of the loop to be modeled. The loop models created by threading are later refined by energy optimization.

Among the available methods to predict loop conformations are:

2.4.1 MODLOOP
MODLOOP (http://alto.compbio.ucsf.edu/modloop/) [49] is a web server for automated modeling of loops in protein structures. The server relies on the loop modeling routine in MODELLER that predicts the loop conformations by satisfaction of spatial restraints.

2.4.2 PLOP
PLOP [57] (http://francisco.compbio.ucsf.edu/~jacobson/plop_manual/plop_overview.htm). (This methodology has been included in PRIME from Schrodinger [51]). This algorithm initially creates a large ensemble of loop conformations using a sampling algorithm that employs a hierarchical approach. In the first stage a large number of loops are generated by growing simultaneously from the N- and C-terminal ends using a dihedral angle-based buildup procedure, followed by an extensive dihedral search using rotamer libraries for backbone dihedral angles. In this fashion, the generated ensemble does not depend on the initial conformation of the loop or the shape of the energy surface. The conformations are processed with a clustering algorithm that selects a representative set of loop candidates to optimize and score with the OPLS_2005 force field and SGB implicit solvent model. The loop prediction algorithm is applied multiple times to each loop and further sampling is done using Cartesian restrains. This algorithm has proven to be remarkably accurate for loops up to eight residues long, showing an intermediate but encouraging performance for loops up to ten residues.

2.4.3 Mehler et al.
Mehler et al. [58] have developed a protocol for the prediction of loops in GPCRs. Using Monte Carlo simulations in a temperature annealing protocol combined with a scaled collective variables technique in the context of a Screen Coulomb Potential- Implicit Solvent model [59] (implemented in CHARMm31b1), the developed method is able to find an absolute minimum free energy ensemble representative of the native funnel. When this method is applied to the EL2 in rhodopsin and D2R, the results are consistent with the crystal structure of rhodopsin and site directed mutagenesis [52]. The method is available for users with a CHARMM license by contacting Dr. Mehler (http://physiology.med.cornell.edu/faculty/mehler/mehler-research.html).

2.4.4 The LOOPSEARCH
The LOOPSEARCH module of the SYBYL package [60] is a threading algorithm used to find good starting conformations for the loops.

2.5 Model refinement
2.5.1 Refinement of proline kinks and other distortions
The initial homology model needs to be optimized to account for the major differences between the target sequence and rhodopsin. Special attention needs to be paid to the presence (or absence) of non-conserved prolines or other distortion-inducing motifs such as Glycines (G, GG, or GXG sequences).

Proline, with its backbone nitrogen embedded in a ring, does not satisfy the characteristic α-helical i to $i-4$ hydrogen bonding pattern. This property makes proline both a good helix initiator when present in the first four positions of an α-helix and a helix breaker when present at the end of a helix. The lack of this hydrogen bond renders the stretch of residues extending from proline (residues i to $i-4$) a particularly flexible motif, that is often associated with the conformational rearrangements needed for receptor activation [8,61]. However proline is not only a point of flexibility; it is also a point for distortion of the characteristic helical axis [8]. Proline's nitrogen substitution introduces bulkiness close to the backbone that can cause a steric clash with the carbonyl moiety of residue $i-3$. To alleviate this clash, the helix needs to bend, inducing a kink in the helix that can adopt a wide variety of values [62], making it difficult to model. Furthermore, the distortion introduced by proline can be modulated by threonine at positions $i-1$ to $i-4$ from proline [63,64]. At these positions, threonine can influence the direction and magnitude of the kink changing its bend and wobble angles [8] by forming a hydrogen bond between the side-chain of threonine and the backbone of the residue $i+3$ from threonine.

The distortions induced by proline in an α-helix can vary significantly, from very small to quite large [8,62]. When an alignment between a target sequence and rhodopsin show a conserved proline in the middle of a transmembrane segment, it is safe to model it using rhodopsin as a template. When there is a non-conserved proline or when rhodopsin contains a proline that is absent in our target receptor, the particular environment surrounding the different residue needs to be studied in order to make a decision about which type of distortion should be modeled. It is possible that our target receptor does not contain a proline present in rhodopsin but that other residues are able to produce the same type of distortions. This has been described as "structural mimicry", where the 3D environment surrounding a group of residues can induce a distortion in an adjacent helix to preserve the general 3D structure [53]. In other cases, the residue aligning with proline and those immediately preceding it, need to be modeled differently to be able to preserve the helical pattern of hydrophobicity and conservation [14].

The modification of the kink can be done (i) manually using a visual interface followed by energy minimization and dynamics; (ii) using a constrained energy based method, like molecular mechanics followed by molecular dynamics simulation with a force constant imposed on the phi and psi angles of residues i to $i-4$ from proline; (iii) using a Monte Carlo method [46,65,66] to find the most energy favorable conformation of the helix in the context of the surrounding helices, (iv) using an energy based function to let the proline adapt to the presence of the ligand. In all cases, the cytoplasmic part of the helix is kept fixed, up to residue $i-4$ from proline (or the residue replacing the proline), and the backbone angles of residues after proline are kept constrained to their initial values to preserve the integrity of the helix at the same time that full backbone flexibility is permitted only to residues i to $i-4$, which are the residues that have been found to absorb the proline distortion in protein crystal structures [62].

When proline kinks are modified, it is important to consider that the transmembrane segments are tethered by loops. There are two main ways to deal with this issue. The first one is to build an initial model of the transmembrane segments, therefore excluding the loops, and construct those only after the modification of the distortions in the transmembrane segments has been done. The second possibility is to build the full model including the loops, and use a combination of constrained molecular mechanics/dynamics simulations to modify the kink, letting the loop adapt to the changes in the helices. In both cases the full model needs to be optimized with an energy based procedure. In section 2.4 we talked in more detail about some of the possible methods available to loop construction.

2.5.2 Ligand-based model refinement

After a full model, including the loops, has been built, the next step in the refinement of the model is to optimize the binding site by docking ligands known to bind the receptor with high activity. In our experience, it is common to find that models based on rhodopsin do not have enough space in the binding site to fit most known ligands. In those cases, a ligand-based optimization is necessary. The ligand can be manually docked in the binding site using site directed mutagenesis information regarding the residues that are involved in ligand binding [15,18,19]. Then the complex can be minimized fixing the position of the ligand and letting the protein adapt to the presence of the ligand in the binding site. There are a number of ways this can be achieved, but in our experience it is advisable to keep the most intracellular half of the receptor constrained (often up to a proline or another distortion in the transmembrane segments) while leaving the rest unconstrained. Another possibility is to maintain the position of the intracellular half of the transmembrane segments using a force constant and in addition maintain the integrity of the helices with a force constant on the dihedral angles of the backbone of the rest, excluding those residues defining proline kinks or other distortions. Using this simple procedure three goals are achieved: (i) the protein adapts to the presence of the ligand by changing only the dihedral angles of the residues involved in proline kinks (or other flexible motifs like G, GG, or GXG motifs), (ii) the cytoplasmic half of the transmembrane segments is not changed, (iii) the segment following a proline toward the extracellular side maintains its dihedral angles, and therefore the regular pattern of hydrogen bonds characterizing an α-helix, avoiding large distortions in residues that are not likely to break their regular α-helical hydrogen pattern. Finally the whole complex should be minimized without constrains. If the complex is energetically stable, the ligand should not undergo large changes after the constraints are released. In some cases, the ligand "flies away" from the binding site, indicating that the position of the ligand in the modeled binding site is not energetically feasible.

Sherman et al. [67] have developed an "Induced-Fit" protocol, involving a protein-ligand docking method that accounts for both ligand and receptor flexibility by iteratively combining rigid receptor docking with protein structure prediction techniques. This technique aims to account for the ligand induced conformational changes in receptor active sites. It uses proprietary software to exhaustively consider possible binding modes and the associated conformational changes within receptor active sites. The Induced Fit protocol is initiated by docking the active ligand in the binding site to generate a diverse ensemble of ligand poses.

In the initial docking phase, the procedure uses a customizable reduced van der Waals radii and an increased Coulomb–vdW cutoff, and has the option to temporarily remove highly flexible side chains mutating them to alanine. For each pose, the side chains of nearby residues are changed to accommodate the ligand. In this step, residues that were mutated to alanine in the initial docking phase are reinstated to their original residue, and the conformation of the side chain is predicted in the context of the whole protein-ligand complex. These residues and the ligand are then minimized. Finally, each ligand is re-docked into its corresponding low energy protein structures and the resulting complexes are ranked according to GlideScore [68].

The last step in the refinement process is to apply an energy based method as molecular mechanics and dynamics simulation to obtain a model that is energetically stable. In the absence of a membrane that keep the helices together, it is advisable to introduce a small force constant to keep the backbone of the transmembrane segments in place and to avoid gross artificial distortions. This force constant can be reduced in subsequent cycles of dynamics simulation (equilibration followed by dynamics) in gradual steps until completely eliminated.

3. AB INITIO METHODS

Given the low sequence identity that most GPCRs have with rhodopsin, with a 19% average [24], some groups opt for ab initio GPCR modeling which does not heavily rely on rhodopsin as structural template. These models have been shown useful for virtual high-throughput screening, giving interesting enrichment factors over random screening. We find that these ab initio methods are particularly useful to model GPCRs that do not belong to the rhodopsin-like family, and can be used to obtain structural inferences that can be tested experimentally. For many years similar procedures were used successfully to study the structure and function of rhodopsin-like GPCRs. In this chapter we will briefly discuss a few methods:

3.1 PREDICT

PREDICT [69] is an ab initio modeling technique that models the structure of the transmembrane domain of GPCRs and other membrane proteins based on their primary sequence of amino acids, without relying on known structures or multiple sequence alignments. PREDICT uses the protein sequence and the physicochemical properties of the membrane environment to create a large number of possible structures that are optimized in cycles of successive refinement steps, from coarse to fine. An energy function ranks the models and gives the best scoring one. This method uses assumptions regarding the overall general structure of rhodopsin and other GPCRs, obtained from previous published knowledge. Among these assumptions are that loops connecting the helices are relatively short, and thus that the helices are packed in an antiparallel orientation, and that TM helices are arranged in a clockwise manner when viewed from the intracellular side. In addition, PREDICT uses experimentally known interactions between non-adjacent helices, such as those found to be correlated mutations, as secondary constraints.

3.2 TASSER

TASSER (Threading, Assembly and Refinement) [24] combines threading and ab initio algorithms, and includes a "hydrophilic" inside potential for predicted transmembrane regions. It builds a final model using low sequence identity templates, with a methodology that consists of template identification, structure assembly, and model selection in an automated manner. Only the sequence is given to TASSER as input, and no other extrinsic knowledge like active site information or other experimental restraints are applied. When TASSER is applied to a group of 38 membrane proteins, the authors report that most proteins with a transmembrane helical topology were well modeled. Moreover when TASSER is applied to rhodopsin (including only those templates with identity <30% as well as bacteriorhodopsin) the authors report a full length RMSD to native of 4.6 Å and 2.1 Å when only transmembrane regions are considered. Skolnick's group has created models of 907 putative GPCRs in the human genome, which are available for noncommercial users at http://cssb.biology.gatech.edu/skolnick/files/gpcr/gpcr.html.

3.3 MEMBSTRUK

MEMBSTRUK [70] uses the Eisenberg hydrophobicity scale to calculate the hydrophobicity profile of a multiple sequence alignment of GPCRs to predict the transmembrane regions and extends the helices at top and bottom analyzing the presence of helix-breaking residues, positively and negatively charged residues to extend the helices up to 6 residues beyond the limits dictated by the hydrophobicity profile. Once the transmembrane regions have been defined, canonical right-handed α-helices are built and are oriented in space according to the electron density map of frog rhodopsin [71] using a similar method to that developed earlier by Filizola et al. [72]. The relative translation and rotation of the helices is optimized with a Monte Carlo-like procedure using rigid right-handed ideal α-helices. Later on, the helices are optimized separately to allow for backbone distortions with a 500 ps molecular dynamic simulation. Finally this bundle is embedded in an explicit membrane formed by 52 molecules of diauroylphophatidylcholine and a rigid-body molecular dynamic simulation of 50 ps is performed. The loops are built and minimized with the transmembrane segments fixed.

4. MODELING THE ACTIVATED STATE

In the simplest model for receptor activation, GPCRs exist in two states, the "activated state" characterized by higher affinity for agonists and capable of binding G-proteins, and a "ground state" characterized by its higher affinity for antagonists [73–75] and its low affinity for G-protein. However the activation picture is much more complex than a simple two state model; there is emerging evidence showing that GPCRs exist not only in one, but in multiple active conformational states, each one possessing differentiated pharmacological properties and G-protein coupling profiles, in some cases being able to even trigger a signaling cascade through an entirely different mechanism, not mediated by G-proteins [76]. Furthermore, some GPCRs have the ability to hetero-oligomerize when co-expressed in the cell membrane, forming complexes with unique pharmacological profiles. This is the case of the oligomerization of the chemokine receptor CCR5 and opioid receptors [77], opiod receptors [78], CCR5 and CCR2b [79], dopamine D1 and D2 receptors [80], the SNSR-4 and the δ-opiod [81], the β2 and β3 adrenergic receptors [82] and the adenosine 1 receptor and the P2Y1 receptor [83] among others.

As a result, agonists can selectively bind not only one, but multiple different activated states, eliciting diverse response profiles depending on the signaling process to which that particular active state is coupled. A deeper understanding of the specific conformational characteristics of each active state and their effector-coupling properties opens the door to the design of finely tuned drugs that stimulate only a partial set of responses, having the potential to elicit fewer side effects. Therefore, it can be necessary to model the conformational changes associated to receptor activation to facilitate rational drug design of agonists. Based on abundant literature data, receptor activation involves: (i) a clockwise rotation and movement of the cytoplasmic end of TM6 apart from TM3 (viewed from the intracellular side) [84–86], (ii) a reduction in the distance between the cytoplasmic ends of TM5 and TM6 [87] and (iii) a switch in the orientation of W6.48 from perpendicular to parallel to the plane of the membrane [38] associated with a concerted change in the conformation of the aromatic cluster of residues in TM6 (F6.44(254), W6.48(258) and F6.52(262)) [10].

We have developed a detailed activated model for the melanocortin MC4 receptor that incorporates the conformational changes that occur upon receptor activation and are likely to affect the shape of the binding site [18]. In a first step the conformation of F6.44, W6.48 and F6.52 was changed to accommodate the concerted change in the conformation of the aromatic cluster of residues in TM6 described above, followed by a reduction of the P6.50 proline kink from an initial 30° to a final, lower kink of 11°. The generated clashes among helices were alleviated by rigid body repositioning of the modified helices keeping the three dimensional arrangement of the key structural motifs described above. A final energy minimization with the CHARMM force field [88] yielded a model of the activated state of the MC4 receptor capable of explaining the binding of endogenous and exogenous agonists.

Bissantz et al. [19] built agonist-bound activated GPCR models by (i) rotating TM6 30° clockwise around its axis (when viewed from the intracellular side), (ii) docking a full agonist into the binding site following experimental information, pose that was later used as reference, (iii) superimposing structurally unrelated agonists onto the reference agonist, (iv) manual adjustment of ligands, (v) multi-ligand receptor minimization with fixed ligands. These authors therefore use a procedure similar to that described above regarding ligand-based receptor refinement, coupled to the introduction of the experimentally observed rotation of TM6 [84–86]. The authors report that the resulting active models are better suited for agonist virtual screening than the inactive ones.

Niv et al. [89] have developed a method to model activated states of GPCRs using the structure-related information for the active state of rhodopsin and other GPCRs (http://physiology.med.cornell.edu/GPCRactivation/gpcrindex.html) as structural constrains for a Simulated Annealing Molecular Dynamics protocol.

It is worth mentioning the recent publication of the crystal structure of a photoactivated intermediate of rhodopsin (2I37.pdb) diffracting at 4.15 Å resolution [7]. This structure suggests that the activation process may involve only minor changes in the structure, in contrast to previous work based on mutagenesis, fluorescence, cys-cross linking experiments that predicted larger structural changes. Given the discrepancy with all the previous experimental data, additional work needs to be done to determine whether this structure is a real representation of an activated rhodopsin applicable to other GPCRs.

5. CONCLUSIONS

Molecular models for GPCRs have proven to be useful tools to understand the molecular structure and mechanism of activation of this family of receptors. GPCR models can be successfully used in structure-based drug design programs, giving reasonable docking hypothesis that in conjunction with experimental testing can support programs in the hit discovery and lead optimization phases, providing a structural framework to rationalize structure based relationships and even serving as tools for virtual high-throughput screening. The success of those models depends on the level of detail employed to refine the initial homology model to account for the important differences between each particular target and rhodopsin. In this paper we have summarized some of the main methods that are available to build reliable models of GPCRs with especial emphasis on modeling the differences with rhodopsin, the only template currently available. In the future more crystal structures of a variety of GPCRs will hopefully be available, and will enormously help to build more robust models to aid in the discovery of pharmaceutically relevant targets.

ACKNOWLEDGEMENT

We thank Scott Rowland for proof reading this manuscript and Mihaly Mezei for the calculation of the proline kinks in rhodopsin.

REFERENCES

1. Takeda, S., Kadowaki, S., Haga, T., Takaesu, H., Mitaku, S. Identification of G protein-coupled receptor genes from the human genome sequence, FEBS Lett. 2002, 520(1–3), 97–101.
2. Fredriksson, R., Lagerstrom, M.C., Lundin, L.G., Schioth, H.B. The G-protein-coupled receptors in the human genome form five main families. Phylogenetic analysis, paralogon groups, and fingerprints, Mol. Pharmacol. 2003, 63(6), 1256–72.
3. Li, J., Edwards, P.C., Burghammer, M., Villa, C., Schertler, G.F. Structure of bovine rhodopsin in a trigonal crystal form, J. Mol. Biol. 2004, 343(5), 1409–38.

4. Teller, D.C., Okada, T., Behnke, C.A., Palczewski, K., Stenkamp, R.E. Advances in determination of a high-resolution three-dimensional structure of rhodopsin, a model of G-protein-coupled receptors (GPCRs), Biochemistry 2001, 40(26), 7761–72.
5. Okada, T., Fujiyoshi, Y., Silow, M., Navarro, J., Landau, E.M., Shichida, Y. Functional role of internal water molecules in rhodopsin revealed by X-ray crystallography, Proc. Natl. Acad. Sci. USA 2002, 99(9), 5982–7. Epub 2002 Apr 23.
6. Okada, T., Sugihara, M., Bondar, A.N., Elstner, M., Entel, P., Buss, V. The retinal conformation and its environment in rhodopsin in light of a new 2.2 A crystal structure, J. Mol. Biol. 2004, 342(2), 571–83.
7. Salom, D., Lodowski, D.T., Stenkamp, R.E., Le Trong, I., Golczak, M., Jastrzebska, B., Harris, T., Ballesteros, J.A., Palczewski, K. Crystal structure of a photoactivated deprotonated intermediate of rhodopsin, Proc. Natl. Acad. Sci. USA 2006, 103(44), 16123–8. Epub 2006 Oct 23.
8. Visiers, I., Braunheim, B.B., Weinstein, H., Prokink: a protocol for numerical evaluation of helix distortions by proline. Protein Eng. 2000, 13(9), 603–6.
9. Ballesteros, J., Kitanovic, S., Guarnieri, F., Davies, P., Fromme, B.J., Konvicka, K., Chi, L., Millar, R.P., Davidson, J.S., Weinstein, H., Sealfon, S.C. Functional microdomains in G-protein-coupled receptors. The conserved arginine-cage motif in the gonadotropin-releasing hormone receptor, J. Biol. Chem. 1998, 273(17), 10445–53.
10. Filizola, M., Visiers, I., Skrabanek, L., Campagne, F., Weinstein, H., Functional mechanisms of GPCRs in a structural context, in Schousboe, A., Bräuner-Osborne, H., editors. Strategies in Molecular Neuropharmacology. Humana Press, 2003.
11. Flanagan, C.A., Zhou, W., Chi, L., Yuen, T., Rodic, V., Robertson, D., Johnson, M., Holland, P., Millar, R.P., Weinstein, H., Mitchell, R., Sealfon, S.C. The functional microdomain in transmembrane helices 2 and 7 regulates expression, activation, and coupling pathways of the gonadotropin-releasing hormone receptor, J. Biol. Chem. 1999, 274(41), 28880–6.
12. Gether, U., Lin, S., Ghanouni, P., Ballesteros, J.A., Weinstein, H., Kobilka, B.K. Agonists induce conformational changes in transmembrane domains III and VI of the beta2 adrenoceptor, Embo J. 1997, 16(22), 6737–47.
13. Visiers, I., Ebersole, B.J., Dracheva, S., Ballesteros, J., Sealfon, S.C., Weinstein, H. Structural motifs as functional microdomains in G-protein-coupled receptors: energetic considerations in the mechanism of activation of the serotonin 5-HT2A receptor by disruption of the ionic lock of the arginine cage, International Journal of Quantum Chemistry 2002, 88(1), 65–75.
14. Visiers, I., Ballesteros, J.A., Weinstein, H. Three-dimensional representations of G protein-coupled receptor structures and mechanisms, Methods Enzymol. 2002, 343, 329–71.
15. Ebersole, B.J., Visiers, I., Weinstein, H., Sealfon, S.C. Molecular basis of partial agonism: orientation of indoleamine ligands in the binding pocket of the human serotonin 5-HT2A receptor determines relative efficacy, Mol. Pharmacol. 2003, 63(1), 36–43.
16. Almaula, N., Ebersole, B.J., Zhang, D., Weinstein, H., Sealfon, S.C. Mapping the binding site pocket of the serotonin 5-hydroxytryptamine2A receptor. Ser3.36(159) provides a second interaction site for the protonated amine of serotonin but not of lysergic acid diethylamide or bufotenin, J. Biol. Chem. 1996, 271(25), 14672–5.
17. Almaula, N., Ebersole, B.J., Ballesteros, J.A., Weinstein, H., Sealfon, S.C. Contribution of a helix 5 locus to selectivity of hallucinogenic and nonhallucinogenic ligands for the human 5-hydroxytryptamine2A and 5-hydroxytryptamine2C receptors: direct and indirect effects on ligand affinity mediated by the same locus, Mol. Pharmacol. 1996, 50(1), 34–42.
18. Hogan, K., Peluso, S., Gould, S., Parsons, I., Ryan, D., Wu, L., Visiers, I. Mapping the binding site of melanocortin 4 receptor agonists: a hydrophobic pocket formed by I3.28(125), I3.32(129), and I7.42(291) is critical for receptor activation, J. Med. Chem. 2006, 49(3), 911–22.
19. Bissantz, C., Bernard, P., Hibert, M., Rognan, D. Protein-based virtual screening of chemical databases. II. Are homology models of G-protein coupled receptors suitable targets?, Proteins 2003, 50(1), 5–25.
20. Evers, A., Hessler, G., Matter, H., Klabunde, T. Virtual screening of biogenic amine-binding G-protein coupled receptors: comparative evaluation of protein- and ligand-based virtual screening protocols, J. Med. Chem. 2005, 48(17), 5448–65.

21. Evers, A., Klabunde, T. Structure-based drug discovery using GPCR homology modeling: successful virtual screening for antagonists of the alpha1A adrenergic receptor, J. Med. Chem. 2005, 48(4), 1088–97.
22. Becker, O.M., Dhanoa, D.S., Marantz, Y., Chen, D., Shacham, S., Cheruku, S., Heifetz, A., Mohanty, P., Fichman, M., Sharadendu, A., Nudelman, R., Kauffman, M., Noiman, S. An integrated in silico 3D model-driven discovery of a novel, potent, and selective amidosulfonamide 5-HT1A agonist (PRX-00023) for the treatment of anxiety and depression, J. Med. Chem. 2006, 49(11), 3116–35.
23. Ballesteros, J.A., Weinstein, H. Integrated methods for the construction of three-dimensional models and computational probing of structure-function relations in G protein-coupled receptors, Methods Neurosci. 1995, 25, 366–428.
24. Zhang, Y., Arakaki, A.K., Skolnick, J. TASSER: an automated method for the prediction of protein tertiary structures in CASP6, Proteins 2005, 61(Suppl. 7), 91–8.
25. Palczewski, K., Kumasaka, T., Hori, T., Behnke, C.A., Motoshima, H., Fox, B.A., Le Trong, I., Teller, D.C., Okada, T., Stenkamp, R.E., Yamamoto, M., Miyano, M. Crystal structure of rhodopsin: A G protein-coupled receptor, Science 2000, 289(5480), 739–45.
26. Perlman, J.H., Colson, A.O., Wang, W., Bence, K., Osman, R., Gershengorn, M.C. Interactions between conserved residues in transmembrane helices 1, 2, and 7 of the thyrotropin-releasing hormone receptor, J. Biol. Chem. 1997, 272(18), 11937–42.
27. Donnelly, D., Maudsley, S., Gent, J.P., Moser, R.N., Hurrell, C.R., Findlay, J.B. Conserved polar residues in the transmembrane domain of the human tachykinin NK2 receptor: functional roles and structural implications, Biochem. J. 1999, 339(Pt 1), 55–61.
28. Donnelly, D., Findlay, J.B., Blundell, T.L. The evolution and structure of aminergic G protein-coupled receptors, Receptors Channels 1994, 2(1), 61–78.
29. Nie, J., Lewis, D.L. Structural domains of the CB1 cannabinoid receptor that contribute to constitutive activity and G-protein sequestration, J. Neurosci. 2001, 21(22), 8758–64.
30. Ballesteros, J.A., Jensen, A.D., Liapakis, G., Rasmussen, S.G., Shi, L., Gether, U., Javitch, J.A. Activation of the beta 2-adrenergic receptor involves disruption of an ionic lock between the cytoplasmic ends of transmembrane segments 3 and 6, J. Biol. Chem. 2001, 276(31), 29171–7.
31. Angelova, K., Fanelli, F., Puett, D. A model for constitutive lutropin receptor activation based on molecular simulation and engineered mutations in transmembrane helices 6 and 7, J. Biol. Chem. 2002, 277(35), 32202–13. Epub 2002 Jun 17.
32. Huang, P., Li, J., Chen, C., Visiers, I., Weinstein, H., Liu-Chen, L.Y. Functional role of a conserved motif in TM6 of the rat mu opioid receptor: constitutively active and inactive receptors result from substitutions of Thr6.34(279) with Lys and Asp., Biochemistry 2001, 40(45), 13501–9.
33. Huang, P., Visiers, I., Weinstein, H., Liu-Chen, L.Y. The local environment at the cytoplasmic end of TM6 of the mu opioid receptor differs from those of rhodopsin and monoamine receptors: introduction of an ionic lock between the cytoplasmic ends of helices 3 and 6 by a L6.30(275)E mutation inactivates the mu opioid receptor and reduces the constitutive activity of its T6.34(279)K mutant, Biochemistry 2002, 41(40), 11972–80.
34. Capra, V., Veltri, A., Foglia, C., Crimaldi, L., Habib, A., Parenti, M., Rovati, G.E. Mutational analysis of the highly conserved ERY motif of the thromboxane A2 receptor: alternative role in G protein-coupled receptor signaling, Mol. Pharmacol. 2004, 66(4), 880–9. Epub 2004 Jun 30.
35. Chung, D.A., Wade, S.M., Fowler, C.B., Woods, D.D., Abada, P.B., Mosberg, H.I., Neubig, R.R. Mutagenesis and peptide analysis of the DRY motif in the alpha2A adrenergic receptor: evidence for alternate mechanisms in G protein-coupled receptors, Biochem. Biophys. Res. Commun. 2002, 293(4), 1233–41.
36. Feng, W., Song, Z.H. Effects of D3.49A, R3.50A, and A6.34E mutations on ligand binding and activation of the cannabinoid-2 (CB2) receptor, Biochem. Pharmacol. 2003, 65(7), 1077–85.
37. Lagane, B., Ballet, S., Planchenault, T., Balabanian, K., Le Poul, E., Blanpain, C., Percherancier, Y., Staropoli, I., Vassart, G., Oppermann, M., Parmentier, M., Bachelerie, F. Mutation of the DRY motif reveals different structural requirements for the CC chemokine receptor 5-mediated signaling and receptor endocytosis, Mol. Pharmacol. 2005, 67(6), 1966–76. Epub 2005 Mar 10.
38. Lin, S.W., Sakmar, T.P. Specific tryptophan UV-absorbance changes are probes of the transition of rhodopsin to its active state, Biochemistry 1996, 35(34), 11149–59.

39. Shi, L., Liapakis, G., Xu, R., Guarnieri, F., Ballesteros, J.A., Javitch, J.A. Beta2 adrenergic receptor activation. Modulation of the proline kink in transmembrane 6 by a rotamer toggle switch, J. Biol. Chem. 2002, 277(43), 40989–96. Epub 2002 Aug 6.
40. Prioleau, C., Visiers, I., Ebersole, B.J., Weinstein, H., Sealfon, S.C. Conserved helix 7 tyrosine acts as a multistate conformational switch in the 5HT2C receptor. Identification of a novel "locked-on" phenotype and double revertant mutations, J. Biol. Chem. 2002, 277(39), 36577–84.
41. Fritze, O., Filipek, S., Kuksa, V., Palczewski, K., Hofmann, K.P., Ernst, O.P. Role of the conserved NPxxY(x)5,6F motif in the rhodopsin ground state and during activation, Proc. Natl. Acad. Sci. USA 2003, 100(5), 2290–5.
42. Lehmann, N., Alexiev, U., Fahmy, K. Linkage between the intramembrane H-bond network around aspartic acid 83 and the cytosolic environment of helix 8 in photoactivated rhodopsin, J. Mol. Biol. 2007, 366(4), 1129–41. Epub 2006 Dec 15.
43. Krishna, A.G., Menon, S.T., Terry, T.J., Sakmar, T.P. Evidence that helix 8 of rhodopsin acts as a membrane-dependent conformational switch, Biochemistry 2002, 41(26), 8298–309.
44. Mielke, T., Alexiev, U., Glasel, M., Otto, H., Heyn, M.P. Light-induced changes in the structure and accessibility of the cytoplasmic loops of rhodopsin in the activated MII state, Biochemistry 2002, 41(25), 7875–84.
45. Swift, S., Leger, A.J., Talavera, J., Zhang, L., Bohm, A., Kuliopulos, A. Role of the PAR1 receptor 8th helix in signaling: the 7-8-1 receptor activation mechanism, J. Biol. Chem. 2006, 281(7), 4109–16. Epub 2005 Dec 13.
46. Konvicka, K., Guarnieri, F., Ballesteros, J.A., Weinstein, H. A proposed structure for transmembrane segment 7 of G protein-coupled receptors incorporating an asn-Pro/Asp-Pro motif, Biophys. J. 1998, 75(2), 601–11.
47. Flanagan, C.A., Rodic, V., Konvicka, K., Yuen, T., Chi, L., Rivier, J.E., Millar, R.P., Weinstein, H., Sealfon, S.C. Multiple interactions of the Asp(2.61(98)) side chain of the gonadotropin-releasing hormone receptor contribute differentially to ligand interaction, Biochemistry 2000, 39(28), 8133–41.
48. Xu, W., Campillo, M., Pardo, L., Kim de Riel, J., Liu-Chen, L.Y. The seventh transmembrane domains of the delta and kappa opioid receptors have different accessibility patterns and interhelical interactions, Biochemistry 2005, 44(49), 16014–25.
49. Eswar, N., John, B., Mirkovic, N., Fiser, A., Ilyin, V.A., Pieper, U., Stuart, A.C., Marti-Renom, M.A., Madhusudhan, M.S., Yerkovich, B., Sali, A. Tools for comparative protein structure modeling and analysis, Nucleic Acids Res. 2003, 31(13), 3375–80.
50. Group, C.C. MOE (The Molecular Operating Environment), Chemical Computing Group. 1010 Sherbrooke Street West, Suite 910, Montreal, Canada H3A 2R7. http://www.chemcomp.com: Montreal, Canada.
51. Schrodinger, Schrodinger LLC, http://www.schrodinger.com/.
52. Shi, L., Javitch, J.A. The second extracellular loop of the dopamine D2 receptor lines the binding-site crevice, Proc. Natl. Acad. Sci. USA 2004, 101(2), 440–5. Epub 2004 Jan 2.
53. Ballesteros, J.A., Shi, L., Javitch, J.A. Structural mimicry in G protein-coupled receptors: implications of the high-resolution structure of rhodopsin for structure-function analysis of rhodopsin-like receptors, Mol. Pharmacol. 2001, 60(1), 1–19.
54. Soderhall, J.A., Polymeropoulos, E.E., Paulini, K., Gunther, E., Kuhne, R. Antagonist and agonist binding models of the human gonadotropin-releasing hormone receptor, Biochem. Biophys. Res. Commun. 2005, 333(2), 568–82.
55. Olah, M.E., Jacobson, K.A., Stiles, G.L. Role of the second extracellular loop of adenosine receptors in agonist and antagonist bonding. Analysis of chimeric A1/A3 adenosine receptors, J. Biol. Chem. 1994, 269(40), 24692–8.
56. Kim, J., Jiang, Q., Glashofer, M., Yehle, S., Wess, J., Jacobson, K.A. Glutamate residues in the second extracellular loop of the human A2a adenosine receptor are required for ligand recognition, Mol. Pharmacol. 1996, 49(4), 683–91.
57. Jacobson, M.P., Pincus, D.L., Rapp, C.S., Day, T.J., Honig, B., Shaw, D.E., Friesner, R.A. A hierarchical approach to all-atom protein loop prediction, Proteins 2004, 55(2), 351–67.
58. Mehler, E.L., Hassan, S.A., Kortagere, S., Weinstein, H. Ab initio computational modeling of loops in G-protein-coupled receptors: lessons from the crystal structure of rhodopsin, Proteins 2006, 64(3), 673–90.

59. Hassan, S.A., Mehler, E.L., Zhang, D., Weinstein, H. Molecular dynamics simulations of peptides and proteins with a continuum electrostatic model based on screened Coulomb potentials, Proteins 2003, 51(1), 109–25.
60. SYBYL, Tripos Inc., St. Louis, MO 63144.
61. Sansom, M.S., Weinstein, H. Hinges, swivels and switches: the role of prolines in signalling via transmembrane alpha-helices, Trends Pharmacol. Sci. 2000, 21(11), 445–51.
62. Sankararamakrishnan, R., Vishveshwara, S. Geometry of proline-containing alpha-helices in proteins, Int. J. Pept. Protein Res. 1992, 39(4), 356–63.
63. Visiers, I., Weinstein, H., Rudnick, G., Stephan, M.M. A second site rescue mutation partially restores functional expression to the serotonin transporter mutant V382P, Biochemistry 2003, 42(22), 6784–93.
64. Govaerts, C., Blanpain, C., Deupi, X., Ballet, S., Ballesteros, J.A., Wodak, S.J., Vassart, G., Pardo, L., Parmentier, M. The TXP motif in the second transmembrane helix of CCR5. A structural determinant of chemokine-induced activation, J. Biol. Chem. 2001, 276(16), 13217–25.
65. Singh, R., Hurst, D.P., Barnett-Norris, J., Lynch, D.L., Reggio, P.H., Guarnieri, F. Activation of the cannabinoid CB1 receptor may involve a W6 48/F3 36 rotamer toggle switch, J. Pept. Res. 2002, 60(6), 357–70.
66. Shi, L., Liapakis, G., Xu, R., Guarnieri, F., Ballesteros, J.A., Javitch, J.A. Beta2 adrenergic receptor activation. Modulation of the proline kink in transmembrane 6 by a rotamer toggle switch, J. Biol. Chem. 2002, 277(43), 40989–96.
67. Sherman, W., Day, T., Jacobson, M.P., Friesner, R.A., Farid, R. Novel procedure for modeling ligand/receptor induced fit effects, J. Med. Chem. 2006, 49(2), 534–53.
68. Halgren, T.A., Murphy, R.B., Friesner, R.A., Beard, H.S., Frye, L.L., Pollard, W.T., Banks, J.L. Glide: a new approach for rapid, accurate docking and scoring. 2. Enrichment factors in database screening, J. Med. Chem. 2004, 47(7), 1750–9.
69. Shacham, S., Marantz, Y., Bar-Haim, S., Kalid, O., Warshaviak, D., Avisar, N., Inbal, B., Heifetz, A., Fichman, M., Topf, M., Naor, Z., Noiman, S., Becker, O.M. PREDICT modeling and in-silico screening for G-protein coupled receptors, Proteins 2004, 57(1), 51–86.
70. Trabanino, R.J., Hall, S.E., Vaidehi, N., Floriano, W.B., Kam, V.W., Goddard, W.A. First principles predictions of the structure and function of g-protein-coupled receptors: validation for bovine rhodopsin, Biophys. J. 2004, 86(4), 1904–21.
71. Schertler, G.F. Structure of rhodopsin, Eye 1998, 12(Pt 3b), 504–10.
72. Filizola, M., Perez, J.J., Carteni-Farina, M. BUNDLE: a program for building the transmembrane domains of G-protein-coupled receptors, J. Comput. Aided Mol. Des. 1998, 12(2), 111–8.
73. Lefkowitz, R.J., Cotecchia, S., Samama, P., Costa, T. Constitutive activity of receptors coupled to guanine nucleotide regulatory proteins, Trends Pharmacol. Sci. 1993, 14(8), 303–7.
74. Samama, P., Cotecchia, S., Costa, T., Lefkowitz, R.J. A mutation-induced activated state of the beta 2-adrenergic receptor. Extending the ternary complex model, J. Biol. Chem. 1993, 268(7), 4625–36.
75. Seifert, R., Wenzel-Seifert, K., Gether, U., Kobilka, B.K. Functional differences between full and partial agonists: evidence for ligand-specific receptor conformations, J. Pharmacol. Exp. Ther. 2001, 297(3), 1218–26.
76. Sun, Y., Huang, J., Xiang, Y., Bastepe, M., Juppner, H., Kobilka, B.K., Zhang, J.J., Huang, X.Y. Dosage-dependent switch from G protein-coupled to G protein-independent signaling by a GPCR, Embo J. 2007, 26(1), 53–64. Epub 2006 Dec 14.
77. Suzuki, S., Chuang, L.F., Yau, P., Doi, R.H., Chuang, R.Y. Interactions of opioid and chemokine receptors: oligomerization of mu, kappa, and delta with CCR5 on immune cells, Exp. Cell Res. 2002, 280(2), 192–200.
78. Gomes, I., Filipovska, J., Devi, L.A. Opioid receptor oligomerization. Detection and functional characterization of interacting receptors, Methods Mol. Med. 2003, 84, 157–83.
79. El-Asmar, L., Springael, J.Y., Ballet, S., Andrieu, E.U., Vassart, G., Parmentier, M. Evidence for negative binding cooperativity within CCR5-CCR2b heterodimers, Mol. Pharmacol. 2005, 67(2), 460–9. Epub 2004 Oct 27.
80. Dziedzicka-Wasylewska, M., Faron-Gorecka, A., Andrecka, J., Polit, A., Kusmider, M., Wasylewski, Z. Fluorescence studies reveal heterodimerization of dopamine D1 and D2 receptors in the plasma membrane, Biochemistry 2006, 45(29), 8751–9.

81. Breit, A., Gagnidze, K., Devi, L.A., Lagace, M., Bouvier, M. Simultaneous activation of the delta opioid receptor (deltaOR)/sensory neuron-specific receptor-4 (SNSR-4) hetero-oligomer by the mixed bivalent agonist bovine adrenal medulla peptide 22 activates SNSR-4 but inhibits deltaOR signaling, Mol. Pharmacol. 2006, 70(2), 686–96. Epub 2006 May 8.
82. Breit, A., Lagace, M., Bouvier, M. Hetero-oligomerization between beta2- and beta3-adrenergic receptors generates a beta-adrenergic signaling unit with distinct functional properties, J. Biol. Chem. 2004, 279(27), 28756–65. Epub 2004 Apr 27.
83. Nakata, H., Yoshioka, K., Kamiya, T., Tsuga, H., Oyanagi, K. Functions of heteromeric association between adenosine and P2Y receptors, J. Mol. Neurosci. 2005, 26(2–3), 233–8.
84. Hubbell, W.L., Cafiso, D.S., Altenbach, C. Identifying conformational changes with site-directed spin labeling, Nat. Struct. Biol. 2000, 7(9), 735–9.
85. Dunham, T.D., Farrens, D.L. Conformational changes in rhodopsin. Movement of helix f detected by site-specific chemical labeling and fluorescence spectroscopy, J. Biol. Chem. 1999, 274(3), 1683–90.
86. Farrens, D.L., Altenbach, C., Yang, K., Hubbell, W.L., Khorana, H.G. Requirement of rigid-body motion of transmembrane helices for light activation of rhodopsin, Science 1996, 274(5288), 768–70.
87. Ghanouni, P., Steenhuis, J.J., Farrens, D.L., Kobilka, B.K. Agonist-induced conformational changes in the G-protein-coupling domain of the beta 2 adrenergic receptor, Proc. Natl. Acad. Sci. USA 2001, 98(11), 5997–6002.
88. Brooks, B.R., Bruccoleri, R.E., Olafson, B.D., States, D.J., Swaminathan, S., Karplus, M. CHARMM: a program for macromolecular energy, minimization, and dynamics calculations, Journal of Computational Chemistry 1983, 4(2), 187–217.
89. Niv, M.Y., Skrabanek, L., Filizola, M., Weinstein, H. Modeling activated states of GPCRs: the rhodopsin template, J. Comput. Aided Mol. Des. 2006, 20(7–8), 437–48. Epub 2006 Nov 11.

INDEX

^{12}C ^{16}O$_2$, 168
3D-QSAR, 67, 71
π–π interactions, 183

ab initio, 215, 219, 220
ACPF, 163
action optimization, 17, 19
activated state, 220–222
active database, 157
Active Thermochemical Tables, 159
adiabatic Jacobi correction (AJC), 158
"alchemical" free energy transformations, 41–53
alignment-independent molecular descriptors, 69
anisotropic polarizability tensors, 180
ANO basis, 201
apparent errors, 196
aromatic cluster, 212, 221
asymmetric top notation, 159
atomistic simulations
    time scale, 15
    transition path methods, 16
    boundary value formulation in length (*see* stochastic difference equation in length)
aug-cc-pVnZ, 198

B-factors, 32, 34, 35
basis sets, 195
Bennett acceptance ratio, 44, 45
benzene dimers, 188
benzene–water, 186
Bessel-DVR, 167
BO approximation, 158
body-fixed frame, 166
bond vector(s), 167, 168
Born–Oppenheimer (BO), 156
Bragg's Law, 89, 90, 97
Breit, 164
Breit term, 163

cc-pCVnZ, 198, 199
cc-pV(n+d)Z, 197
cc-pVnZ, 196, 199, 202

cc-pVnZ-DK, 200, 202
cc-pVnZ-PP, 201, 202
cc-pwCVnZ, 198, 199
CCSD(T), 160
CH$_2$ radical, 156
charge transfer interactions, 180
Chemical Markup Language (CML), 116, 126
CO$_2$, 162, 168
cold shock proteins (CSP), 24
complete basis set, 196
complete basis set (CBS) full configuration interaction (FCI), 156
conformational change(s), 32–36
conformational restraints, 49, 50
conformational sampling, 48, 49
continuum solvation models, 181
core correlation, 198, 203
core-valence, 199, 202
correlating functions, 197
correlation-consistent, 160, 196
CPHMD, 6
Crooks relationship, 45
cross validation
    leave-group-out, 67
    leave-one-out, 67
Crystallographic Courseware, 96
Cu, Zn superoxide dismutase (SOD), 24, 25
curvilinear, 166
CVRQD, 161–164
CYP inhibitor, 65, 71
CYP substrate, 65, 71
cytochrome c, 22
cytochrome P450, 63, 64

D/ERY motif, 211
D2.50, 211
database, 169
Davidson correction, 163
DBOC, 160, 163
DEWE, 168
diagonal Born–Oppenheimer corrections (DBOC), 158
digital repository, 103, 107, 108, 125, 129

229

Index

digital-object identifier (DOI), 99, 102, 103, 109, 110, 115, 119, 121, 124, 129
dipole polarizability, 179
discrete path sampling (DPS), 16
discrete variable representation (DVR), 166
displacement coordinates, 168
dissipative MD, 139
distributed multipolar expansion, 179
DKH, 200
DMS, 156
DMSs, 163, 165
DOPI, 166, 168
drug discovery, 64
drug-drug interactions, 63
Dublin-core metadata (DC), 104, 107, 108, 125
DVR, 167

E6.30, 211
Eckart–Watson Hamiltonians, 167
effective core potentials, 200
effective fragment potential (EFP), 178
EFP, 178, 190
    induced dipoles, 181
    EFP-QM, 182
    EFP/PCM, 181
elastic network model(s), 31–37
electrostatic interaction, 179
empirical PESs, 164
Euler angles, 168
exchange repulsion, 179, 180
exponential damping functions, 180
extensible metadata platform (XMP), 104, 107, 109–111

FCI, 160
few-body systems, 158
few-electron systems, 156
first-principles thermochemistry, 160
FIS3, 161, 162, 164
FKBP, 52
focal-point approach (FPA), 160
folding intermediate states, 9
force fields, 162
FPA, 160
free energy calculations, 41–53
functional microdomains, 211

G-protein coupled receptors (GPCRs), 209
GAMESS, 190
Gaussian quadratures, 166
generalized finite basis representation (GFBR), 167
GRAFS, 210

[H,C,N], 163
H_2, 158
H_2^+-like systems, 158
$H_2^{16}O$, 160, 164
$H_2^{17}O$, 159, 160, 164
$H_2^{18}O$, 164
H_2O, 162, 163, 168
H_2S, 163
H_2^+, 158
Hartree–Fock (HF), 160
HEAT (High-accuracy Extrapolated *Ab initio* Thermochemistry), 160
HF limit, 197
high-resolution spectra, 157
homology models, 211

ICMRCI, 163
implicit solvent, 5
Induced Fit, 218
information triple, 109, 110, 128, 131
internal coordinates, 166
International Chemical Identifier (InChI), 108–110, 120, 125–128
intrinsic errors, 196

Jacobi coordinates, 158
Jarzynski relationship, 45, 46
Jmol, 99, 113–117, 119–121, 125, 126

Lamb-shift, 163, 164
lambda dynamics, 6
Lanczos technique, 166
Langevin, 140, 144, 145
level density, 156
ligand binding, 42, 43, 51
LOOPSEARCH, 216

MARVEL, 157–162, 165
MaxFlux, 16
maximum likelihood methods, 44
MEMBSTRUCK, 220
metadata, 99, 101–112, 116, 118, 119, 121, 123–126, 131
metal surface, 137
Miller indices h, k, l, 91
MLR, 67
model applicability domain, 68, 74
MODELLER, 213
MODLOOP, 216
MOE, 214
MOIL, 19
molecular descriptors, 66
molecular dynamics, 140
    atomistic models, 143
    coarse-grained, 138, 144
    with electronic friction, 143

molecular interaction field, 66
multicanonical methods, 48
MULTIMODE, 166
multiple sequence alignment, 211–213
MV, 163
MVD1, 164
MVD2, 163

n-mode representation, 167
N_2O, 162
N1.50, 211
N7.49, 211, 212
nonadiabatic, 158
nonequilibrium work, 45, 46
normal coordinates, 163, 167, 168
normal mode, 159
NPXXY motif, 212
nuclear motion computations, 166
nuclear-motion, 169
nudged-elastic-band (NEB) method, 16

objectives for teaching crystallography, 86–89
OMTKY3, 189
Onsager-Machlup action, 17, 18
optical interference, 96
orthogonal coordinates, 166

partial least squares (PLS), 67
PCM induced charges, 181
periodic boundary conditions, 181
perturbation theory (PT), 156
pH-coupled molecular dynamics, 4
pH-modulated helix-coil transitions, 9
pK_a, 4, 188
pK_a prediction, 4
PLOP, 216
Podcast, 99, 118–121, 131
point group symmetry, 94
polarization consistent, 196
portable-document-format (PDF, Acrobat), 99, 101–107, 110, 116, 118, 131
potential energy surface (PES), 156, 158–160, 162–167
PREDICT, 219
PRIME, 214
projective models, 144
proline, 213, 216, 221
protein A, 22
protein folding, 22
protein misfolding and aggregation, 9
pseudopotentials, 200

QED, 158, 163
QM/EFP/PCM, 181
QM/MM, 182, 188, 190

QSAR, 66
quantum electrodynamics (QED), 155
quantum number, 164
QZVPP, 197

rare event, 140
reaction kinetics, 158
receptor activation, 221
refinement, 216, 218, 219
relativity, 200
replica-exchange, 7
resource description framework (RDF), 99, 107–112, 116, 123, 124, 126, 128
rotational–vibrational
    energy levels, 159
    spectra, 169
    transitions, 159
rovibrational eigenvalues, 157

SDEL (see stochastic difference equation in length)
semantic Wiki, 110, 123, 126–128, 131
sextic force fields, 162
small molecule solvation, 50
"soft core" Lennard-Jones interactions, 47
space group symmetry, 94
spectroscopic accuracy, 157
spectroscopic network (SN), 159
spherical harmonics, 167
standard pK_a, 4
standard uncertainty (su), 87
Steepest Descent Path (SDP), 19
stochastic difference equation in length (SDEL), 17–19
    advantages, 20
    disavantages, 20
stochastic difference equation in time (SDET), 17
string method, 16
structural mimicry, 217
structural motifs, 211
surface diffusion, 138, 140
Sutcliffe–Tennyson triatomic rovibrational Hamiltonian, 167

T4 lysozyme, 52
TASSER, 220
tautomeric interconversion, 7
thermochemistry, 158
thermodynamic integration (TI), 44, 45
time-independent Schrödinger equation, 167
toggle switch, 212
transition path sampling (TPS), 16
transition path theory, 16

transition state theory, 141
two-electron integrals, 182

vibrational
    band origins (VBOs), 164, 168
    energy levels, 161
    states, 160
VPT2, 163

water dimer, 188
water–benzene dimer, 186, 188

Web 2.0, 100, 111, 122, 124, 131
weighted histogram analysis method (WHAM), 44, 45, 47
Wigner rotation functions, 166
Wiki, 99, 103, 108, 110, 117, 121–131
Wikipedia, 99, 112, 122, 124, 129, 131
Wn (Weizmann-n), 160

Zwanzig relationship, 43, 44

CUMULATIVE INDEX VOLS 1–2

ab initio modelling, 1, 187–8
ab initio thermochemical methods, 1, 33, 37, 45
absorption
    intestinal, 1, 137–8
    see also ADMET properties
accelerated molecular dynamics, 2, 230
active transport, 1, 139–40
acyl carrier protein synthase (AcpS), 1, 179
adenosine triphosphate (ATP) site recognition, 1, 187–8
adiabatic approximations, 1, 20, 25, 27
ADMET properties
    active transport, 1, 139–40
    aqueous solubility, 1, 135–7, 162
    blood–brain barrier permeation, 1, 140–2
    computational prediction, 1, 133–51
    cytochrome P450 interactions, 1, 143–4
    drug discovery, 1, 159–62
    efflux by P-glycoprotein, 1, 140, 160–1
    intestinal absorption, 1, 137–8
    intestinal permeability, 1, 134–5, 161
    metabolic stability 142–3, 162
    oral bioavailability, 1, 134, 138–9, 159–60
    plasma protein binding, 1, 142
    toxicity, 1, 144
AGC group of kinases, 1, 196
agrochemicals, 1, 163
AK peptide, 2, 91
AMBER force fields, 1, 92, 94–7, 99, 119–21
AMBER, 2, 91
angular wavefunctions, 1, 225–8
applicability domain, 2, 113, 118, 120, 123, 125
aqueous solubility, 1, 135–7, 162
atomic orbital representations, 1, 225–8
atomistic simulation
    boundary conditions, 1, 80
    experimental agreement, 1, 77–8
    force fields, 1, 77, 79, 80–2
    methodological advances, 1, 79
    nucleic acids, 1, 75–89
    predictive insights, 1, 78–9
    sampling limitations, 1, 80–2
ATP *see* adenosine triphosphate
AUTODOCK, 1, 122–3; 2, 184

B3LYP functional, 1, 32, 48–50
back-propagation neural networks (BPNN), 1, 136–7
Bad, 2, 197, 203
bagging, 2, 136
Bak, 2, 197–8, 203–5
barrier heights, 2, 64, 73
base pair opening, 1, 77
basis set superposition errors (BSSE), 2, 68, 74, 76, 78
basis sets, 1, 13–15, 32–3
Bax, 2, 197–8, 203–4
Bayes model, 2, 157
Bayesian methods, 2, 132
Bcl-2, 2, 197–8, 201, 203–6
Bcl-xL, 2, 197, 203–6
Betanova, 1, 248–9
Bethe–Salpeter equation, 1, 27
bias potential, 2, 224–6, 229–30
Bid, 2, 197, 203, 205
Bim, 2, 197, 203
binding affinities, 1, 78
binding free energy
    calculating, 1, 114–19
    protein–ligand interactions, 1, 113–30
    scoring functions, 1, 119–26
bioavailability, 1, 134, 138–9, 159–60
bio-molecular simulation
    atomistic simulation, 1, 75–82
    nonequilibrium approaches, 1, 108
    protein force fields, 1, 91–102
    protein–ligand interactions, 1, 113–30
    water models, 1, 59–74
biospectrum similarity, 2, 150
Bleep, 2, 162
blood–brain barrier permeation, 1, 140–2, 160–1
bond breaking
    configuration interaction, 1, 51
    coupled cluster methods, 1, 52–3
    generalized valence bond method, 1, 47–8
    Hartree–Fock theory, 1, 46, 48–51
    multireference methods, 1, 51–3

perturbation theory, 1, 51–2
potential energy surface, 1, 54
quantum mechanics, 1, 45–56
self-consistent field methods, 1, 46–7, 53
spin-flip methods, 1, 53
boost energy, 2, 225–7
boosting, 2, 136, 151
Born–Oppenheimer approximation, 1, 3, 54
Boss program, 2, 264
boundary conditions, 1, 80
Boyer Commission, 1, 206–7
BPNN see back-propagation neural networks
Bridgman tables, 1, 224
BSSE see basis set superposition errors

CAMK group of kinases, 1, 186, 196
Carnegie Foundation, 1, 206–7
casein kinase 2 (CK2), 1, 197
Casida's equations, 1, 21–2, 25
caspase-3, 2, 206
caspase-9, 2, 206, 208
CASSCF see complete-active-space self-consistent field
CATS3D, 2, 149
CBS-n methods, 1, 36–7
CC see coupled cluster
CD see circular dichroism
CDKs see cyclin-dependent kinases
central nervous system (CNS) drugs, 1, 160–1
chance correlations, 2, 153
charge transfer (CT), 1, 26
CHARMM force fields, 1, 77, 79, 92–5, 97–9, 119–20
chemical amplification, 2, 11
Chemical Kinetics Simulator, 2, 4
chemical space (size of), 2, 143
chemical vapor deposition (CVD), 1, 232–3
ChemScore, 2, 162
CI see configurational interaction
circular dichroism (CD) spectra, 1, 22–4
circular fingerprints, 2, 144
cis-trans isomerization, 2, 228, 229
cluster-based computing, 1, 113
CMAP see correction maps
CMGC group of kinases, 1, 186, 192–4
CMS see central nervous system
combinatorial QSAR, 2, 113, 120
CoMFA, 2, 152
Compartmentalization, 2, 11
complete-active-space self-consistent field (CASSCF) method, 1, 47, 53
compound equity, 1, 171
computational protein design (CPD), 1, 245–53
degrees of freedom, 1, 246
energy function, 1, 246–7

examples, 1, 248–50
search methods, 1, 247–8
solvation and patterning, 1, 247
target structures, 1, 246
computational thermochemistry
ab initio methods, 1, 33, 37, 45
CBS-n methods, 1, 36–7
density functional theory, 1, 32–3
empirical corrections, 1, 34–6
explicitly correlated methods, 1, 39
G1, G2, G3 theory, 1, 34–6
hybrid extrapolation/correction, 1, 36–7
isodesmic/isogyric reactions, 1, 34
nonempirical extrapolation, 1, 37–9
quantum mechanics, 1, 31–43
semi-empirical methods, 1, 31–2
Weizmann-n theory, 1, 37–9
configurational interaction (CI), 1, 9–10, 48, 51
configurational space, 2, 84
conformational changes, substrate induced P450, 2, 173
conformational flexibility, 1, 173
Conformational flooding, 2, 221, 223–4
Conformational Transitions, 2, 221–2, 227
consensus approaches, 1, 145
consensus scoring, 2, 158
correction maps (CMAP), 1, 95, 96, 98
correlation energy, 2, 53–4, 59–62, 64–71, 73–4, 76
correlation methods, 1, 8–11
Council for Chemical Research, 1, 240
Council on Undergraduate Research (CUR), 1, 206–7, 208
coupled cluster (CC) methods, 1, 10–11, 37–40, 48–50, 52–3
CPD see computational protein design
cross-validation, 2, 153–4
CT see charge transfer
CUR see Council on Undergraduate Research
current density, 1, 27
CVD see chemical vapor deposition
cyclin-dependent kinases (CDKs), 1, 186, 192–4
cytochrome P450, 2, 171
3A4, 2, 172
BM-3, 2, 174
2C5, 2, 172
2C9, 2, 172
eryF, 2, 174
terp, 2, 174
cytochrome P450 interactions, 1, 143–4

3D QSAR, 2, 182
D&C see divide and conquer

DA *see* discriminant analysis
database mining, 2, 114, 121–5
databases
    drug-likeness, 1, 155–6
    ligand-based screening, 1, 172–5
    self-extracting, 1, 223, 225
    symbolic computation engines, 1, 223, 224–5
dead-end elimination (DEE), 1, 247–8, 249
degrees of freedom, 1, 246
de novo protein design, 1, 245
density fitting, 2, 55, 74, 77
density functional theory (DFT)
    bond breaking, 1, 48–9
    computational thermochemistry, 1, 32–3
    protein–ligand interactions, 1, 116
    state of the art, 1, 4, 11–12, 13–15
    time-dependent, 1, 20–30
descriptor binarization effect, 2, 152
DEZYMER algorithm, 1, 249
DF-LCCSD(T), 2, 55
DF-LMP2, 2, 55, 73, 75
DFT *see* density functional theory
    discriminant analysis (DA), 1, 138
distant pairs, 2, 54, 62–3
distributed computing, 1, 113
distribution *see* ADMET properties
divide and conquer (D&C) algorithm, 1, 116–17
DMA gyrase, 2, 280
DOCK program, 1, 173–4, 177, 178, 189
DOCK, 2, 157, 159, 161, 179, 184–6, 299–303, 308, 314–7, 319–20
docking, 1, 79, 114, 119, 121, 155, 169, 172–4, 178, 189–96; 2, 141, 145, 157, 159, 161–2, 284, 297–303, 305–7, 309, 311, 313–21, 323
DockIt, 2, 299–300, 317
DockScore, 2, 161
DockVision, 2, 299–300, 315–17
domain approximation, 2, 53, 64, 73–6, 78
domain extensions, 2, 54, 59, 62–3, 77
drug discovery, 1, 155–68
    agrochemicals, 1, 163
    aqueous solubility, 1, 162
    chemistry quality, 1, 157
    CMS drugs, 1, 160–1
    databases, 1, 155–6
    drug-likeness, 1, 155–7
    intestinal permeability, 1, 161
    lead-likeness, 1, 159
    metabolic stability, 1, 162
    oral drug activity, 1, 159–60
    positive desirable chemistry filters, 1, 158–9
    promiscuous compounds, 1, 162–3
drug-likeness, 1, 155–7; 2, 160

DrugScore, 2, 161–2
D-Score, 2, 161

education
    research-based experiences, 1, 205–14
    stochastic models, 1, 215–20
    symbolic computation engines, 1, 221–35
efflux by P-glycoprotein, 1, 140, 160–1
EFP, 2, 267
electron correlation methods, 1, 8–11
electronic embedding, 2, 37
electronic Schrödinger equation, 1, 3–15
empirical force fields, 1, 91–102
empirical scoring functions, 1, 122–3
energy function, 1, 246–7
enrichment, 2, 297, 302–3, 305–9, 313–19
essential dynamics, 2, 233, 236, 242–4, 247
Ewald summation techniques, 1, 59, 62, 75
Ewald summation, 2, 265
exact exchange, 1, 26–7
excited state structure/dynamics, 1, 24
excretion *see* ADMET properties
explicit solvent, 2, 98–9, 101–2, 104–6
extended systems, 1, 26

feature selection, 2, 151, 153
FEP *see* free energy perturbation
FEPOPS, 2, 146
Fingal, 2, 148
FlexX, 1, 173, 178, 189; 2, 157, 159, 184, 186, 299–300, 308, 313–19
Flo+, 299–300, 317
FLO99, 1, 178
Florida Memorial College, 1, 212
fluctuation theorem, 1, 109
fluid properties, 1, 239–44
focal-point approach, 1, 39
force fields
    molecular simulations, 1, 239–40
    nucleic acids, 1, 77, 79, 80–2
    protein–ligand interactions, 1, 116, 119–21
    proteins, 1, 91–102
    structure-based lead optimization, 1, 177
fragment positioning, 1, 175–7
FRED, 2, 148, 161, 299–300, 313–14, 317, 319
free energy, 1, 96, 103–11, 113–30
free energy perturbation (FEP), 1, 104, 106; 2, 265
F-Score, 2, 161
fuzzy clustering, 2, 160
fuzzy logic, 1, 218

G1, G2, G3 theory, 1, 34–6
Gaussian Geminal Methods, 2, 25
GB-1 beta hairpin, 2, 91–2

generalized Born, 2, 222
generalized conductor-like screening model (GCOSMO), 2, 266
generalized gradient approximation (GGA), 1, 12
generalized valence bond (GVB) method, 1, 47–8
Ghose/Crippen descriptors, 2, 160
Glide, 2, 161, 299–300, 302–3, 313–19
global matrices, 1, 116–17
glutathione peroxidase, 2, 47
GOLD, 2, 161–2, 184–6, 299–300, 313–19
graphical representations, 1, 225–8, 232–3
GRID, 2, 148–9
GRIND, 2, 148
GROMACS, 2, 89, 91
GROMOS force fields, 1, 97
GROMOS, 2, 91
G-Score, 1, 123; 2, 161
GVB *see* generalized valence bond

Hartree–Fock (HF) method, 1, 4, 5–11, 13–15, 20–1, 46, 48–51
HDM2, 2, 209
Hellmann–Feynman theorem, 1, 21
hierarchical protein design, 1, 245
high throughput docking (HTD), 2, 298–302, 304–6, 308, 309, 317–20
high-throughput screening (HTS), 1, 171–2
HINT, 2, 162
Hohenberg–Kohn (HK) theorem, 1, 11, 20
homodesmotic reactions, 1, 34
homology models, 1, 170, 188–9
HTD *see* high throughput docking
HTS *see* high-throughput screening
HTS data analysis, 2, 156
HTS Data Mining and Docking Competition, 2, 159
hybrid quantum and molecular mechanical simulation (QM/MM), 2, 263–8
hybrid solvent, 2, 106
hybridization, structure-based, 1, 191–2
hydration free energies, 1, 103
Hylleraas Method, 2, 21
Hylleraas-CI method, 2, 24
Hyperdynamics, 2, 221, 224–5

IAPs, 2, 206
ICM, 2, 299–300, 308, 313–14, 318–19
IL-2, 2, 214
implicit solvent, 2, 99–100
intermolecular potential functions, 1, 241–2
intestinal absorption, 1, 137–8
intestinal permeability, 1, 134–5, 161
iron chelation, modeling of, 2, 185

isodesmic/isogyric reactions, 1, 34

Jarzynski relationship, 1, 103, 104–10

Kemp decarboxylation, 2, 263–4, 271–3, 275
kinome targeting, 1, 185–202
    applications, 1, 192–7
    ATP site recognition, 1, 187–8
    homology models, 1, 188–9
    kinase family, 1, 186–7
    methodology, 1, 188–92
    selectivity, 1, 190–1
    structure-based hybridization, 1, 191–2
    virtual screening, 1, 189–90
knowledge-based scoring functions, 1, 123–5
Kohn–Sham (KS) equations, 1, 11, 20–2, 25
Kohonen maps, 2, 181
Kriging, 2, 151

laboratory course modules, 1, 7
LCCSD(T), 1, 54, 62, 71, 78
LCCSD(TO), 1, 64
lead optimization *see* structure-based lead optimization
lead-likeness, 1, 159
Lennard–Jones (LJ) potential, 1, 93–4, 116, 121
LES *see* locally enhanced sampling
library enumeration, 1, 178
ligand binding, 1, 103
ligand-based screening, 1, 172–5, 178–9
LigandFit, 2, 299–300, 302–3, 315–17, 319
LigScore2, 2, 161
linear interaction energy, 1, 117
Linear R12 methods, 2, 28
linear scaling, 2, 54–5, 62, 64, 77
LINGO, 2, 146
link atoms, 2, 37
LJ *see* Lennard–Jones
LMP2, 2, 55, 60–78
local correlation, 2, 53, 77
local coupled cluster, 2, 54
local spin density approximation, 1, 11–12
localized orbitals, 2, 53–4, 57
locally enhanced sampling (LES), 1, 79
LUDI scoring function, 1, 123, 173
lysozyme, 2, 199

many-body perturbation theory, 1, 10
Maple, 1, 228, 230–2
master equations, 1, 115–16, 119–20
Mathematical Association of America, 1, 215–16
maximum common substructure, 2, 160
MC *see* Monte Carlo

MCSCF *see* multi-configurational self-consistent field
MCSS program, 1, 173–4, 177
MD *see* molecular dynamics
MDM2, 2, 197, 200, 209–11
mechanical embedding, 2, 37
Menshutkin reaction, 2, 263, 265–8, 275
metabolic stability, 1, 142–3, 162
    see also ADMET properties
MLR *see* multiple linear regression
MM *see* molecular mechanics
Model scope, 2, 155
MOEDock, 2, 299–300, 317
molecular descriptors, 2, 141, 144–6, 151
molecular dynamics (MD) simulation, 1, 75–8, 217, 239, 242
molecular dynamics, 2, 98–9, 221–4, 227–30, 233–8, 243–4, 246–7
molecular mechanics (MM), 1, 119, 120–2
molecular modeling, 1, 59–130
    atomistic simulation of nucleic acids, 1, 75–89
    free energy, 1, 103–11, 113–30
    nonequilibrium approaches, 1, 103–11
    protein force fields, 1, 91–102
    protein–ligand interactions, 1, 113–30
    water models, 1, 59–74
        TIP4P, 1, 62–4, 69–72
        TIP4P-EW, 1, 64–5, 69–72
        TIP5P, 1, 65–7, 69–72
        TIP5P-E, 1, 67–72
molecular orbital representation, 1, 229–31
molecular similarity, 2, 141
molecular simulations, 1, 177–8, 239–44
Møller–Plesset form, 1, 10, 48–50
MOLPRINT 2D, 2, 145
Monte Carlo methods, 1, 216–18, 239, 242, 247–8
Monte Carlo simulation (MC), 2, 263–8, 270–1, 273, 275
multi-configurational self-consistent field (MCSCF) method, 1, 9–10, 46–7
multiple excitations, 1, 25
multiple linear regression (MLR), 1, 136
multipole approximations, 2, 62 multireference methods, 1, 51–3

National Science Foundation (NSF), 1, 206–7, 209
neural networks, 2, 181
nonequilibrium approaches
    computational uses, 1, 109
    experimental applications, 1, 108
    free energy calculations, 1, 103–11
    Jarzynski relationship, 1, 103, 104–10

theoretical developments, 1, 108–9
nonlinear models, 2, 152
NR, 2, 211
NSF *see* National Science Foundation
nuclear hormone receptor, 2, 211
nucleic acids, 1, 75–89
nucleophilic aromatic substitution (S$_N$Ar), 2, 263–4
nuisance compounds, 1, 162–3, 190

ONIOM, 2, 35
OPLS/AA force fields, 1, 92–4, 97
OPLS-VA/VA force fields, 2, 265, 273
oral bioavailability, 1, 134, 138–9, 159–60
oral drug activity, 1, 159–60
orbital domains, 2, 58–9, 61–3
orbital representations, 1, 225–8, 229–31
oscillating systems, 1, 232–3
overfilling, 2, 154

p53, 2, 197, 200, 209–11
PAO, 2, 53–62, 68
parallel computing, 1, 242
PARAM force fields, 1, 97
partial least squares (PLS) analysis, 1, 134–5, 138
patterning, 1, 247
PB *see* Poisson–Boltzmann
PCM, 2, 266, 271, 275
PDB *see* Protein Data Bank
PDBbind, 2, 161
PDDG/PM3, 2, 263–5, 267–8, 273–5
PDF inhibitor, 2, 288
permeability, intestinal, 1, 134–5, 161
perturbation theory, 1, 10, 51–2
PES *see* potential energy surface
p-glycoprotein, 1, 140, 160–1
pharmaceutical chemicals
    ADMET properties, 1, 133–51
    drug discovery, 1, 155–68
    structure-based lead optimization, 1, 169–83
    virtual screening protocols, 1, 114, 120, 125
pharmacophore models, 1, 172–4
pharmacophores, 2, 182–3
PhDOCK, 1, 173–4, 177
physical chemistry, 1, 215–17
Pipek–Mezey localization, 2, 56, 68
plasma protein binding (PPB), 1, 142
PLP2, 2, 161
PLS *see* partial least squares
PMF, 2, 161–2, 263, 266
PMFScore, 1, 124–5
Poisson–Boltzmann (PB) equation, 1, 117–19, 120–2

polarizable continuum model (PCM), 2, 264, 266, 271
poly(organo)silanes, 1, 232–3
polymer-source chemical vapor deposition (PS-CVD), 1, 232–3
positive desirable chemistry filters, 1, 158–9
PostDOCK, 2, 157
potential energy landscape, 2, 221–4, 227, 229–30
potential energy surface (PES), 1, 3–4, 54
potential functions, 1, 241–2
Potential of mean force (PMF), 2, 263–8
PPB *see* plasma protein binding
predictive modeling, 1, 133–51, 240
principal component analysis, 2, 233, 235–6
privileged structures, 1, 158
probabilistic protein design, 1, 249–50
problem-solving templates, 1, 228
process design, 1, 231–2
projected atomic orbitals, 2, 53
promiscuous compounds, 1, 162–3, 190
Protein Data Bank (PDB), 1, 113, 117, 123–4
protein design, 1, 245–53
    degrees of freedom, 1, 246
    energy function, 1, 246–7
    examples, 1, 248–50
    search methods, 1, 247–8
    solvation and patterning, 1, 247
    target structures, 1, 246
protein force fields, 1, 91–102
    condensed-phase, 1, 94–6
    free energies of aqueous solvation, 1, 96
    gas-phase, 1, 94–6
    optimization, 1, 96–9
    united-atom, 1, 97
protein kinases *see* kinome targeting
protein–ligand interactions, 1, 113–30
protein–protein interaction, 2, 197–8, 200, 202–3, 205, 211, 214–15
PS-CVD *see* polymer-source chemical vapor deposition

QM/MM, 2, 35, 263–8, 270–1, 273–5
QSAR/QSPR models, 1, 133–51
quantum–classical enzymatic calculations, 1, 103
quantum mechanics, 1, 3–56
    basis sets, 1, 13–15, 32–3
    bond breaking, 1, 45–56
    computational thermochemistry, 1, 31–43
    configurational interaction, 1, 9–10, 48, 51
    coupled cluster methods, 1, 10–11, 37–40, 48–50, 52–3
    density functional theory, 1, 4, 11–12, 13–15, 32–3, 48–9
    electron correlation methods, 1, 8–11
    generalized valence bond method, 1, 47–8
    Hartree–Fock method, 1, 4, 5–11, 13–15, 20–1, 46, 48–51
    perturbation theory, 1, 10, 51–2
    potential energy surface, 1, 3–4, 54
    self-consistent field methods, 1, 6–8, 9–10, 37, 46–7, 53
    semi-empirical methods, 1, 12–13, 15
    symbolic computation engines, 1, 225–8
    time-dependent density functional theory, 1, 20–30
quasi-static (QS) transformations, 1, 105, 133–51

random forest, 2, 136, 151
RASSCF *see* restricted-active-space self-consistent field
reaction energies, 2, 53–4, 64, 71, 74–5, 77
REMD *see* Replica Exchange Molecular Dynamics
re-parameterizations, 1, 59, 60–1, 67, 72
Replica Exchange Molecular Dynamics, 2, 83, 85, 87, 89–91, 93, 95, 222
Replica exchange with solute tempering (REST), 2, 86
Research Experiences for Undergraduates (REU), 1, 209
research institutions, 1, 205–14
restrained electrostatic potential, 1, 92–3
restricted Hartree–Fock (RHF), 1, 46, 48–50
restricted-active-space self-consistent field (RASSCF) method, 1, 47
REU *see* Research Experiences for Undergraduates
R-group descriptor, 2, 147
RHF *see* restricted Hartree–Fock
RISM, 2, 266–7
ROC curve, 2, 297, 306–7, 315
ROCS, 2, 318
Roothaan–Hall equations, 1, 6–8
$Ru(bpy)_3^{2+}$, 7
Runge–Gross theorem, 1, 27

$S_N A$, 2, 270–1
$S_N Ar$, 2, 268–70, 275
sampling barriers, 1, 242–3
SAR *see* structure–activity relationships
scads, 1, 250
scaling methods, 1, 6–8
Schrödinger equation, 1, 3–15; 2, 297–9, 313–14, 316, 318–20
scoring functions, 1, 119–26
scoring functions, quality, 2, 161–2

self-consistent field (SCF) methods, 1, 6–10, 37, 46–7, 53
self-consistent reaction field (SCRF), 1, 118, 121
self-extracting databases, 1, 223, 225
semi-empirical methods, 1, 12–13, 15, 31–2
  PDDG/PM3, 2, 264, 265, 267, 268, 272, 274, 276
SHAKE algorithm, 2, 222
signal trafficking *see* kinome targeting
similar property principle, 2, 141
Slater geminal methods, 2, 28, 30
Smac, 2, 206, 208–9
solubility, 1, 135–7
solvation, 1, 117–19, 247
spin-flip methods, 1, 53
standard domains, 2, 53, 57, 59, 64, 68–9, 71, 73–6
statistical computational assisted design strategy (scads), 1, 250
Stochastic Gradient Boosting, 2, 137
stochastic models, 1, 215–20
storage capacity, 1, 224–5
strong pairs, 2, 59, 62–3, 68–9, 71, 73, 75, 77
structure–activity relationships (SAR), 1, 91, 133–51
structure-based design, 2, 197, 202, 205, 209
structure-based drug design, 1, 114, 120, 125
structure-based hybridization, 1, 191–2
structure-based lead optimization, 1, 169–83
  application to specific targets, 1, 179
  compound equity, 1, 171
  discovery, 1, 171–5
  fragment positioning, 1, 175–7
  high-throughput screening, 1, 171–2
  library enumeration, 1, 178
  ligand–target complex evaluation, 1, 178–9
  modification, 1, 175–9
  molecular simulation, 1, 177–8
  structure visualization, 1, 175
  virtual screening, 1, 169, 172–5
structure-based ligand design, 2, 184
structure-based virtual screening, 2, 284
structure-property relationships, 2, 142
substrate access, P450, 2, 178
substrate prediction, P450, 2, 172
support vector machines, 1, 137, 145; 2, 128, 149
Surflex, 2, 161
symbolic computation engines (SCE), 1, 221–35
  advanced application-specific procedures, 1, 229–31
  computation power, 1, 228–9
  emulation of professional software, 1, 229–31

  graphical representations, 1, 225–8, 232–3
  process design, 1, 231–2
  quantification, 1, 225, 231–3
  self-extracting databases, 1, 223
  specialized procedures, 1, 228–9
  storage capacity, 1, 224–5

target structures, 1, 246
TC5b, 2, 89
TDDFT *see* time-dependent density functional theory
temperature programmed-desorption, 2, 6
template approach, 1, 228–9
thermal conductivity, 1, 242–3
thermochemistry, computational, 1, 31–43
thermodynamics
  integration method, 1, 104
  nonequilibrium approaches, 1, 103–11
  protein–ligand interactions, 1, 113–30
  symbolic computation engines, 1, 224–5
  water models, 1, 59–72
thermogravimetric analysis, 2, 6
thyroid hormone, 2, 197, 201, 211
time-dependent density functional theory (TDDFT), 1, 20–30
  computational aspects, 1, 21–2
  developments, 1, 26–8
  electronic excitations, 1, 20–1
  exact exchange, 1, 26–7
  performance, 1, 22–4
  qualitative limitations, 1, 25–6
time-dependent Hamiltonian operators, 1, 104
TIP3P, 2, 86, 89, 266
TIP4P, 1, 62–4, 69–72; 2, 265–7
TIP4P-EW, 1, 64–5, 69–72
TIP5P, 1, 65–7, 69–72
TIP5P-E, 1, 67–72
TKL *see* tyrosine kinase-like
TKs *see* tyrosine kinases
Top7, 1, 249
toxicity, 1, 144, 190
  see also ADMET properties
TR, 2, 212
transamination, 1, 232–3
transferable intermolecular potential (TIP) water molecules, 1, 59–74
transition state theory, 2, 224, 229
Trp-cage, 2, 89–90, 93
Turbo Similarity Searching, 2, 153
two-electron integrals, 1, 6–7, 12–13
tyrosine kinase-like (TKL) group of kinases, 1, 186, 196–7
tyrosine kinases (TKs), 1, 186, 194–5

UHF *see* unrestricted Hartree–Fock
umbrella potential, 2, 223
umbrella sampling, 2, 221, 223–4, 228, 230
undergraduate research, 1, 205–14
Undergraduate Research Programs (URPs), 1, 208, 209–12
united-atom protein force fields, 1, 97
university research, 1, 205–14
unrestricted Hartree–Fock (UHF), 1, 46, 50–1
URPs *see* Undergraduate Research Programs

van't Hoff reactions, 1, 228–9
vertical excitation, 1, 22–4
virtual database screening, 2, 201
virtual screening, 1, 169, 172–5, 189–90; 2, 158
   high throughput, 1, 120
   protocols, 1, 114, 120, 125
Virtual Screening, performance assessment of algorithms, 2, 144
viscosity, 1, 242–3
visualization, 1, 175, 225–8, 232–3

water models, 1, 59–74; 2, 98, 102
   bio-molecular simulation, 1, 59–61
   effective fragment potential (EFP), 2, 267
   five-site, 1, 65–72
   four-site, 1, 62–5, 69–72
   generalized conductor-like screening model (GCOSMO), 2, 266
   methods, 1, 61–2
   rerference interaction site model (RISM), 2, 267, 268
   TIP3P, 2, 266, 267
   TIP4P, 1, 62–4, 69–72; 2, 265, 266, 267
   TIP4P-EW, 1, 64–5, 69–72
   TIP5P, 1, 65–7, 69–72
   TIP5P-E, 1, 67–72
wavefunctions, 1, 225–8
weak pairs, 2, 62–3, 68
Weighted Probe Interaction Energy Method, 2, 147
Weizmann-*n* theory, 1, 37–9

XED, 2, 159
XIAP, 2, 206, 208–9
XScore, 1, 123; 2, 161–2

Z-factor equation, 1, 22
zeolites, 2, 45